Introduction to Crops of India

2nd Revised & Enlarged Edition

Prof. N.R. DAS

M.Sc. (Ag.), Ph. D., D.W.P. (CIMMYT-MEXICO), FIBR (INDIA)

Ex - Professor of Agronomy
Bidhan Chandra Krishi Viswavidyalaya,
Mohanpur, Nadia, West Bengal

Published by
SCIENTIFIC PUBLISHERS (INDIA)

Jodhpur –

5 A, New Pali Road
P.O. Box 91
Jodhpur - 342 001 INDIA

Delhi –

4806/24, Ansari Road
Daryaganj
New Delhi - 110 002 INDIA

ISBN: 978-93-86652-29-4

Visit the Scientific Publishers (India) website at
http://www.scientificpub.com

Printed in India

Dedicated to my

Son Partha,
Daughter Aparna
and
Wife Arati

PREFACE (2nd Enlarged Edition)

Information on crops and other are the continuous process, with the advancement of research and development. In view, of this, the book entitled, 'Introduction to Crops of India' has been revised. Besides this, one chapter [**Chapter 13 : Crops in Water** (in different seasons)], has been added under the new **'Part Three : Water-Crops'**.

Chapter 13 deals with different water-crops that grow in waterbodies, which mostly have not yet been utilized commercially. With the increase of human and cattle population in India, these crops should be utilized fully for different purposes (such as, foods/fruits/medicines/ fodders / vegetables / fuels / flowers / beautification / soil-binders / organic matters / industries etc.) in different seasons. Some lowlying areas are there all over the country, particularly in eastern India, wherefrom these water-crops are available.

In this chapter, the same format, as earlier, has been used in the description of the water-crops, omitting 'water management' and adding 'peoples' response for use', with continuous serial numbers in crops. In this chapter, there are 74 crops (527 to 600), giving total number of 600 crops.

Other than the addition of chapter, the entire book has been revised, fully with the addition or deletion of materials other than correction of words/scientific names etc.

I am very much thankful to my colleagues of the Uttarbanga Krishi Viswavidyalaya / Calcutta University / Viswa Bharati University (Palli Siksha Bhavan) / Bidhan Chandra Krishi Viswavidyalaya and Kalyani University, for their appreciation and encouragement. At the end, I am thankful to my family members (wife Arati, son Partha and daughter Aparna) for their helps directly and indirectly.

N.R. Das

Subuj Niketan, Kalyani, W.B.
April, 28 2017

PREFACE (1st Edition)

While I was a student in agriculture, I felt some need of the hand-book with all crops along with brief and ready information. Again, while I was working as an agricultural teacher in the University, I felt the same need, because the students of both undergraduate and post-graduate, were demanding the hand-book, having some ready-made information of crops of all seasons in most of the areas of the country. Then, I thought of writing such a new book (in the later part of my life, after 35 years of teaching experience) named : "*Crops of India*", to meet the requirement of the students of all agricultural universities of the country.

It is, of course, very difficult to cover all the crops occurring in India in one hande book, containing some major information. In this book, therefore, some of the major crops of India, have been recorded with salient information, to keep it handy. The crops have been divided into two parts i.e. in part one, `*Field crops*' and in part two '*Plantation crops*'. Field crops, mostly seasonal, have been divided into cereals, pulses, oil seeds, fibres, tubers, sugars, vegetables, fodders, green-manuring crops, medicinal plants, spices, fruits and flowers (with succulents and ornamentals), in different seasons (*rabi*/ pre *kharif* and *kharif*). The plantation crops, have been divided into fruits, medicinal plants, timbers, tree-fodders, beverage and narcotics, spices, industrial crops and fuel crops.

As far as possible, English, Hindi and local names of the crops have been given along with scientific names, families and uses.

Climatologically, India is divided into major 5 ecosystems and these ecosystems are: (i) Arid Ecosystem, (ii) Semiarid

Ecosystem, (iii) Sub-humid Ecosystem, (iv) Humid Ecosystem and (v) Coastal Ecosystem. All the crops are within these ecosystems.

I will be very happy, if this handy agricultural book, containing crops of India, is fully utilized by agricultural students of both undergraduate and post graduate courses of the agricultural universities all over India and by agricultural personnel including, teachers, researchers, extension workers, planners, farmers and common people and also all agricultural organizations connected with teaching, research and extension activities, for crop production in the country.

Graduate and post-graduate students of general universities of the country will also be benefited with this book, for their syllabus on economic plants in Botany subject.

I am very thankful to my colleagues of Tamil Nadu Agricultural University, T.N.; Calcutta University, W.B.; Palli Siksha Bhavan, Viswa Bharati University, W.B., Bidhan Chandra Krishi Viswavidyalaya, W.B., Kalyani University, W.B. and North Bengal Agricultural University, W.B., for their encouragement.

N.R. Das

Sabuj Niketan, Kalyani, W.B.
21 Oct, 2006
Deepawali

CONTENTS

PART ONE : FIELD CROPS

Part 1 :
FIELD CROPS

Chapter - 1

WINTER / *RABI* SEASON CROPS

INTRODUCTION

Chapter 1 deals with different crops in cereals, pulses, oil seeds, fibres, tubers/rhizomes, sugar crops, vegetables, fodders, green manuring crops, medicinal plans, spices, fruits, flowers (alongwith succulents and ornamentals), beverage and narcotics, in winter (*rabi*) season.

Winter season in India changes from place-to-place. In the northern part, winter is severe (temperature 0 to 8ºC), where it is mild in the east, central and western tracts (7 to 14ºC), while it is nominal in the southern states.

During the winter season in India, rainfall is very low, in general, but in the southern parts, rainfall is received due to 'north-east monsoon', slight amount of rainfall is obtained; also small amount of rainfall is received in other parts of the country particularly in the eastern parts.

So crop cultivation during this period, is either with irrigation water or under rainfed conditions. In the eastern parts of the country i.e. West Bengal, Assam, Meghalaya, Tripura, Mizoram etc., unlike other areas, high soil moisture is observed, and as a result green vegetation is available.

1. CEREALS

1. ***CROP* — WHEAT / Gehu / *Gam***

   Scientific Name — Triticum aestivum L. emend. Thell.

   Family — Gramineae

   Herb / Shrub / Tree — Herb / 50-60 cm height

   Annual / Biennial / Perennial — Annual

   Used as — Food (Carbohydrate) / fodder

   1. *Climatic requirements* — Minimum temperature ranges from 5 to 16ºC, dry condition, winter season, For ripening it should be 35-36ºC.

2. *Soil requirements* — Alluvial / loamy / sandy loam / clayey loam soils
3. *Required land situations* — Flat land better; can also be grown in hills
4. *Importance of crop* — Major food in India; normally awn present in it
5. *Fertilizer management* — N, P_2O_5 and K_2O @ 100, 50 and 50 kg/ha respectively
6. *Water management* — 4-5 irrigations at 3 weeks intervals with 6-7 cm of water depth
7. *Duration of crop* — 4-5 months in plains and 5-6 months in hills
8. *Parts used* — Grain for human beings and straw for cattle
9. *Where grown in India* — All the states in India, except South India
10. *Export possibility* — Very high specially *durum* wheat
11. *Economic yields (q / ha)* — 25-35
12. *Economic value (Rs. / ha)* — 25,000 – 35,000
13. *By-product (If any)* — Straw and bran
14. *Use of by-product* — Cattle feed / mulching / fuel / organic matter / thatching for house / paper board

2. *CROP* — RICE / Chauval / Dhan / dhanya / Anna / Chawal

Scientific Name* — *Oryza sativa L.

Family — Gramineae

Herb / Shrub / Tree — Herb / 50-60 cm height

Annual / Biennial / Perennial — Annual

Used as — Food (Carbohydrate) / fodder

1. *Climatic requirements* — High temperature, humidity and water stagnation needed (transplantation in cool conditions)
2. *Soil requirements* — Clay / clayey / loamy / clay loam soils
3. *Required land situations* — Low / medium / flat land
4. *Importance of crop* — Major foods (carbohydrate)
5. *Fertilizer management* — N, P_2O_5 and K_2O @ 80-100, 40-50 and 40-50 kg/ha respectively
6. *Water management* — Water stagnated, if irrigated, 5-6 times irrigation, 8-10 cm depth
7. *Duration of crop* — 4-6 Months
8. *Parts used* — Seeds for human beings and straw for cattle

9. *Where grown in India* — All the states in India, except western dry land
10. *Export possibility* — Very high with quality rice (Basmati etc.)
11. *Economic yields (q / ha)* — 25-60
12. *Economic value (Rs. / ha)* — 30,000 to 80,000
13. *By-product (If any)* — Straw and bran
14. *Use of by-product* — Cattle feed / roofing / mulching / organic matter / fuel / paper board

3. *CROP* — MAIZE / Corn / Indian corn / *Makai / Bhutta*

Scientific Name* — *Zea mays L.

***Family* —** Gramineae

***Herb / Shrub / Tree* —** Herb / 2-4 m height

***Annual / Biennial / Perennial* —** Annual

***Used as* —** Food / fodder

1. *Climatic requirements* — Medium to high temperature, medium to high humidity and rainfall
2. *Soil requirements* — Loamy / alluvial / sandy loam soils
3. *Required land situations* — High land in plains and flat lands in hills
4. *Importance of crop* — Major food and for industrial products
5. *Fertilizer management* — N, P_2O_5 and K_2O @ 120, 60 and 60 kg/ha respectively
6. *Water management* — In winter season, 5-6 irrigations; in summer and wet seasons 3-4 irrigations
7. *Duration of crop* — 4-6 months
8. *Parts used* — Seeds for human consumption and stover for cattle feed / trailing vine crops.
9. *Where grown in India* — In all the states of India, except low lands, deserts/ snowy areas
10. *Export possibility* — Very high after processing
11. *Economic yields (q / ha)* — 20-25
12. *Economic value (Rs. / ha)* — 30,000 to 80,000
13. *By-product (If any)* — Stover
14. *Use of by-product* — Cattle feed and leaves for trailing vine crops in croppings / fuel / pegs

4. *CROP* — BARLEY / Jau / Jav / *Jab*

Scientific Name* — *Hordeum vulgare L. emend. Bowden

Family — Gramineae

Herb / Shrub / Tree — Herb / 60-70 cm height

Annual / Biennial / Perennial — Annual

Used as — Food / fodder

1. *Climatic requirements* — Low / medium temperature; dry weather
2. *Soil requirements* — Loamy / sandy loam / loamy / clayey soils
3. *Required land situations* — Flat lands of plains / hills
4. *Importance of crop* — Minor food but major forage
5. *Fertilizer management* — N, P_2O_5 and K_2O @ 80, 40 and 40 kg/ha respectively
6. *Water management* — Normally rainfed; if irrigated only one or two irrigations at crown root initiation and flowering
7. *Duration of crop* — 4-5 months.
8. *Parts used* — Seeds for human consumption and straw for cattle
9. *Where grown in India* — All states of India, except southern states
10. *Export possibility* — High after processing
11. *Economic yields (q / ha)* — 10-15.
12. *Economic value (Rs. / ha)* — 10,000 to 25,000
13. *By-product (If any)* — Straw
14. *Use of by-product* — Cattle feed / fuel / organic mater / roofing

5. *CROP* — **OAT / Jai / Ot**

Scientific Name — Avena sativa L.

Family — Gramineae

Herb / Shrub / Tree — Herb / 70-80 cm height

Annual / Biennial / Perennial — Annual

Used as — Food / fodder

1. *Climatic requirements* — Low temperature, dry condition
2. *Soil requirements* — Alluvial / loamy / saline soil
3. *Required land situations* — Flat lands in plains and hills
4. *Importance of crop* — Food / fodder can be obtained
5. *Fertilizer management* — N, P_2O_5 and K_2O @ 80, 30 and 30 kg/ha respectively
6. *Water management* — Normally rainfed; 1-2 irrigations give high yields
7. *Duration of crop* — 4-5 months
8. *Parts used* — Seeds for man and animal and straw for cattle feeds

9. *Where grown in India* — All states except South India
10. *Export possibility* — High, if processed
11. *Economic yields (q / ha)* — 15-20
12. *Economic value (Rs. / ha)* — 10,000 to 25,000
13. *By-product (If any)* — Straw
14. *Use of by-product* — Cattle feed / roofing / organic matter / fuel / mulching

2. PULSES

6. *CROP* — GRASSPEA / Chickling pea / *Khesari* / Jarosse

Scientific Name* — *Lathyrus sativus L.

Family — Papilionaceae

Herb / Shrub / Tree — Herb / 60-70 cm length and 50 cm height (as it lies on soil)

Annual / Biennial / Perennial — Annual

Used as — Pulse and Fodder

1. *Climatic requirements* — Medium to low temperature, dry condition
2. *Soil requirements* — Clay silt / clayey / loamy / alluvial soils
3. *Required land situations* — Low land in plains and hills
4. *Importance of crop* — Minor dal (pulse) / fodder / soil-moisture enriching crop/ cover crop / green manuring crop
5. *Fertilizer management* — Practically no fertilizer is needed
6. *Water management* — Grown as rainfed crop, hence no irrigation
7. *Duration of crop* — 4-5 months.
8. *Parts used* — Seed as pulse and stover as cattle feed
9. *Where grown in India* — All states where low temperature ranges between 7°-15°C in low lying areas (mainly in eastern states)
10. *Export possibility* — High with particular variety (Nirmal 1 & 2)
11. *Economic yields (q / ha)* — 10-15
12. *Economic value (Rs. / ha)* — 7,000 to 25,000
13. *By-product (If any)* — Stover / *bhushi*
14. *Use of by-product* — Cattle feeds / organic matter / fuel / mulching

7. *CROP* — LENTIL / Mossor / Musur / Masar

Scientific Name* — *Lens culinaris Medik

Family — Papilionaceae

Herb / Shrub / Tree — Herb / 30-40 cm height

Annual / Biennial / Perennial — Annual

Used as — Pulse food

1. *Climatic requirements* — Medium or low temperature and dry conditions
2. *Soil requirements* — Loamy / clay loam soils
3. *Required land situations* — Flat lands of plains / hills
4. *Importance of crop* — Minor pulse in north east India
5. *Fertilizer management* — N, P_2O_5 and K_2O @ 40, 60 and 30 kg/ha respectively
6. *Water management* — Grown under rainfed condition; only one irrigation may be given at flowering, if irrigated
7. *Duration of crop* — 4-5 months.
8. *Parts used* — Seeds of pulse for human beings and stover / *bhushi* for cattle
9. *Where grown in India* — All states of India, except southern states
10. *Export possibility* — High after processing
11. *Economic yields (q / ha)* — 10-15.
12. *Economic value (Rs. / ha)* — 25,000 to 40,000
13. *By-product (If any)* — Stover / *bhushi*
14. *Use of by-product* — Cattle feed / fuel / organic matter

8. *CROP* **— PEA / Matar / Motor / Karaisuti**

Scientific Name — Pisum sativum L.

Family — Papilionaceae

Herb / Shrub / Tree — Herb / 50-60 cm length and 40 cm height

Annual / Biennial / Perennial — Annual

Used as — Pulse food / vegetables / forage

1. *Climatic requirements* — Low or medium temperature; dry weather
2. *Soil requirements* — Sandy / Loamy / Clayey loam soils
3. *Required land situations* — Flat / undulated lands in plains and hills
4. *Importance of crop* — Major pulse crop / vegetable crop / creeper
5. *Fertilizer management* — N, P_2O_5 and K_2O @ 40, 60 and 40 kg/ha respectively

6. *Water management* — Rainfed, two irrigations at flowering, if irrigated
7. *Duration of crop* — 4-5 months.
8. *Parts used* — Seeds as pulse, green pod as vegetable, green plants as forage and green manure
9. *Where grown in India* — All states in India except southern states
10. *Export possibility* — Very high after processing
11. *Economic yields (q / ha)* — Seed-10; Green pod 50 and biomass 150
12. *Economic value (Rs. / ha)* — Seed-20,000; Green pod 25,000 and biomass 20,000
13. *By-product (If any)* — Stover / *bhushi*
14. *Use of by-product* — Cattle feed / fuel / organic mater / mulching

9. *CROP* — PIGEONPEA / ARAHAR / Tur / Red gram / *Rahar*

Scientific Name — ***Cajanus cajan*** (L.) Mill sp.

Family — Papilionaceae

Herb / Shrub / Tree — Shrub / 1-3 m. height

Annual / Biennial / Perennial — Annual

Used as — Seed as pulse, stover as fuel, twigs as forages

1. *Climatic requirements* — Medium temperature; dry / wet condition needed.
2. *Soil requirements* — Alluvial / loamy / clayey soils
3. *Required land situations* — High land in plains and medium land in hills
4. *Importance of crop* — Major pulse with yellow flower
5. *Fertilizer management* — N, P_2O_5 and K_2O @ 40, 60 and 40 kg/ha respectively
6. *Water management* — Normally rainfed; if necessary, one irrigation at flowering should be given.
7. *Duration of crop* — 4-8 months.
8. *Parts used* — Seed for pulse / stover for fuel
9. *Where grown in India* — All the states except Tamil Nadu and Kerala
10. *Export possibility* — High potential crop for export
11. *Economic yields (q / ha)* — 15-20
12. *Economic value (Rs. / ha)* — 20,000 to 30,000
13. *By-product (If any)* — Stover and *busha*
14. *Use of by-product* — Good cattle feed / pigs / vine trailing / fuel

10. *CROP* — **GRAM / Chick-pea / Chana /** ***Chhola / but***

Scientific Name — Cicer arietinum L.

Family — Papilionaceae

Herb / Shrub / Tree — Herb / 30-50 cm height

Annual / Biennial / Perennial — Annual

Used as — Pulse foods and food-products

1. *Climatic requirements* — Low or medium temperature; dry weather
2. *Soil requirements* — Loamy / sandy loam soils
3. *Required land situations* — Flat land of plains and hills
4. *Importance of crop* — Major pulse, bushy with branches
5. *Fertilizer management* — N, P_2O_5 and K_2O @ 40, 60 and 60 kg/ha respectively
6. *Water management* — Normally rainfed crop; one irrigation at flowering may be given, if water available.
7. *Duration of crop* — 5-6 months.
8. *Parts used* — Seed for human beings or horses and *bhusa* for cattle
9. *Where grown in India* — All over India except southern states
10. *Export possibility* — Very high, if properly processed and packed
11. *Economic yields (q / ha)* — 15-20
12. *Economic value (Rs. / ha)* — 20,000 to 30,000
13. *By-product (If any)* — Stover / *bhusa*
14. *Use of by-product* — Cattle feed / fuel / organic matter / mulching

11. *CROP* — **BLACK GRAM / Urd /** *Maskalai* / *Biuri dal*

Scientific Name — Vigna mungo L. Hepper

Family — Papilionaceae

Herb / Shrub / Tree — Herb / 60-70 cm length and 40 cm height

Annual / Biennial / Perennial — Annual

Used as — Food / pulse / fodder / green manure

1. *Climatic requirements* — High temperature, humidity and rainfall
2. *Soil requirements* — Alluvial / loamy soil
3. *Required land situations* — Flat land / undulated hills
4. *Importance of crop* — Major pulse, slimy in nature, good for *'dosa'*.
5. *Fertilizer management* — N, P_2O_5 and K_2O @ 20, 60 and 40 kg/ha respectively

6. *Water management* — Normally grown as rainfed crop but one irrigation at flowering stage, if irrigated.
7. *Duration of crop* — 4-5 months
8. *Parts used* — Seed as pulse, stover / *bhusa* as cattle feed / green biomass as green manure and fodder
9. *Where grown in India* — All states except drier north and western states in India
10. *Export possibility* — High with proper processing
11. *Economic yields (q / ha)* — 10-15
12. *Economic value (Rs. / ha)* — 15,000 to 30,000
13. *By-product (If any)* — Dry stover and *bhusa* / green biomass
14. *Use of by-product* — Cattle feed / fodder or forage / organic matter / fuel / green manure.

12. *CROP* **— RAJMA / KIDNEY BEAN / Vilaiti sim / Faraz bean / *Rajma***

Scientific Name* — *Phaseolus vulgaris L.

Family — Papilionaceae

Herb / Shrub / Tree — Herb / 30-40 cm height

Annual / Biennial / Perennial — Annual

Used as — Seed as pulse / stover as cattle feed / vegetable (Pod)

1. *Climatic requirements* — Low medium temperature, less humidity and rainfall.
2. *Soil requirements* — Alluvial / loamy soil
3. *Required land situations* — High flat land in plains and flat land in hills
4. *Importance of crop* — Minor pulse / bushy creeper
5. *Fertilizer management* — N, P_2O_5 and K_2O @ 60, 60 and 60 kg/ha respectively
6. *Water management* — Normally rainfed but 1 or 2 irrigations when grown for green pod for vegetables.
7. *Duration of crop* — 4-5 months.
8. *Parts used* — Seed as pulse / stover and other as cattle feed / fuel / organic matter
9. *Where grown in India* — All the states except southern states
10. *Export possibility* — Very high, if processed
11. *Economic yields (q / ha)* — 10-15
12. *Economic value (Rs. / ha)* — 25,000 to 30,000
13. *By-product (If any)* — Stover / green biomass after green pods

14. *Use of by-product* — Cattle feed / organic matter / fuel / green manure / vegetable

3. OIL SEEDS

13. ***CROP*** **— RAPE / Indian rape / Toria / Indian Mustard /** ***Sarse / Sarisha*** **/ Rai**

Scientific Name **—** ***Brassica campestris*** L. var ***toria*** Duth and Full. & ***B. juncea*** (L.) Czern & Cross

Family — Cruciferae

Herb / Shrub / Tree — Herb / 30-50 cm height

Annual / Biennial / Perennial — Annual

Used as — Oil / cattle feeds

1. *Climatic requirements* — Low or medium temperature with dry conditions.
2. *Soil requirements* — Loamy / alluvial soils
3. *Required land situations* — Flat lands in plains and hills
4. *Importance of crop* — Major oilseed crop, with yellow flower
5. *Fertilizer management* — N, P_2O_5 and K_2O @ 80, 60 and 60 kg/ha respectively
6. *Water management* — Normally rainfed irrigation may be given before and after flowering, twice
7. *Duration of crop* — 3-4 months.
8. *Parts used* — Seed for oils and cake for cattle feed and manure and stover for cattle feeds
9. *Where grown in India* — All the states except Tamil Nadu, Kerala and Andhra Pradesh
10. *Export possibility* — Very high (as pickle) after processing
11. *Economic yields (q / ha)* — 10-15
12. *Economic value (Rs. / ha)* — 20,000 to 30,000
13. *By-product (If any)* — Oil cake and stover
14. *Use of by-product* — Cattle feeds / manure / fuel / organic matter

Note — 13(a) Crop – Mustard / Rai / Sarisha / Sarshe is also another annual herb for oilseeds, the scientific name of which is *Brassica juncea* Czern and Cross and is cultivated for oils, in whole of North India in winter. The seed yield, 10-25 quintal/ha, economic value of which is Rs. 15,000 – 40,000. Other characters are like Rape/Indian rape, though plant is bigger in size.

14. *CROP* — **LINSEED / Flax / Alsi / *Tisi***

Scientific Name* — *Linum usitatissimum L.

Family — Linaceae

Herb / Shrub / Tree — Herb / 30-50 cm height

Annual / Biennial / Perennial — Annual

Used as — Varnish oil / perfume base / oil

1. *Climatic requirements* — Low / medium temperature with dry weather
2. *Soil requirements* — Loamy / sandy / alluvial soils
3. *Required land situations* — Flat lands of plains and hills
4. *Importance of crop* — Minor oil (may be used as *paira* crop)
5. *Fertilizer management* — N, P_2O_5 and K_2O @ 60, 30 and 30 kg/ha respectively
6. *Water management* — Normally rainfed, if irrigated, one irrigation before flowering
7. *Duration of crop* — 4-5 months.
8. *Parts used* — Seed for oil (for varnish)
9. *Where grown in India* — All states except Tamil Nadu, Kerala and A.P.
10. *Export possibility* — Very high for furniture varnishing and also for dilution of painting materials.
11. *Economic yields (q / ha)* — 8-10
12. *Economic value (Rs. / ha)* — 10,000 to 15,000 (Sometimes even 50,000)
13. *By-product (If any)* — Oil cake and stover
14. *Use of by-product* — Cattle feed and flax (fibre) / manures / fuel / organic matter.

15. *CROP* — **NIGER / Rantil / Kalatil / Ramtil**

Scientific Name* — *Guizotia abyssinica Cass

Family — Composiae

Herb / Shrub / Tree — Herb / 60-90 cm height

Annual / Biennial / Perennial — Annual

Used as — Oil / condiments / varnish / soap

1. *Climatic requirements* — Low or medium temperature with dry weather
2. *Soil requirements* — Loamy / alluvial soils
3. *Required land situations* — Flat lands of plains and hills

4. *Importance of crop* — Minor oils
5. *Fertilizer management* — N, P_2O_5 and K_2O @ 60, 30 and 30 kg/ha respectively
6. *Water management* — Normally rainfed, one irrigation at flowering, if irrigated.
7. *Duration of crop* — 4-5 months.
8. *Parts used* — Seeds for human / cattle and stover for fuel
9. *Where grown in India* — Except southern states, all over India
10. *Export possibility* — High with processing for pickles / varnish
11. *Economic yields (q / ha)* — 15-20
12. *Economic value (Rs. / ha)* — 25,000 to 40,000
13. *By-product (If any)* — Oil cake and stover
14. *Use of by-product* — Cattle feed / manure / fuel / organic matter.

16. *CROP* — **SAFFLOWER /** ***KUSUM***

Scientific Name — Carthamus tinctorius L.

Family — Compositae

Herb / Shrub / Tree — Herb / Height 60-80 cm.

Annual / Biennial / Perennial — Annual

Used as — Minor oils / Perfumes / varnish / Linoleum

1. *Climatic requirements* — Low / medium temperature with dry weather
2. *Soil requirements* — Sandy loam / alluvial soils
3. *Required land situations* — Flat lands of plains and hills
4. *Importance of crop* — Minor edible oils / base perfume materials / spines in plant
5. *Fertilizer management* — N, P_2O_5 and K_2O @ 60, 30 and 30 kg/ha respectively
6. *Water management* — Normally rainfed, if irrigated apply one irrigation at flowering
7. *Duration of crop* — 4-5 months.
8. *Parts used* — Seeds for oils / perfumes, stover for fuel, flower for dye
9. *Where grown in India* — All states in dry areas except Tamil Nadu, Kerala and Andhra Pradesh
10. *Export possibility* — Very high after processing (i.e. perfume)
11. *Economic yields (q / ha)* — 10-12
12. *Economic value (Rs. / ha)* — 10,000 to 25,000

13. *By-product (If any)* — Stalk (thorny)
14. *Use of by-product* — Fuel / organic manure / mulching

4. FIBRES

17. *CROP* — FLAX / Alsi / *Tisi*

Scientific Name* — *Linum usitatissimum L.

Family — Linaceae

Herb / Shrub / Tree — Herb / Height 50-70 cm

Annual / Biennial / Perennial — Annual

Used as — Fibre / oil seeds

1. *Climatic requirements* — Low / medium temperature with dry weather
2. *Soil requirements* — Sandy loam / loam / alluvial soils
3. *Required land situations* — Flat lands of plains and hills
4. *Importance of crop* — Minor fibre crop with less importance
5. *Fertilizer management* — N, P_2O_5 and K_2O @ 60, 30 and 30 kg/ha respectively
6. *Water management* — Normally rainfed, if irrigated, one irrigation should be applied at 45 Days after sowing
7. *Duration of crop* — 4-5 months.
8. *Parts used* — Bast fibre / stover
9. *Where grown in India* — All the states, except the southern part
10. *Export possibility* — High after processing
11. *Economic yields (q / ha)* — 8-12
12. *Economic value (Rs. / ha)* — 10,000 to 25,000
13. *By-product (If any)* — Stover or stick
14. *Use of by-product* — Paper / board / fuel / organic matter / mulching

5. TUBERS / RHIZOMES

18. *CROP* — POTATO / Àlu / Golalu

Scientific Name* — *Solanum tuberosum L.

Family — Solanaceae

Herb / Shrub / Tree — Herb, height 40-50 cm

Annual / Biennial / Perennial — Annual

Used as — Major starchy food / vegetable

1. *Climatic requirements* — Low / medium temperature with dry weather
2. *Soil requirements* — Loamy / alluvial soils
3. *Required land situations* — Flat lands of plains and hills
4. *Importance of crop* — Major tuber crop / tuber grows underground
5. *Fertilizer management* — N, P_2O_5 and K_2O @ 120, 60 and 60 kg/ha respectively
6. *Water management* — Irrigated crop, 5-8 irrigations, at 10 days interval and next at 20 days interval
7. *Duration of crop* — 4-5 months.
8. *Parts used* — Tuber as food / vegetable; haulm for cattle feed / organic matter / mulch
9. *Where grown in India* — Except southern states, all over India
10. *Export possibility* — Very high after processing
11. *Economic yields (q / ha)* — 200 to 250 (raw / green)
12. *Economic value (Rs. / ha)* — 30,000 to 40,000
13. *By-product (If any)* — Haulm and leaves
14. *Use of by-product* — Catle feed / organic manure / mulch / fuel.

6. SUGARS

19. ***CROP* — BEET / Beet root / Chukandar / *Bit Sugar***

Scientific Name — Beta vulgaris L.

Family — Chenopodiaceae

Herb / Shrub / Tree — Herb, height 50-70 cm

Annual / Biennial / Perennial — Biennial / Annual (for beet)

Used as — Sugar/ vegetable

1. *Climatic requirements* — Low or medium temperature with dry weather
2. *Soil requirements* — Loamy / alluvial / saline soils
3. *Required land situations* — Flat lands of plains and hills
4. *Importance of crop* — Major sugar / vegetable can withstand salinity
5. *Fertilizer management* — N, P_2O_5 and K_2O @ 120, 60 and 60 kg/ha respectively

6. *Water management* — 3-4 irrigations, 20 / 25 days interval
7. *Duration of crop* — 4-5 months.
8. *Parts used* — Swollen root for sugar / vegetable; foliage as cattle feed
9. *Where grown in India* — Except southern states, all over India
10. *Export possibility* — Very high after processing
11. *Economic yields (q / ha)* — 150-200 (green)
12. *Economic value (Rs. / ha)* — 20,000 to 40,000
13. *By-product (If any)* — Green foliage
14. *Use of by-product* — Cattle feed / organic manure / fuel / mulch

20. *CROP* — YAM BEAN / *Sakalu*

Scientific Name* — *Pachyrhizus erosus (L.) Urb.

Family — Papilionaceae

Herb / Shrub / Tree — Herb (Vine) – Very tall

Annual / Biennial / Perennial — Perennial

Used as — Sugar food / starchy food / vegetable (pods)

1. *Climatic requirements* — Higher temperature with high rainfall and humidity
2. *Soil requirements* — Loamy / alluvial / sandy soils
3. *Required land situations* — High lands
4. *Importance of crop* — Roots are swollen, eaten as such and white in colour
5. *Fertilizer management* — N, P_2O_5 and K_2O @ 40, 60 and 50 kg/ha respectively
6. *Water management* — Normally rainfed, 1-2 irrigation, if irrigated
7. *Duration of crop* — 120-140 days or 1-2 years
8. *Parts used* — Root (tuberous) eaten as such raw
9. *Where grown in India* — Assam, West Bengal, Bihar, Orissa, Andhra Pradesh, etc.
10. *Export possibility* — High
11. *Economic yields (q / ha)* — 150-200 (green)
12. *Economic value (Rs. / ha)* — 20,000 to 40,000
13. *By-product (If any)* — Twig with leaves and vine
14. *Use of by-product* — Cattle feed / fuel / mulch / organic matter

7. VEGETABLES (Annual)

21. *CROP* — **RED GOURD / Red Pumkin / *Sitaphal / Misti Kumro***

Scientific Name — *Cucurbita maxima* Duch

Family — Cucurbitaceae

Herb / Shrub / Tree — Herb / 3-5 m length, 50 cm height, trailing type

Annual / Biennial / Perennial — Annual

Used as — Fruit as vegetable / vine and leaves for leafy vegetable

1. *Climatic requirements* — In all climates it grows, except heavy dry and cold conditions
2. *Soil requirements* — All types of soils
3. *Required land situations* — Plains are suitable
4. *Importance of crop* — Major vegetable / trailing type
5. *Fertilizer management* — N, P_2O_5 and K_2O @ 80, 40 and 40 kg/ha respectively
6. *Water management* — Both under rainfed or irrigated condition it can be grown. Apply 2 irrigations before and after flowering, if irrigated
7. *Duration of crop* — 4-5 months.
8. *Parts used* — Fruit, vine and leaves as vegetable
9. *Where grown in India* — All over the country except extreme dry / wet and cold in northern parts
10. *Export possibility* — High, if processed
11. *Economic yields (q / ha)* — 150 to 200 (green fruit)
12. *Economic value (Rs. / ha)* — 15,000 to 20,000
13. *By-product (If any)* — Vines and leaves
14. *Use of by-product* — Vegetable / cattle feed / fuel / organic matter.

22. *CROP* — **PUMKIN / Marrow / Vilaiti kaddu / Lau**

Scientific Name — *Cucurbita pepo* L.

Family — Cucurbitaceae

Herb / Shrub / Tree — Herb / 3-5 m length and 50 cm height

Annual / Biennial / Perennial — Annual (creeper)

Used as — Fruit, vine and leaves used as vegetables

1. *Climatic requirements* — High or medium/low temperature, with dry weather
2. *Soil requirements* — Loamy / alluvial soils

3. *Required land situations* — Except very high altitude, all lands are suitable
4. *Importance of crop* — Major vegetable / trailing type
5. *Fertilizer management* — N, P_2O_5 and K_2O @ 80, 40 and 40 kg/ha respectively
6. *Water management* — Spot irrigation, 4-5 times
7. *Duration of crop* — 4-5 months.
8. *Parts used* — Fruit, vine and leaves as vegetables
9. *Where grown in India* — All over the country, except too dry and too cold conditions
10. *Export possibility* — High if properly protected
11. *Economic yields (q / ha)* — 150-200 (green)
12. *Economic value (Rs. / ha)* — 20,000 to 30,000
13. *By-product (If any)* — Vine / leaves
14. *Use of by-product* — Vegetable / cattle feed / fuel / mulch / organic matter.

23. *CROP* — **DRUM STICK / Horse-radish / Sahinjna /** *Soanjna / Sajna*

Scientific Name — Moringa oleifera Lamk

Family — Moringaceae

Herb / Shrub / Tree — Tree (small and soft)

Annual / Biennial / Perennial — Perennial

Used as — Green stick as vegetable / leaves or twig as fodder / vegetable

1. *Climatic requirements* — All types of climates, except extreme dry and cold
2. *Soil requirements* — All types of soils
3. *Required land situations* — Lands, except hills and low-lands
4. *Importance of crop* — Major vegetable / very long pods
5. *Fertilizer management* — N, P_2O_5 and K_2O @ 60, 30 and 30 kg/ha respectively
6. *Water management* — Spot irrigation, 4-5 times a year
7. *Duration of crop* — 5 years
8. *Parts used* — Tender fruit as vegetable, leaves as fodder / fuel / vegetable and flower as vegetables
9. *Where grown in India* — All over India except extreme hot and cold conditions
10. *Export possibility* — High, if properly packed

11. *Economic yields (q / ha)* — 100-200 (green)
12. *Economic value (Rs. / ha)* — 25,000 to 30,000
13. *By-product (If any)* — Soft branch, leaves and flower
14. *Use of by-product* — Cattle feeds / match boxes / fuel / organic matter / vegetable

24. ***CROP* — TOMATO / *Tomator* / *Tamato* / Vilaiti begun**

Scientific Name* — *Lycopersicon esculentum L.

Family — Solanaceae

Herb / Shrub / Tree — Herb / Height 60-70 cm

Annual / Biennial / Perennial — Annual

Used as — Vegetable / sauce / salad / fruit

1. *Climatic requirements* — Low / medium temperature, dry weather
2. *Soil requirements* — Loamy / alluvial soils
3. *Required land situations* — Flat lands in plains / hills
4. *Importance of crop* — Major vegetable / food processing
5. *Fertilizer management* — N, P_2O_5 and K_2O @ 100, 50 and 50 kg/ha respectively
6. *Water management* — If irrigated, 3-4 irrigations, one at 20 days after sowing and two before and after flowering
7. *Duration of crop* — 4-5 months.
8. *Parts used* — Fruit as vegetable / sauce, haulm as cattle feeds
9. *Where grown in India* — All over India, except extreme dry and cold conditions
10. *Export possibility* — Very high, if properly processed for sauce
11. *Economic yields (q / ha)* — 150-200
12. *Economic value (Rs. / ha)* — 30,000 to 40,000
13. *By-product (If any)* — Haulm
14. *Use of by-product* — Cattle feed / manure / fuel / organic matter

25. *CROP* — **CABBAGE / Bandhgobi / Bandhakapi**

Scientific Name* — *Brassica oleraceae L. var. ***capitata***

Family — Cruciferae

Herb / Shrub / Tree — Herb / Height 40-50 cm

Annual / Biennial / Perennial — Annual (round head) / Biennial

Used as — Head as vegetable / dry vegetable

1. *Climatic requirements* — Low or medium temperature with dry weather

2. *Soil requirements* — All types of soil except sandy one
3. *Required land situations* — Flat land in plains and hills
4. *Importance of crop* — Major leafy vegetable in winter
5. *Fertilizer management* — N, P_2O_5 and K_2O @ 120, 60 and 60 kg/ha respectively
6. *Water management* — After transplantation, 4 irrigations at 20 days interval
7. *Duration of crop* — 4-5 months.
8. *Parts used* — Head as leafy vegetable
9. *Where grown in India* — All states except southern states
10. *Export possibility* — Very high with packaging
11. *Economic yields (q / ha)* — 200-300 (green)
12. *Economic value (Rs. / ha)* — 30,000 to 50,000
13. *By-product (If any)* — Stocks
14. *Use of by-product* — Cattle feeds / fuel / organic matter

26. *CROP* — CAULIFLOWER / Phulgobi / *Phulkapi*

Scientific Name* — *Brassica oleracea L. var. ***botrytis***

Family — Cruciferae

Herb / Shrub / Tree — Herb / 60-70 cm height

Annual / Biennial / Perennial — Annual / Biennial

Used as — Used as vegetable

1. *Climatic requirements* — Low or medium temperature with dry weather
2. *Soil requirements* — Except sandy soil, all type of soils
3. *Required land situations* — Flat land in plains and hills
4. *Importance of crop* — Major vegetable / Head is inflorescence
5. *Fertilizer management* — N, P_2O_5 and K_2O @ 120, 60 and 60 kg/ha respectively
6. *Water management* — 4-5 irrigations, 20 days interval after transplantation
7. *Duration of crop* — 4-5 months.
8. *Parts used* — Inflorescence as vegetable, leaf as cattle feed or manure / fuel
9. *Where grown in India* — All over India except southern states
10. *Export possibility* — High, if processed well or protected well
11. *Economic yields (q / ha)* — 200-300 (green)
12. *Economic value (Rs. / ha)* — 30,000 to 50,000

13. *By-product (If any)* — Leaves and stalk
14. *Use of by-product* — Cattle feed / fuel / organic matter / mulch

27. ***CROP* — RADISH / Muli / *Mula / Mulo***

Scientific Name* — *Raphanus sativus L.

Family — Cruciferae

Herb / Shrub / Tree — Herb / Height 40-50 cm

Annual / Biennial / Perennial — Annual

Used as — Root vegetable

1. *Climatic requirements* — Low or medium temperature with dry weather
2. *Soil requirements* — Except heavy, all types of soils
3. *Required land situations* — Flat land in plains and hills
4. *Importance of crop* — Major vegetable in winter / root above soil
5. *Fertilizer management* — N, P_2O_5 and K_2O @ 60, 80 and 60 kg/ha respectively
6. *Water management* — 4-5 irrigations at interval of 20 days after sowing
7. *Duration of crop* — 4-5 months.
8. *Parts used* — Roots and leaves as vegetable
9. *Where grown in India* — All over India except too high or too low temperature conditions
10. *Export possibility* — High if proper packaging is done
11. *Economic yields (q / ha)* — 200-300 (green)
12. *Economic value (Rs. / ha)* — 20,000 to 30,000
13. *By-product (If any)* — Leaf
14. *Use of by-product* — Vegetable / cattle feed / fuel / organic matter.

28. ***CROP* — KNOL-KHOL / Kohlrabi / Ganthgobi / *Nalkhal***

Scientific Name* — *Brassica oleracea L. var. ***gongylodes*** L.

Family — Cruciferae

Herb / Shrub / Tree — Herb / 10-15 cm height

Annual / Biennial / Perennial — Annual (round ball)

Used as — Vegetable

1. *Climatic requirements* — Low or medium temperature with dry weather
2. *Soil requirements* — All soils without stone, sand and heavy clay
3. *Required land situations* — Flat land in plains and hills

4. *Importance of crop* — Major vegetable in winter / round ball above soil
5. *Fertilizer management* — N, P_2O_5 and K_2O @ 50, 30 and 30 kg/ha respectively
6. *Water management* — 4-5 irrigations, if irrigated, at 20 days interval
7. *Duration of crop* — 4-5 months
8. *Parts used* — Swollen stem as vegetable and foliage as cattle feed
9. *Where grown in India* — Except southern states all over the country
10. *Export possibility* — High if packed
11. *Economic yields (q / ha)* — 100-200 (green)
12. *Economic value (Rs. / ha)* — 10,000 to 20,000
13. *By-product (If any)* — Leaves
14. *Use of by-product* — Cattle feed / manure / organic / fuel / mulch.

29. ***CROP* — SPINACH / PALAK / *Palong***

Scientific Name — Spinacia oleracea L.

Family — Chenopodiace

Herb / Shrub / Tree — Herb / Height 30-50 cm

Annual / Biennial / Perennial — Annual

Used as — Leaf as vegetable (whole plant)

1. *Climatic requirements* — Low or medium temperature with dry weather
2. *Soil requirements* — Except sandy, all soils with organic matter
3. *Required land situations* — Flat land in plains and hills
4. *Importance of crop* — Major leafy vegetable in *rabi*
5. *Fertilizer management* — N, P_2O_5 and K_2O @ 100, 50 and 50 kg/ha respectively
6. *Water management* — 2-3 irrigations for leafy vegetable, at 20 days interval
7. *Duration of crop* — 4-5 months.
8. *Parts used* — Leaf as vegetable
9. *Where grown in India* — All states except in southern states in India
10. *Export possibility* — High if properly packed
11. *Economic yields (q / ha)* — 200-250 (green)
12. *Economic value (Rs. / ha)* — 30,000 to 50,000

13. *By-product (If any)* — Stem (when seed is taken)
14. *Use of by-product* — Cattle feed / mulch / organic matter.

30. *CROP* — BEET ROOT / Garden beet / Chukandar / *Bit*

Scientific Name* — *Beta vulgaris L.

Family — Chenopodiaceae

Herb / Shrub / Tree — Herb / 30-40 cm height

Annual / Biennial / Perennial — Biennial (if grown for seeds, otherwise annual)

Used as — Swollen root as vegetable/sugar

1. *Climatic requirements* — Low or medium temperature with dry weather
2. *Soil requirements* — All soil except sandy and clay soils
3. *Required land situations* — Flat lands in plains and hills
4. *Importance of crop* — Major vegetable in winter (round ball)
5. *Fertilizer management* — N, P_2O_5 and K_2O @ 60, 30 and 30 kg/ha respectively
6. *Water management* — 3-4 irrigations, at 20 days interval
7. *Duration of crop* — 4-5 months.
8. *Parts used* — Swollen root as vegetable, foliage as cattle feed/ manure
9. *Where grown in India* — Except southern states, all over the country
10. *Export possibility* — High, if properly processed
11. *Economic yields (q / ha)* — 200-250 (green)
12. *Economic value (Rs. / ha)* — 20,000 to 30,000
13. *By-product (If any)* — Leaves
14. *Use of by-product* — Cattle feed / manure / fuel / organic matter

31. *CROP* — CARROT / *Gajor*

Scientific Name* — *Daucus carota L.

Family — Umbelliferae

Herb / Shrub / Tree — Herb / Height 30-40 cm

Annual / Biennial / Perennial — Annual / Biennial (Annual, when grown for vegetable)

Used as — Fleshy root as vegetable

1. *Climatic requirements* — Low or medium temperature with dry weather

2. *Soil requirements* — Except sandy or clay soil, grown in all other soils
3. *Required land situations* — Flat lands of plains and hills
4. *Importance of crop* — Minor vegetable in winter, yellow in colour
5. *Fertilizer management* — N, P_2O_5 and K_2O @ 60, 30 and 30 kg/ha respectively
6. *Water management* — 3-4 irrigations at 20 days interval
7. *Duration of crop* — 4-5 months.
8. *Parts used* — Swollen tap root as vegetable and foliage as cattle feed
9. *Where grown in India* — Except the southern states, all over the country
10. *Export possibility* — Very high, if packed and processed properly
11. *Economic yields (q / ha)* — 200-300 (green)
12. *Economic value (Rs. / ha)* — 25,000 to 35,000
13. *By-product (If any)* — Leaves
14. *Use of by-product* — Cattle feed / manure / fuel / mulching

32. *CROP* — **COWPEA (*Barbati*) / Lobia / Chowli / *Lalsa***

Scientific Name — Vigna unguiculata L. Walp

Family — Papilionaceae

Herb / Shrub / Tree — Herb (Creeper) / 2-7 m. length but 50 cm height

Annual / Biennial / Perennial — Annual

Used as — Green pod as vegetable foliage as fodder; seed as pulse

1. *Climatic requirements* — Mostly all climatic conditions except too hot and too cold
2. *Soil requirements* — All types of soils except sandy one
3. *Required land situations* — Flat lands of plains and hills
4. *Importance of crop* — Major vegetable / vine type
5. *Fertilizer management* — N, P_2O_5 and K_2O @ 40, 80 and 40 kg/ha respectively
6. *Water management* — 2-3 irrigations before flowering and onwards at 20 days interval, if land is irrigated
7. *Duration of crop* — 4-6 months.
8. *Parts used* — Green pod as vegetable, green plants as fodder / green manure crop
9. *Where grown in India* — All over the country except too hot and too cold conditions

10. *Export possibility* — High, if processed properly
11. *Economic yields (q / ha)* — 250-300 (green)
12. *Economic value (Rs. / ha)* — 30,000 to 40,000
13. *By-product (If any)* — Vine and leaves
14. *Use of by-product* — Cattle feed / green manure / fuel / mulch / organic matter

33. *CROP* — **POTATO / Alu / *Golalu***

Scientific Name — Solanum tuberosum L.

Family — Solanaceae

Herb / Shrub / Tree — Herb / Bushy 40-50 cm height

Annual / Biennial / Perennial — Annual

Used as — Tuber (potato) as vegetable

1. *Climatic requirements* — Low or medium temperature with dry weather
2. *Soil requirements* — Loamy / alluvial soils
3. *Required land situations* — Flat lands of plains and hills
4. *Importance of crop* — Major vegetable / tuber grows in side soils
5. *Fertilizer management* — N, P_2O_5 and K_2O @ 120, 60 and 60 kg/ha respectively
6. *Water management* — 5-8 irrigations, first 2 at 10 days interval and next at 20 days interval
7. *Duration of crop* — 4-5 months.
8. *Parts used* — Tuber (potato) as vegetable, haulm and leaves as cattle feed and manure / mulch
9. *Where grown in India* — All over India, except the southern states
10. *Export possibility* — Very high if processed / packed properly
11. *Economic yields (q / ha)* — 200-250 (green)
12. *Economic value (Rs. / ha)* — 50,000 to 80,000
13. *By-product (If any)* — Haulm and leaves
14. *Use of by-product* — Cattle feed / manure / mulch / organic matter.

34. *CROP* — **ONION / Piaz / Piaj / Pez**

Scientific Name — Allium cepa L.

Family — Alliaceae

Herb / Shrub / Tree — Herb / 20-30 cm height

Annual / Biennial / Perennial — Biennial (Cultivated as annual)

Used as — Pungent bulb is used as vegetable / medicine / spice

1. *Climatic requirements* — Medium temperature with dry weather
2. *Soil requirements* — Loamy / alluvial soils / black soils
3. *Required land situations* — Flat lands of plains / hills
4. *Importance of crop* — Major vegetable, bulb remains in soils
5. *Fertilizer management* — N, P_2O_5 and K_2O @ 100, 50 and 50 kg/ha respectively
6. *Water management* — 4-5 irrigations at 20 days interval
7. *Duration of crop* — 4-5 months
8. *Parts used* — Bulbs, inflorescence with-stick and leaves used as vegetables
9. *Where grown in India* — All over India except too hot and too cold climate
10. *Export possibility* — Very high, as such, with proper packaging
11. *Economic yields (q / ha)* — 150-200 (green)
12. *Economic value (Rs. / ha)* — 40,000 to 60,000
13. *By-product (If any)* — Inflorescence stick and leaves
14. *Use of by-product* — As vegetable / cattle feed

35. *CROP* — **GARLIC / Loosun / Lasun / *Rasun* / Lahsun**

Scientific Name — Allium sativum L.

Family — Alliaceae

Herb / Shrub / Tree — Herb / 15-20 cm height

Annual / Biennial / Perennial — Annual

Used as — Pungent bulb used as vegetable / medicine / spice

1. *Climatic requirements* — Low or medium temperature with dry weather
2. *Soil requirements* — Loamy / alluvial soils / black soils
3. *Required land situations* — Flat lands of plains and hills of lower height
4. *Importance of crop* — Major condiments / spice / vegetable / medicine
5. *Fertilizer management* — N, P_2O_5 and K_2O @ 80, 50 and 50 kg/ha respectively
6. *Water management* — 3-4 irrigations at 25 days after sowing
7. *Duration of crop* — 4-5 months.
8. *Parts used* — Bulbs used as spice / condiment / medicine / leaves as vegetables or cattle feed

9. *Where grown in India* — All over the country except too high or too low temperature
10. *Export possibility* — Very very high, with proper packaging and processing
11. *Economic yields (q / ha)* — 50-80 (green)
12. *Economic value (Rs. / ha)* — 60,000 to 90,000
13. *By-product (If any)* — Leaf
14. *Use of by-product* — Vegetable / mulch / fuel medicine

36. *CROP* — **TURNIPS / Shaljam / Shalgam**

Scientific Name — Brassica rapa L.

Family — Cruciferae

Herb / Shrub / Tree — Herb (30-40 cm height)

Annual / Biennial / Perennial — Annual

Used as — Swollen root as vegetable

1. *Climatic requirements* — Low or medium temperature with dry climate
2. *Soil requirements* — Loamy / alluvial soils
3. *Required land situations* — Flat lands in plains or low hills
4. *Importance of crop* — Minor vegetable / small round-ball above soil
5. *Fertilizer management* — N, P_2O_5 and K_2O kg/ha @ 60, 40 and 40 kg/ha respectively
6. *Water management* — 3-4 irrigations, 25 days interval
7. *Duration of crop* — 4-5 months.
8. *Parts used* — Swollen roots as vegetable, foliage as cattle feed / manure
9. *Where grown in India* — All over India during winter season
10. *Export possibility* — High
11. *Economic yields (q / ha)* — 100-300 (green)
12. *Economic value (Rs. / ha)* — 20,000 to 40,000
13. *By-product (If any)* — Foliage
14. *Use of by-product* — Cattle feed / manure / fuel / vegetable.

37. *CROP* — **AGATI SESBANIA / *Bakful* / Basna / Agastoya**

Scientific Name — Sesbania grandiflora (L.) Poir

Family — Papilionaceae

Herb / Shrub / Tree — Tree, small quick growing

Annual / Biennial / Perennial — Perennial

Used as — Flower / tender fruits used as vegetable, tree fodder

1. *Climatic requirements* — Medium temperature with wet conditions
2. *Soil requirements* — Loamy / alluvial soils
3. *Required land situations* — Flat lands in plains and low hills
4. *Importance of crop* — Minor (flower as vegetable) vegetable
5. *Fertilizer management* — N, P_2O_5 and K_2O @ 50, 60 and 30 kg/ha respectively
6. *Water management* — 2-3 irrigations in dry season of the year
7. *Duration of crop* — 3-5 years
8. *Parts used* — Flower and tender fruit as vegetables, Leaf as manure, wood as stick (match) preparation
9. *Where grown in India* — Where mild temperature prevails
10. *Export possibility* — Less
11. *Economic yields (q / ha)* — 15-20 (Flower)
12. *Economic value (Rs. / ha)* — (i) Flower 5,000 to 10,000 (ii) Wood 5,000 to 10,000 Total 15,000 to 30,000
13. *By-product (If any)* — Leaf and wood
14. *Use of by-product* — Manures / fuel / match box preparation / tree fodder

38. *CROP* — **BROCCOLE / Borecole / Kale /** ***Karam sag***

Scientific Name — Brassica oleracea L. var. ***acephala*** D.C.

Family — Cruciferae

Herb / Shrub / Tree — Herb /Height 30-40 cm

Annual / Biennial / Perennial — Annual for vegetable, biennial for seed

Used as — Vegetable

1. *Climatic requirements* — Low or medium temperature with dry weather
2. *Soil requirements* — Loamy / alluvial soil
3. *Required land situations* — Flat land of plains and low hills
4. *Importance of crop* — Minor vegetable
5. *Fertilizer management* — N, P_2O_5 and K_2O @ 80, 40 and 40 kg/ha respectively
6. *Water management* — 3-4 irrigations, 25 days intervals
7. *Duration of crop* — 4-5 months.

8. *Parts used* — Tender inflorescence / bud for vegetable / stock portion for cattle feed
9. *Where grown in India* — All over the country except too hot or too cold areas
10. *Export possibility* — Very high, with packaging and processing
11. *Economic yields (q / ha)* — 200-300 (green)
12. *Economic value (Rs. / ha)* — 20,000 to 40,000
13. *By-product (If any)* — Stock
14. *Use of by-product* — Cattle feed / manure

39. *CROP* — **BRUSSELS SPROUTS / Bud bearing cabbage / buttamgobhi / *Botam-kapi***

Scientific Name — Brassica oleraceae L. var. ***gemmifera*** D.C.

Family — Cruciferae

Herb / Shrub / Tree — Herb / 40-50 cm height

Annual / Biennial / Perennial — Annual (for sprouts)/Biennial (for seed)

Used as — Vegetable

1. *Climatic requirements* — Low or medium temperature with dry weather
2. *Soil requirements* — Loamy / alluvial soils
3. *Required land situations* — Flat lands in plains and low hills
4. *Importance of crop* — Minor vegetable (Head and Sprouts)
5. *Fertilizer management* — N, P_2O_5 and K_2O @ 100, 50 and 50 kg/ha respectively
6. *Water management* — 3-4 irrigations, at 25 days interval
7. *Duration of crop* — 4-5 months.
8. *Parts used* — Head and Sprouts as vegetable
9. *Where grown in India* — All over the country except extreme hot and cold areas
10. *Export possibility* — Very high, if packaging done properly
11. *Economic yields (q / ha)* — 200-300 (green)
12. *Economic value (Rs. / ha)* — 40,000 to 50,000
13. *By-product (If any)* — Stock and leaves
14. *Use of by-product* — Cattle feed / organic manure / mulch / fuel.

40. *CROP* — **MOTH BEAN / Moth / Mot / Bhringga**

Scientific Name — Vigna aconitifolia Marechal

Family — Papilionaceae

Herb / Shrub / Tree — Herb (vine)

Annual / Biennial / Perennial — Annual

Used as — Pod as vegetable, Vine for fodder

1. *Climatic requirements* — High hot, humid conditions
2. *Soil requirements* — Loamy / alluvial / sandy soils
3. *Required land situations* — High land in plains
4. *Importance of crop* — Vine can be used as green manure crop
5. *Fertilizer management* — N, P_2O_5 and K_2O @ 40, 60 and 40 kg/ha respectively
6. *Water management* — Normally rainfed, 1-2 times irrigations, if irrigated
7. *Duration of crop* — 110-130 days
8. *Parts used* — Pod as vegetable and vine as fodder
9. *Where grown in India* — All over India except too hot and too cold areas
10. *Export possibility* — Less
11. *Economic yields (q / ha)* — 100-200 (green)
12. *Economic value (Rs. / ha)* — 10,000 to 20,000
13. *By-product (If any)* — Twigs, stalk and leaves
14. *Use of by-product* — Cattle feed / organic matter / fuel.

41. *CROP* — CHINEESE CABBAGE / Shantung cabbage / *China kapi*

Scientific Name — Brassica pekinensis Ruper

Family — Cruciferae

Herb / Shrub / Tree — Herb / 30-40 cm height

Annual / Biennial / Perennial — Annual (for head) / Biennial (for seed)

Used as — Vegetable

1. *Climatic requirements* — Low and medium temperature with dry weather
2. *Soil requirements* — Loamy / alluvial soils
3. *Required land situations* — Flat land in plains or low hills
4. *Importance of crop* — Minor vegetable (small round head)
5. *Fertilizer management* — N, P_2O_5 and K_2O @ 100, 50 and 50 kg/ha respectively
6. *Water management* — 3-4 irrigations, at 25 days interval
7. *Duration of crop* — 4 – 5 months

8. *Parts used* — Head for vegetable and stock for cattle feed
9. *Where grown in India* — All over the country with moderate winter
10. *Export possibility* — High, if properly packed
11. *Economic yields (q / ha)* — 200-300 (green)
12. *Economic value (Rs. / ha)* — 30,000 to 40,000
13. *By-product (If any)* — Stock
14. *Use of by-product* — Cattle feed / organic matter / fuel / mulch.

42. *CROP* — **GARDEN PEAS / Mator / *Karaisuti***

Scientific Name* — *Pisum sativum L. var. ***arvense*** L.

Family — Papilionaceae

Herb / Shrub / Tree — Herb / 60 to 80 length and 50 cm height

Annual / Biennial / Perennial — Annual

Used as — Green pod is used as vegetable, dry seed for pulse

1. *Climatic requirements* — Low or medium temperature with dry weather
2. *Soil requirements* — Loamy / alluvial soils
3. *Required land situations* — Flat land in plains and lower hills
4. *Importance of crop* — Minor vegetable / trailing type
5. *Fertilizer management* — N, P_2O_5 and K_2O @ 40, 80 and 40 kg/ha respectively
6. *Water management* — 3-4 irrigations, at 25 days interval
7. *Duration of crop* — 4 – 5 months
8. *Parts used* — Tender pod / seeds as vegetable. Plant as cattle feed
9. *Where grown in India* — All over India with moderate winter
10. *Export possibility* — Very high, with package
11. *Economic yields (q / ha)* — 200-300 (green)
12. *Economic value (Rs. / ha)* — 40,000 to 50,000
13. *By-product (If any)* — Leaf and stem
14. *Use of by-product* — Cattle feed / manure / organic matter / mulch.

43. *CROP* — **FRENCH BEAN / Kidney bean / Vilaiti sim / *Rajma* / Bin**

Scientific Name* — *Phaseolus vulgaris L.

Family — Papilionaceae

Herb / Shrub / Tree — Herb / Height 40-60 cm

Annual / Biennial / Perennial — Annual

Used as — Green pod as vegetable / seed as pulse

1. *Climatic requirements* — Low and medium temperature with dry weather
2. *Soil requirements* — Loamy / alluvial soils
3. *Required land situations* — Flat land of plains and low hills
4. *Importance of crop* — Major vegetable; plant bushy
5. *Fertilizer management* — N, P_2O_5 and K_2O @ 80, 80 and 40 kg/ha respectively
6. *Water management* — 3-4 irrigations, at 25 days interval
7. *Duration of crop* — 4 – 5 months
8. *Parts used* — Green pod as vegetable, plant as cattle feed, seed as pulse
9. *Where grown in India* — All over the country where *rabi* winter season prevails
10. *Export possibility* — High, if properly packed
11. *Economic yields (q / ha)* — 200-300 (green)
12. *Economic value (Rs. / ha)* — 30,000 to 40,000
13. *By-product (If any)* — Leaves and stock (plains)
14. *Use of by-product* — Cattle feed / manure / fuel / mulch.

44. *CROP* **— CLUSTER BEAN /** ***GUAR*** **/ Gawar /** ***Goar***

Scientific Name — Cyamopsis tetragonoloba L. Taub.

Family — Papilionaceae

Herb / Shrub / Tree — Herb / 70-80 cm height

Annual / Biennial / Perennial — Annual

Used as — Green pod as vegetable, plants as green manure

1. *Climatic requirements* — Medium temperature with dry weather
2. *Soil requirements* — Loamy / alluvial soils
3. *Required land situations* — Flat lands in plains and low hills
4. *Importance of crop* — Minor vegetable / green manure crop
5. *Fertilizer management* — N, P_2O_5 and K_2O @ 40, 80 and 40 kg/ha respectively
6. *Water management* — 2-3 irrigations, at 30 days interval or in rainfed areas
7. *Duration of crop* — 4 – 5 months
8. *Parts used* — Green pod as vegetable and plants (leaves and plants) as cattle feed / green manure

9. *Where grown in India* — All over the country except too hot or too low temperature areas
10. *Export possibility* — High, if properly packed
11. *Economic yields (q / ha)* — 100-200 (green)
12. *Economic value (Rs. / ha)* — 20,000 to 30,000
13. *By-product (If any)* — Leaves and stems
14. *Use of by-product* — Cattle feed / green manure / fuel / mulch.

45. *CROP* — **LABLAB BEAN / SEEM / Hyacinth bean / Sem / Sim**

Scientific Name* — *Lablab vulgaris L. / ***Dolichos lablab*** L.

Family — Papilionaceae

Herb / Shrub / Tree — Herb / 2-3 m. length and 50 cm height

Annual / Biennial / Perennial — Annual

Used as — Green pod as vegetable, stem and leaves cattle feed

1. *Climatic requirements* — Low or medium temperature with dry weather
2. *Soil requirements* — Loamy / alluvial soils
3. *Required land situations* — Flat land of plains and low hills
4. *Importance of crop* — Minor vegetable in winter season
5. *Fertilizer management* — N, P_2O_5 and K_2O @ 40, 80 and 50 kg/ha respectively
6. *Water management* — 3-4 irrigations, at 30 days interval
7. *Duration of crop* — 4 – 5 months
8. *Parts used* — Green pod as vegetable and stem and leaves as cattle feed
9. *Where grown in India* — All over the country, where *rabi* (winter) season prevails
10. *Export possibility* — High, if properly packed
11. *Economic yields (q / ha)* — 100-200 (green pod)
12. *Economic value (Rs. / ha)* — 20,000 to 30,000
13. *By-product (If any)* — Steams and leaves
14. *Use of by-product* — Cattle feed / green manure / fuel / mulch.

46. *CROP* — **PIG WEED / *BATHUA* / Lambs – quarters / *Bethosag***

Scientific Name* — *Chenopodium album L.

Family — Chenopodiaceae

Herb / Shrub / Tree — Herb / Height 60-70 cm

Annual / Biennial / Perennial — Annual

Used as — Leaf as vegetable

1. *Climatic requirements* — Low or medium temperature with dry weather
2. *Soil requirements* — Loamy / alluvial soils
3. *Required land situations* — Flat land of plains and low hills
4. *Importance of crop* — Minor leafy vegetable
5. *Fertilizer management* — N, P_2O_5 and K_2O @ 60, 60 and 30 kg/ha respectively
6. *Water management* — 3-4 irrigations at 30 days interval
7. *Duration of crop* — 4 – 5 months
8. *Parts used* — Leaf as vegetable and stalk as cattle feed
9. *Where grown in India* — All over the country whenever winter season prevails
10. *Export possibility* — Less
11. *Economic yields (q / ha)* — 100-200 (green)
12. *Economic value (Rs. / ha)* — 10,000 to 20,000
13. *By-product (If any)* — Steams and lower leaves
14. *Use of by-product* — Cattle feed / green manure / fuel / mulch.

8. FODDER CROPS (GRASSY)

47. *CROP* — **MAIZE / CORN / Indian Corn / Makai /** ***Bhutta***

Scientific Name — Zea mays L.

Family — Gramineae

Herb / Shrub / Tree — Herb / Height 2-3 m.

Annual / Biennial / Perennial — Annual

Used as — Seeds / cob as food / food products (flankes), stock-cattle feed

1. *Climatic requirements* — All climatic conditions
2. *Soil requirements* — Loamy / alluvial soils
3. *Required land situations* — Flat land of plains and low hills
4. *Importance of crop* — Grain or dough cob; silk and tursel present
5. *Fertilizer management* — N, P_2O_5 and K_2O @ 120, 60 and 60 kg/ha respectively
6. *Water management* — 4-5 irrigations, at 20 days interval
7. *Duration of crop* — 4 – 5 months
8. *Parts used* — Seeds, cobs, stock and leaves

9. *Where grown in India* — All over the country except too hot or too cold areas
10. *Export possibility* — High after proper processing or cobs or food products
11. *Economic yields (q / ha)* — 30-40 (grain) / 100-150 (cob)
12. *Economic value (Rs. / ha)* — 50,000 to 80,000
13. *By-product (If any)* — Stalk and leaves
14. *Use of by-product* — Cattle feed or used for other creeper crops.

48. *CROP* — **OAT / Jai / Ot**

Scientific Name* — *Avena sativa L.

Family — Gramineae

Herb / Shrub / Tree — Herb / 50-60cm Height

Annual / Biennial / Perennial — Annual

Used as — Plant as fodder / Seed as food for man and animal

1. *Climatic requirements* — Low or medium temperature with dry weather
2. *Soil requirements* — Loamy / alluvial soils
3. *Required land situations* — Flat lands in plains and low hills
4. *Importance of crop* — Main fodder / long ear head
5. *Fertilizer management* — N, P_2O_5 and K_2O @ 80, 50 and 50 kg/ha respectively
6. *Water management* — 1-2 irrigation at 25 days intervals. It can also be raised as rainfed crop
7. *Duration of crop* — 4-5 months
8. *Parts used* — Plant as fodder and seed as human consumption or cattle feed
9. *Where grown in India* — All over the country where *rabi* (winter)seasons prevails
10. *Export possibility* — High, if processed properly
11. *Economic yields (q / ha)* — 20-25
12. *Economic value (Rs. / ha)* — 15,000 to 25,000
13. *By-product (If any)* — No by-product, if cut as fodder
14. *Use of by-product* — When seed is taken, by product is stover and used as fodder.

49. *CROP* — **TEOSINTE / *Ban bhutta***

Scientific Name* — *Euchlaena mexicana Schrad or ***Zea mexicana*** Reeves

Family — Gramineae

Herb / Shrub / Tree — Herb / Height 2-3 m.

Annual / Biennial / Perennial — Annual

Used as — Fodder

1. *Climatic requirements* — Medium / High temperature with high humidity and rainfall
2. *Soil requirements* — Loamy / alluvial soils
3. *Required land situations* — Flat lands in plains and low hills
4. *Importance of crop* — Minor fodder (seeds scattered in cobs)
5. *Fertilizer management* — N, P_2O_5 and K_2O @ 80, 40 and 40 kg/ha respectively
6. *Water management* — 1-2 irrigations, when soil is dry
7. *Duration of crop* — 4-5 months
8. *Parts used* — Entire plant as fodder
9. *Where grown in India* — All over the country except too dry and too cold areas
10. *Export possibility* — Less
11. *Economic yields (q / ha)* — 200-300
12. *Economic value (Rs. / ha)* — 10,000 to 20,000
13. *By-product (If any)* — No by-product, if cut for fodder
14. *Use of by-product* — Nil.

50. *CROP* — **BARLEY / Jau / Jav / Jab**

Scientific Name — *Hordeum vulgare* L., emend. Bowden

Family — Gramineae

Herb / Shrub / Tree — Herb / Height 60-70 cm.

Annual / Biennial / Perennial — Annual

Used as — Fodder / food / organic matter / fuel / mulch

1. *Climatic requirements* — Low or medium temperature with dry weather
2. *Soil requirements* — Loamy / alluvial soils / sandy
3. *Required land situations* — Flat lands in plains and low hills
4. *Importance of crop* — Fodder (entire plant); can be grown in saline tract
5. *Fertilizer management* — N, P_2O_5 and K_2O @ 60, 30 and 30 kg/ha respectively
6. *Water management* — 1-2 irrigations before / after flowering. This is grown also under rainfed condition

7. *Duration of crop* — 4-5 months
8. *Parts used* — Entire plant for fodder
9. *Where grown in India* — Can be grown during winter season
10. *Export possibility* — Less
11. *Economic yields (q / ha)* — 200-300
12. *Economic value (Rs. / ha)* — 15,000 to 25,000
13. *By-product (If any)* — Nil
14. *Use of by-product* — Nil.

NON-GRASSY

51. *CROP* — PURPLE VETCH / Lal hetka

Scientific Name — Vicia benghalensis L.

Family — Papilionaceae

Herb / Shrub / Tree — Herb (50-80 cm length, but 50 cm height)

Annual / Biennial / Perennial — Annual (creeper)

Used as — Pod as vegetable / plant as fodder

1. *Climatic requirements* — High humid condition, low temperature with moderate rainfall
2. *Soil requirements* — Loamy / alluvial / sandy soils
3. *Required land situations* — High / medium / low lands
4. *Importance of crop* — This may be used as fodder also; Flower purple
5. *Fertilizer management* — N, P_2O_5 and K_2O @ 20, 40 and 40 kg/ha respectively
6. *Water management* — Normally rainfed
7. *Duration of crop* — 110-130 days
8. *Parts used* — Pod for vegetable (may be used as fodder / green manure crop)
9. *Where grown in India* — Assam, W.B., Bihar, U.P., Orissa etc.
10. *Export possibility* — Less
11. *Economic yields (q / ha)* — 100-200 (green)
12. *Economic value (Rs. / ha)* — 10,000 to 20,000
13. *By-product (If any)* — Stalk /leaves
14. *Use of by-product* — Cattle feed / organic matter / fuel / mulch

52. *CROP* — **GRASSPEA / *KHESARI* / Chickling pea / Jarosse**

***Scientific Name* — *Lathyrus sativus* L.**

Family — Papilionaceae

Herb / Shrub / Tree — Herb, 80-90 cm length but 50 cm height

Annual / Biennial / Perennial — Annual

Used as — Fodder / pulse / green manure / cover crop

1. *Climatic requirements* — Low or medium temperature with dry weather
2. *Soil requirements* — Clayey, loamy soils or sandy soils
3. *Required land situations* — Flat lands of plains or foothills
4. *Importance of crop* — Winter fodder with blue / red flower
5. *Fertilizer management* — N, P_2O_5 and K_2O @ 20, 40 and 40 kg/ha respectively (Practically no fertilizer needed)
6. *Water management* — Rainfed crop – no irrigation needed
7. *Duration of crop* — 4–5 months
8. *Parts used* — Entire plant as fodder / green manure
9. *Where grown in India* — All over the country except too hot or too cold areas
10. *Export possibility* — Less
11. *Economic yields (q / ha)* — 10-15 (variety Normal 1 and 2)
12. *Economic value (Rs. / ha)* — 10,000 to 20,000
13. *By-product (If any)* — Nil
14. *Use of by-product* — Nil.

53. *CROP* — **GOBISARSON / BLACK MUSTARD / Kali rai / Kalosarse**

***Scientific Name* — *Brassica nigra* L. Koch**

Family — Cruciferae

Herb / Shrub / Tree — Herb / Height 60-80 cm

Annual / Biennial / Perennial — Annual / Biennial

Used as — Cattle feed / *Sag*

1. *Climatic requirements* — Medium temperature with humidity
2. *Soil requirements* — Loamy / alluvial soils
3. *Required land situations* — Flat lands in plains and food hills
4. *Importance of crop* — Cattle feed (Minor) / fodder
5. *Fertilizer management* — N, P_2O_5 and K_2O @ 80, 40 and 40 kg/ha respectively
6. *Water management* — 2 irrigations before and after flowering

7. *Duration of crop* — 4–5 months
8. *Parts used* — Leaf and stem for cattle feed
9. *Where grown in India* — All over the country except too hot and too cold areas
10. *Export possibility* — Less
11. *Economic yields (q / ha)* — 200-250 (green)
12. *Economic value (Rs. / ha)* — 20,000 to 30,000
13. *By-product (If any)* — Nil
14. *Use of by-product* — Nil.

54. *CROP* — **BERSEEM / EGYPTIAN CLOVER / *Barsim***

Scientific Name* — *Trifolium alexandrinum L.

Family — Papilionaceae

Herb / Shrub / Tree — Herb / Height 50-60 cm

Annual / Biennial / Perennial — Annual

Used as — Fodder / green manure / cover crop

1. *Climatic requirements* — Low or medium temperature with dry weather
2. *Soil requirements* — Loamy / Alluvial soils
3. *Required land situations* — Flat land of plains and foot hills
4. *Importance of crop* — Major fodder
5. *Fertilizer management* — N, P_2O_5 and K_2O @ 40, 80 and 60 kg/ha respectively
6. *Water management* — 3-4 irrigations, at 25 days interval
7. *Duration of crop* — 4–5 months
8. *Parts used* — Entire plants for fodder / Hay
9. *Where grown in India* — All over the country except too hot and too cold areas
10. *Export possibility* — Less (seed – high)
11. *Economic yields (q / ha)* — 200-300 (green)
12. *Economic value (Rs. / ha)* — 25,000 to 35,000
13. *By-product (If any)* — Nil
14. *Use of by-product* — Nil.

55. *CROP* — **LUCERNE / ALFALFA / Vilaiti-gawuth / Hol**

Scientific Name* — *Medicago sativa L.

Family — Papilionaceae

Herb / Shrub / Tree — Herb / Height 60-70 cm

Annual / Biennial / Perennial — Perennial

Used as — Fodder / Hay

1. *Climatic requirements* — Low or medium temperature with dry weather
2. *Soil requirements* — Loamy / clayey soils
3. *Required land situations* — Flat land of plains and foot hills
4. *Importance of crop* — Major fodder crop
5. *Fertilizer management* — N, P_2O_5 and K_2O @ 40, 80 and 80 kg/ha respectively
6. *Water management* — 3-4 irrigations, at 25 days interval
7. *Duration of crop* — 4–5 months
8. *Parts used* — Entire plant as fodder
9. *Where grown in India* — All over the country where *rabi* (winter) season prevails
10. *Export possibility* — Less
11. *Economic yields (q / ha)* — 250-300 (Green)
12. *Economic value (Rs. / ha)* — 25,000 to 40,000
13. *By-product (If any)* — Nil
14. *Use of by-product* — Nil.

56. *CROP* — **SENJI / SWEETCLOVER / Banmethi**

Scientific Name — Melilotus indica L.

Family — Papilionaceae

Herb / Shrub / Tree — Herb / Height 60-70 cm

Annual / Biennial / Perennial — Annual

Used as — Green fodder

1. *Climatic requirements* — Low or medium temperature with dry weather
2. *Soil requirements* — Loamy / clayey / alluvial soils
3. *Required land situations* — Flat land of plains and foot hills
4. *Importance of crop* — A prominent weed in *Lathyrus* crop
5. *Fertilizer management* — N, P_2O_5 and K_2O @ 40, 80 and 80 kg/ha respectively
6. *Water management* — 2-3 irrigations, at 20 days interval
7. *Duration of crop* — 4–5 months
8. *Parts used* — Entire plant for fodder
9. *Where grown in India* — All over the country whenever winter season prevails

10. *Export possibility* — Less
11. *Economic yields (q / ha)* — 200-250 (green)
12. *Economic value (Rs. / ha)* — 25,000 to 30,000
13. *By-product (If any)* — Nil
14. *Use of by-product* — Nil.

57. *CROP* — **LABLAB BEAN / HYACINTH BEAN / Sem / *Sim***

Scientific Name* — *Lablab purpureus L. / ***Dolichos lablab*** L.

Family — Papilionaceae

Herb / Shrub / Tree — Twining herb / 3-4 m. length, 50 cm height (Creeper)

Annual / Biennial / Perennial — Annual

Used as — Plant as fodder

1. *Climatic requirements* — Low or medium temperature with dry weather
2. *Soil requirements* — Loamy / alluvial soil
3. *Required land situations* — Flat lands in plains and foot hills
4. *Importance of crop* — Winter fodder
5. *Fertilizer management* — N, P_2O_5 and K_2O @ 40, 80 and 40 kg/ha respectively
6. *Water management* — 3-4 irrigations, at 20 days interval, it can also be grown as rainfed crop
7. *Duration of crop* — 4–5 months
8. *Parts used* — Pods as vegetable and other parts as cattle feed / manure
9. *Where grown in India* — All over the country where winter prevails
10. *Export possibility* — High, if properly packed
11. *Economic yields (q / ha)* — 100-200 (green)
12. *Economic value (Rs. / ha)* — 25,000 to 40,000
13. *By-product (If any)* — Nil
14. *Use of by-product* — Nil.

58. *CROP* — **BORECOLE / KALE MUSTARD / KARAMSAG / *Bara Sarisha***

Scientific Name* — *Brassica oleraceae L. var. ***acephala*** D.C

Family — Cruciferae

Herb / Shrub / Tree — Herb / Height 70-80cm

Annual / Biennial / Perennial — Annual / Biennial

Used as — Fodder

1. *Climatic requirements* — Low or medium temperature with dry weather
2. *Soil requirements* — Loamy / alluvial soils
3. *Required land situations* — Flat land in plains and foot hills
4. *Importance of crop* — Minor fodder
5. *Fertilizer management* — N, P_2O_5 and K_2O @ 100, 50 and 50 kg/ha respectively
6. *Water management* — 3-4 irrigations, at 20 days interval
7. *Duration of crop* — 4–5 months
8. *Parts used* — Fodder
9. *Where grown in India* — All over the country whenever winter prevails
10. *Export possibility* — Less
11. *Economic yields (q / ha)* — 200-300 (green)
12. *Economic value (Rs. / ha)* — 15,000 to 20,000
13. *By-product (If any)* — Nil
14. *Use of by-product* — Nil.

59. *CROP* — **TURNIP / SHALJAM / *Shalgam***

Scientific Name* — *Brassica rapa L. var. ***rapifera*** Metz.

Family — Cruciferae

Herb / Shrub / Tree — Herb / Height 30-40 cm

Annual / Biennial / Perennial — Annual / Biennial

Used as — Fodder / vegetable

1. *Climatic requirements* — Low or medium temperature with dry weather
2. *Soil requirements* — Loamy / alluvial soils
3. *Required land situations* — Flat lands in plains or foot hills
4. *Importance of crop* — Winter fodder (small ball-like)
5. *Fertilizer management* — N, P_2O_5 and K_2O @ 100, 50 and 50 kg/ha respectively
6. *Water management* — 3-4 irrigations
7. *Duration of crop* — 4–5 months
8. *Parts used* — Entire plant as fodder
9. *Where grown in India* — All over India where winter prevails

10. *Export possibility* — Less
11. *Economic yields (q / ha)* — 100-200 (Green)
12. *Economic value (Rs. / ha)* — 10,000 to 25,000
13. *By-product (If any)* — Entire plant
14. *Use of by-product* — Nil.

9. GREEN MANURE CROPS

60. *CROP* — **GRASS PEA / Chikling pea / *Khesari***

Scientific Name* — *Lathyrus sativus L.

Family — Papilionaceae

Herb / Shrub / Tree — Herb, 50-90 cm length, but height 50 cm

Annual / Biennial / Perennial — Annual

Used as — Green manure crop / fodder / cover crop

1. *Climatic requirements* — Low or medium temperature with dry weather
2. *Soil requirements* — Loamy or clayey soils
3. *Required land situations* — Flat lands of plains or foot hills
4. *Importance of crop* — Minor green manure / fodder crop
5. *Fertilizer management* — N, P_2O_5 and K_2O @ 0, 40 and 40 kg/ha; as rainfed *paira* crop, N, P_2O_5 and K_2O not needed
6. *Water management* — Rainfed
7. *Duration of crop* — 4–5 months
8. *Parts used* — Entire plant as green manure / fodder
9. *Where grown in India* — All over the country whenever *rabi* (winter) season is there (except too hot / too cold areas)
10. *Export possibility* — Less
11. *Economic yields (q / ha)* — 200-250 (green)
12. *Economic value (Rs. / ha)* — 20,000 to 80,000
13. *By-product (If any)* — Nil
14. *Use of by-product* — Nil.

61. *CROP* — **SENJI / WHITE CLOVER / Huban clover**

Scientific Name* — *Melilotus alba Medik

Family — Papilionaceae

Herb / Shrub / Tree — Herb / Height 60-80 cm

Annual / Biennial / Perennial — Annual

Used as — Green manure (in winter) / fodder

1. *Climatic requirements* — Low or medium temperature in dry weather
2. *Soil requirements* — Loamy / dry alluvial soil
3. *Required land situations* — Flat lands of plains and foot hills
4. *Importance of crop* — Minor green manure / fodder crop
5. *Fertilizer management* — N, P_2O_5 and K_2O @ 20, 80 and 60 kg/ha respectively
6. *Water management* — Normally irrigation is not given
7. *Duration of crop* — 4–5 months
8. *Parts used* — Entire plant
9. *Where grown in India* — All over the country where winter season prevails
10. *Export possibility* — Less
11. *Economic yields (q / ha)* — 100-200 (Green)
12. *Economic value (Rs. / ha)* — 10,000 to 20,000
13. *By-product (If any)* — Nil
14. *Use of by-product* — Nil.

NON-LEGUMES

62. *CROP* — **OAT / Jai**

Scientific Name — Avena sativa L.

Family — Gramineae

Herb / Shrub / Tree — Herb / Height 6-80 cm

Annual / Biennial / Perennial — Annual

Used as — Green manure / fodder / food grain

1. *Climatic requirements* — Low or medium temperature with dry weather
2. *Soil requirements* — Loamy / alluvial soils
3. *Required land situations* — Flat lands of plains and low hills
4. *Importance of crop* — Minor green manure crop / fodder crop
5. *Fertilizer management* — Fertilizer uneconomical
6. *Water management* — Normally rainfed
7. *Duration of crop* — 4–5 months
8. *Parts used* — Entire plant as green manure / fodder

9. *Where grown in India* — All over the country where winter prevails
10. *Export possibility* — Less
11. *Economic yields (q / ha)* — 100-150 (green)
12. *Economic value (Rs. / ha)* — 10,000 to 15,000
13. *By-product (If any)* — Nil
14. *Use of by-product* — Nil.

63. *CROP* — **BARLEY / Jau / Jav**

Scientific Name — Hordeum vulgare L. emend. Bowden

Family — Gramineae

Herb / Shrub / Tree — Herb / Height 60-80 cm

Annual / Biennial / Perennial — Annual

Used as — Fodder / grain crop

1. *Climatic requirements* — Low or medium temperature with dry weather
2. *Soil requirements* — Loamy / clay / alluvial soils
3. *Required land situations* — Flat land of plains or low hills
4. *Importance of crop* — Suitable crop for saline soils.
5. *Fertilizer management* — Fertilizer is not economical (50, 30 and 30 kg/ha of N, P_2O_5 and K_2O)
6. *Water management* — Rainfed crop
7. *Duration of crop* — 4–5 months
8. *Parts used* — Entire plant
9. *Where grown in India* — All over the country whenever winter season prevails
10. *Export possibility* — Less
11. *Economic yields (q / ha)* — 100-200 (green)
12. *Economic value (Rs. / ha)* — 20,000 to 25,000
13. *By-product (If any)* — Nil
14. *Use of by-product* — Nil.

10. MEDICINAL PLANTS

64. *CROP* — **OPIUM POPPY / POST / Afim / Posto**

Scientific Name — Papaver somniferum L.

Family — Papaveraceae

Herb / Shrub / Tree — Herb / Height 60-70 cm

Annual / Biennial / Perennial — Annual

Used as — Medicinal plant with good flavour

1. *Climatic requirements* — Low or medium temperature with dry weather
2. *Soil requirements* — Sandy loam / loamy / alluvial soils
3. *Required land situations* — Flat lands of plains or foot hills
4. *Importance of crop* — Major medicinal plant with good flavour
5. *Fertilizer management* — N, P_2O_5 and K_2O @ 80, 80 and 40 kg/ha respectively
6. *Water management* — 3-4 irrigations at 25 days interval
7. *Duration of crop* — 4–5 months
8. *Parts used* — Mature fruits posto / lalex (afim)
9. *Where grown in India* — Most of mild / strong winter regions of the country
10. *Export possibility* — Very high with processing
11. *Economic yields (q / ha)* — 1 to 2
12. *Economic value (Rs. / ha)* — 40,000 to 60,000
13. *By-product (If any)* — Entire plant
14. *Use of by-product* — Cattle feed / fuel / mulch / organic matter

65. *CROP* — **ERGOT (Dried *Sclerotium* in Rye crop)**

Scientific Name — Claviceps purpurea Tul

Family — Clavicipitaceae

Herb / Shrub / Tree — Fungus parasite on Rye grass or other grasses

Annual / Biennial / Perennial — Annual

Used as — Medicine

1. *Climatic requirements* — Humid condition with low temperature
2. *Soil requirements* — Harbour on grass particularly rye or triticale
3. *Required land situations* — High hills
4. *Importance of crop* — Parasite fungus on rye / triticale
5. *Fertilizer management* — 5 gm N, 5 gm P_2O_5 and 5gm K_2O per square metre for the harbour plants
6. *Water management* — Water spraying every 7 days on plants
7. *Duration of crop* — 100-120 days
8. *Parts used* — *Sclerotium* (dried)
9. *Where grown in India* — Hilly stations of India

10. *Export possibility* — Very high
11. *Economic yields (q / ha)* — 10-20 kg/ha
12. *Economic value (Rs. / ha)* — 40,000 to 80,000
13. *By-product (If any)* — Plant on it grows
14. *Use of by-product* — Organic matter / fuel / mulch (kept for next year)

66. *CROP* — **DIOSCOREA / WHITE YAM / *KHAMALU* / Asiatic yam**

Scientific Name* — *Dioscorea alata L.

Family — Dioscoreaceae

Herb / Shrub / Tree — Climbing shrub / length 10-15m

Annual / Biennial / Perennial — Perennial

Used as — Tubers / bulbil are used as medicine / vegetables

1. *Climatic requirements* — Medium temperature with humidity and rainfall
2. *Soil requirements* — Loamy / alluvial soils
3. *Required land situations* — Flat lands of plains or foot hills
4. *Importance of crop* — Medicinal / vegetable / Vines
5. *Fertilizer management* — N, P_2O_5 and K_2O @ 100, 50 and 50 g/plant annually, during monsoon respectively
6. *Water management* — 1-2 irrigations annually
7. *Duration of crop* — 2-3 years
8. *Parts used* — Tubers inside the soils are used as medicine / vegetable and plant for cattle feed. Bulbils on vine as vegetable
9. *Where grown in India* — In most of the states where rainfall is high
10. *Export possibility* — Very high, if processed properly
11. *Economic yields (q / ha)* — 100-200 (green)
12. *Economic value (Rs. / ha)* — 50,000 to 70,000
13. *By-product (If any)* — Entire plants
14. *Use of by-product* — Cattle feed / mulch / organic matter / fuel.

67. *CROP* — **BLONDE PSYLLIUM / *ISOPHGOL* / PSYLLIUM BLACK/ Black Isophgol**

Scientific Name* — *Plantago ovata Rosk / ***P. psyllium*** (L)

Family — Plantaginaceae

Herb / Shrub / Tree — Herb / 40-60 cm height

Annual / Biennial / Perennial — Annual

Used as — Dry seeds as medicine (for constipation / dysentery)

1. *Climatic requirements* — Low temperature but dry condition
2. *Soil requirements* — Sandy loam / loam / alluvial soils
3. *Required land situations* — Flat lands of plains or foot hills
4. *Importance of crop* — Medicine / fodder
5. *Fertilizer management* — N, P_2O_5 and K_2O @ 100, 50 and 50 kg/ha respectively
6. *Water management* — 3-4 irrigations, 25 days interval
7. *Duration of crop* — 4-5 months
8. *Parts used* — Dry seeds and outer coating as medicine
9. *Where grown in India* — Most of the dry but low temperature (in winter) areas
10. *Export possibility* — Very high, if properly processed
11. *Economic yields (q / ha)* — 10-20
12. *Economic value (Rs. / ha)* — 30,000 to 40,000
13. *By-product (If any)* — Entire plant
14. *Use of by-product* — Cattle feed / fuel / organic matter / mulch

68. *CROP* — **BELLADONNA / ANGURSHEFA / Sag-angur /** ***Beladona***

Scientific Name — Atropa belladonna L.

Family — Solanaceae

Herb / Shrub / Tree — Herb / Height 60-80 cm

Annual / Biennial / Perennial — Annual

Used as — Roots and leaves as medicine

1. *Climatic requirements* — Low temperature with dry weather
2. *Soil requirements* — Loamy / gravelly soils with organic matter
3. *Required land situations* — High hills (plain lands)
4. *Importance of crop* — Medicinal crop (Ophthalmic treatments)
5. *Fertilizer management* — N, P_2O_5 and K_2O @ 80, 40 and 40 g/plant, during monsoon, respectively
6. *Water management* — 3-4 irrigations, 25 days interval
7. *Duration of crop* — 4-5 months
8. *Parts used* — Root and leaves for medicine
9. *Where grown in India* — Kashmir and other high hills
10. *Export possibility* — Very high, if properly processed
11. *Economic yields (q / ha)* — 50-70
12. *Economic value (Rs. / ha)* — 40,000 to 60,000

13. *By-product (If any)* — Plant stems
14. *Use of by-product* — Fuel / cattle feeds

69. *CROP* — **KING-OF-BITTERS / KALMEGH / KIRYAT / *Kalomegh***

Scientific Name — *Andrographis paniculata* Burm / ***Justicia paniculata*** Burm

Family — Acanthaceae

Herb / Shrub / Tree — Herb / Height 40-60 cm

Annual / Biennial / Perennial — Annual

Used as — Stem, leaves and inflorescence as medicines for worms

1. *Climatic requirements* — Medium or high temperature with humidity and rainfall
2. *Soil requirements* — Loam / alluvial soils
3. *Required land situations* — Flat lands of plains and foot hills
4. *Importance of crop* — Medicine plant, very bitter (leaf and stem)
5. *Fertilizer management* — N, P_2O_5 and K_2O @ 80, 40 and 40 kg/ha respectively
6. *Water management* — 3-4 irrigations, in dry weather
7. *Duration of crop* — 5-6 months
8. *Parts used* — Stem, leaves and inflorescence
9. *Where grown in India* — All over the country except too high or too low temperature
10. *Export possibility* — Very high, if processed properly
11. *Economic yields (q / ha)* — 100-200 (green)
12. *Economic value (Rs. / ha)* — 40,000 to 60,000
13. *By-product (If any)* — Nil
14. *Use of by-product* — Nil

70. *CROP* — **INDIAN DILL / *SOWA / SAJ* / SOYA**

Scientific Name — Anethum sowa Kurz

Family — Umbelliferae

Herb / Shrub / Tree — Herb / Height 50-80 cm

Annual / Biennial / Perennial — Annual

Used as — Leaf / seed as spice / medicine

1. *Climatic requirements* — Low / medium temperature with medium humidity and rainfall
2. *Soil requirements* — Loamy / alluvial soils
3. *Required land situations* — Flat lands of plains and foot hill

4. *Importance of crop* — Spice and medicine; seed emits smell
5. *Fertilizer management* — N, P_2O_5 and K_2O @ 80, 40 and 40 q/ha respectively
6. *Water management* — 1-2 irrigations at 40-50 days interval
7. *Duration of crop* — 5-6 months
8. *Parts used* — Leaves and seeds as medicine / spices
9. *Where grown in India* — Whenever medium temperature with humidity present in the country
10. *Export possibility* — Very high, if processed properly
11. *Economic yields (q / ha)* — 10-40
12. *Economic value (Rs. / ha)* — 30,000 to 50,000
13. *By-product (If any)* — Stem
14. *Use of by-product* — Cattle feed / fuel / organic matter / mulch

71. *CROP* — GARDEN CELERY / *SALARI* / AJMUD / KARAS

Scientific Name — Apium graveolens L.

Family — Umbelliferae

Herb / Shrub / Tree — Herb / Height 60-80 cm

Annual / Biennial / Perennial — Annual

Used as — Vegetable (salad) / spice / medicine

1. *Climatic requirements* — Low or medium temperature with dry weather
2. *Soil requirements* — Loamy soil with organic matter
3. *Required land situations* — Flat lands of hills
4. *Importance of crop* — Vegetable (salad)
5. *Fertilizer management* — N, P_2O_5 and K_2O @ 80, 40 and 40 q/ha respectively
6. *Water management* — 3-4 irrigations, at 25 days interval
7. *Duration of crop* — 4-5 months
8. *Parts used* — Leaf and stalk as vegetable, dry seed as spices / oil as perfumes
9. *Where grown in India* — In hilly regions, all over the country
10. *Export possibility* — Very high, if properly processed
11. *Economic yields (q / ha)* — 100-200 (green)
12. *Economic value (Rs. / ha)* — 30,000 to 50,000
13. *By-product (If any)* — Stems
14. *Use of by-product* — Cattle feed / fuel

11. SPICES

72. *CROP* — **BLACK CUMIN / *KALAJIRA* / Kalenji / Mungrela**

Scientific Name — Nigella sativa L

Family — Ranunculaceae

Herb / Shrub / Tree — Herb 50-60 cm height

Annual / Biennial / Perennial — Annual

Used as — Spice / medicine

1. *Climatic requirements* — Low or medium temperature with dry weather
2. *Soil requirements* — Loamy / alluvial soils
3. *Required land situations* — Flat lands of plains or foot hills
4. *Importance of crop* — Spice / medicine. Can be grown as *paira* crop
5. *Fertilizer management* — N, P_2O_5 and K_2O @ 60, 30 and 30 kg/ha respectively
6. *Water management* — 1-2 irrigations, at 40 days interval
7. *Duration of crop* — 4-5 months
8. *Parts used* — Seed as spice / medicine
9. *Where grown in India* — All over the country where winter season prevails
10. *Export possibility* — Very high, if properly processed
11. *Economic yields (q / ha)* — 5-10
12. *Economic value (Rs. / ha)* — 20,000 to 30,000
13. *By-product (If any)* — Stalk / stover
14. *Use of by-product* — Fuel / cattle feed / mulch / organic matter.

73. *CROP* — **CORIANDER / DHANIA/ Dhane**

Scientific Name — Coriandrum sativum (L.)

Family — Umbelliferae

Herb / Shrub / Tree — Herb / Height 50-70 cm

Annual / Biennial / Perennial — Annual

Used as — Spice / vegetable / medicine

1. *Climatic requirements* — Low or medium temperature with dry weather
2. *Soil requirements* — Loamy / alluvial soil
3. *Required land situations* — Flat lands in plains or foot hills
4. *Importance of crop* — Spice / medicine / condiment

5. *Fertilizer management* — N, P_2O_5 and K_2O @ 80, 40 and 40 kg/ha respectively
6. *Water management* — 1-2 irrigations, 40 days interval
7. *Duration of crop* — 4-5 months
8. *Parts used* — Seed as spice, leave and seeds as medicine
9. *Where grown in India* — All over the country where *rabi* (winter) season prevails
10. *Export possibility* — Very high, if properly processed
11. *Economic yields (q / ha)* — 10-15
12. *Economic value (Rs. / ha)* — 40,000 to 50,000
13. *By-product (If any)* — Dry stalk
14. *Use of by-product* — Cattle feed / fuel / mulch / organic matter

74. *CROP* — FENUGREEK / *METHI*

Scientific Name — Trigonella foenum - graecum (L)

Family — Papilionaceae

Herb / Shrub / Tree — Herb / 20 – 40 cm height

Annual / Biennial / Perennial — Annual

Used as — Seed as spice / medicine

1. *Climatic requirements* — Low or medium temperature with dry weather
2. *Soil requirements* — Loamy / alluvial soil
3. *Required land situations* — Flat lands of plains or foot hills
4. *Importance of crop* — Spice / vegetable. Smaller plant
5. *Fertilizer management* — N, P_2O_5 and K_2O @ 40, 80 and 40 kg/ha respectively
6. *Water management* — 3-4 irrigations, 25 days interval
7. *Duration of crop* — 4-5 months
8. *Parts used* — Seeds for spice, leaves as vegetable / fodder
9. *Where grown in India* — All over India except southern India
10. *Export possibility* — Very high, if properly processed
11. *Economic yields (q / ha)* — 5-10 (seeds)
12. *Economic value (Rs. / ha)* — 40,000 to 60,000
13. *By-product (If any)* — Dry stalk
14. *Use of by-product* — Cattle feed / fuel

75. *CROP* — **ONION / *PIYAZ* / Piaj**

Scientific Name* — *Allium cepa (L)

Family — Alliaceae

Herb / Shrub / Tree — Herb / Height 30-40 cm

Annual / Biennial / Perennial — Biennial /Annual

Used as — Bulb as spice / medicine, inflorescence as vegetable

1. *Climatic requirements* — Low / medium temperature with dry weather
2. *Soil requirements* — Loamy / sandy loam / alluvial soils
3. *Required land situations* — Flat land in plains / foot hills
4. *Importance of crop* — Spice / vegetable / medicine. Bulb white/red
5. *Fertilizer management* — N, P_2O_5 and K_2O @ 100, 50 and 50 kg/ha respectively
6. *Water management* — 3-4 irrigations, 40 days interval
7. *Duration of crop* — 4-5 months
8. *Parts used* — Bulb as spice / vegetable / medicine and leaves and inflorescence as vegetable
9. *Where grown in India* — Wherever mild winter season or low temperature prevails, this can be grown all over the country
10. *Export possibility* — Very high, as such or with proper processing
11. *Economic yields (q / ha)* — 100-200 (green)
12. *Economic value (Rs. / ha)* — 30,000 to 50,000
13. *By-product (If any)* — Nil
14. *Use of by-product* — Nil

76. *CROP* — **GARLIC / *LASUN* or *RASUN* /Lahsan / *Rosun***

Scientific Name* — *Allium sativum (L.)

Family — Alliaceae

Herb / Shrub / Tree — Herb / 30-40 cm height

Annual / Biennial / Perennial — Annual / Biennial

Used as — Bulbs as spice/medicine and leaves and stem as vegetable

1. *Climatic requirements* — Low or medium temperature with dry weather
2. *Soil requirements* — Loamy / sandy loam / alluvial soils
3. *Required land situations* — Flat land in plains / foot hills
4. *Importance of crop* — Spice / vegetable / medicine. Bulb white
5. *Fertilizer management* — N, P_2O_5 and K_2O @ 80, 40 and 40 kg/ha respectively

6. *Water management* — 3-4 irrigations, 25 days interval
7. *Duration of crop* — 4-5 months
8. *Parts used* — Bulbs as spice / medicine and leaves and stem as vegetable
9. *Where grown in India* — All over the country where winter prevails
10. *Export possibility* — Very high, if properly processed
11. *Economic yields (q / ha)* — 100-200 (green)
12. *Economic value (Rs. / ha)* — 30,000 to 50,000
13. *By-product (If any)* — Nil
14. *Use of by-product* — Nil

77. *CROP* — **FENNEL / *Saunf* / Sonp / *Mauri***

Scientific Name* — *Foeniculum vulgare Mill

Family — Umbelliferae

Herb / Shrub / Tree — Herb / Height 50-140 cm

Annual / Biennial / Perennial — Annual / biennial

Used as — Seeds as spice / medicine

1. *Climatic requirements* — Low or medium temperature with dry weather
2. *Soil requirements* — Loamy / alluvial soils
3. *Required land situations* — Flat land in plains / foot hills
4. *Importance of crop* — Spice crop (smell in seed and leaf)
5. *Fertilizer management* — N, P_2O_5 and K_2O @ 80, 40 and 40 kg/ha respectively
6. *Water management* — 3-4 irrigations, 20/25 days interval
7. *Duration of crop* — 5-6 months
8. *Parts used* — Seed as spice / medicine, leaves as vegetable and stalk as fodder
9. *Where grown in India* — All over the country where winter season prevails
10. *Export possibility* — Very high, if properly processed
11. *Economic yields (q / ha)* — 5-10
12. *Economic value (Rs. / ha)* — 40,000 to 60,000 (All)
13. *By-product (If any)* — Stalk
14. *Use of by-product* — Cattle feed / fuel / mulching

78. *CROP* — **RAPE and MUSTARD / *TORIA* and *RAI* /*Sarisa***

Scientific Name — Brassia campestris L. var. ***toria* / *B. juncea*** L.

Family — Cruciferae

Herb / Shrub / Tree — Herb / Height 40-80 cm

Annual / Biennial / Perennial — Annual

Used as — Seeds as spice

1. *Climatic requirements* — Low or medium temperature with dry weather
2. *Soil requirements* — Loamy / alluvial soils
3. *Required land situations* — Flat land in plains and foot hills
4. *Importance of crop* — Oil crop / spice. Yellow flower
5. *Fertilizer management* — N, P_2O_5 and K_2O @ 100, 60 and 60 kg/ha respectively
6. *Water management* — 1-2 irrigations, before or after flowering
7. *Duration of crop* — 4-5 months
8. *Parts used* — Seeds as oil / seeds, leaves as vegetable / spice
9. *Where grown in India* — All over the country where winter season prevails
10. *Export possibility* — Very high, if properly processed
11. *Economic yields (q / ha)* — 10-20
12. *Economic value (Rs. / ha)* — 40,000 to 60,000
13. *By-product (If any)* — Leaves / stalk
14. *Use of by-product* — Cattle feed / fuel / organic matter

79. *CROP* — **CHILLIES / LAL MIRCHA / Red pepper / *Marich* / *Lanka***

Scientific Name — Capsicum annuum L.

Family — Solanaceae

Herb / Shrub / Tree — Herb / Height 40-90 cm

Annual / Biennial / Perennial — Perennial / Annual

Used as — Fruit (green and ripe) as spice / condiments / medicine / vegetable

1. *Climatic requirements* — All types of climatic conditions
2. *Soil requirements* — All types of soils
3. *Required land situations* — Both in plains and hills
4. *Importance of crop* — Spice crop (green / ripe). Fruit very bitter / pungent

5. *Fertilizer management* — N, P_2O_5 and K_2O @ 50, 100, 25-50 and 25-50 kg/ha respectively
6. *Water management* — 3-4 irrigations, 20-25 days interval; rainfed crop in some areas
7. *Duration of crop* — All the year round (1-3 years)
8. *Parts used* — Fruits (green and ripe) as spice
9. *Where grown in India* — All over the country, but intensity varies
10. *Export possibility* — Very high, if properly processed
11. *Economic yields (q / ha)* — 100-200
12. *Economic value (Rs. / ha)* — 50,000 to 60,000
13. *By-product (If any)* — Stalk (with leaves)
14. *Use of by-product* — Fuel / organic matter

12. FRUITS (ANNUAL)

80. *CROP* — **CUCUMBER / *KHIRA / SHASA***

Scientific Name* — *Cucumis sativus L.

Family — Cucurbitaceae

Herb/Shrub/Tree — Herb (trailing, length 304 m but height 50-60 cm)

Annual / Biennial / Perennial — Annual

Used as — Fruits / vegetables

1. *Climatic requirements* — High temperature with high humidity and rainfall
2. *Soil requirements* — Loamy / alluvial soils
3. *Required land situations* — Flat land in plains / foot hills
4. *Importance of crop* — Fruit crop, vine type
5. *Fertilizer management* — N, P_2O_5 and K_2O @ 100, 50 and 50 kg/ha respectively
6. *Water management* — 1-2 irrigations at flowering and after flowering. It can also be raised as rainfed crop in high rainfall areas.
7. *Duration of crop* — 4-5 months
8. *Parts used* — Fruit / vegetable
9. *Where grown in India* — All over India except too low and too high temperature
10. *Export possibility* — High, if properly processed
11. *Economic yields (q / ha)* — 100-200 (green)
12. *Economic value (Rs. / ha)* — 20,000 to 40,000

13. *By-product (If any)* — Foliage and vine
14. *Use of by-product* — Cattle feed / fuel / mulching / organic matter

81. *CROP* — **BANANA / KELA or KALA / Kadali / Rambha**

Scientific Name* — *Musa paradisiaca L. / ***M. sapientum*** L.

Family — Musaceae

Herb / Shrub / Tree — Shrub / Height 2-4 m

Annual / Biennial / Perennial — Annual

Used as — Fruits / vegetables

1. *Climatic requirements* — High temperature with high humidity and high rainfall
2. *Soil requirements* — Loamy / alluvial soils
3. *Required land situations* — Undulated / flatland of plains or foot hills
4. *Importance of crop* — Succers (for propagation) come out of it
5. *Fertilizer management* — N, P_2O_5 and K_2O @ 120, 60 and 80 kg/ha respectively
6. *Water management* — 3-4 irrigations, at 20-25 days interval
7. *Duration of crop* — 8-12 months
8. *Parts used* — Fruits as fruit / green fruit as vegetable / stem for medicine, leaves as plate
9. *Where grown in India* — All over the country where humid condition prevails with high rainfall and high temperature
10. *Export possibility* — Very high, if properly processed
11. *Economic yields (q / ha)* — 200-300 (green)
12. *Economic value (Rs. / ha)* — 40,000 to 60,000
13. *By-product (If any)* — Stem / leaves
14. *Use of by-product* — Cattle feed / fuel / plate / vegetable

82. *CROP* — **MUSKMELON / Kharbuja / Kharmaj**

Scientific Name* — *Cucumis melo L.

Family — Cucurbitaceae

Herb / Shrub / Tree — Herb /length 304 m; height 50-60 cm (trailing type)

Annual / Biennial / Perennial — Annual

Used as — Fruits

1. *Climatic requirements* — High temperature with dry weather
2. *Soil requirements* — Loamy / sandy loam / alluvial / sandy soil

3. *Required land situations* — Flat lands in plains / foot hills
4. *Importance of crop* — Fruits (in dry period)
5. *Fertilizer management* — N, P_2O_5 and K_2O @ 80, 40 and 40 kg/ha respectively
6. *Water management* — 1-2 irrigations, before and after flowering
7. *Duration of crop* — 4-5 months
8. *Parts used* — Fruit as fruits
9. *Where grown in India* — All over the country in hot areas
10. *Export possibility* — High, if properly packed
11. *Economic yields (q / ha)* — 100-200 (green)
12. *Economic value (Rs. / ha)* — 20,000 to 40,000
13. *By-product (If any)* — Foliage and vine
14. *Use of by-product* — Cattle feed / fuel / mulch.

83. *CROP* — **WATERMELON / *Tarmuz / Tarmuj* / Tarbooz**

Scientific Name* — *Citrullus lanatus (Thunb) Mansf.

Family — Cucurbitaceae

Herb / Shrub / Tree — Herb / length 3-4 m; height 60 cm (trailing)

Annual / Biennial / Perennial — Annual

Used as — Fruits

1. *Climatic requirements* — Hot and dry weather
2. *Soil requirements* — Sandy loam / loam / alluvial soils
3. *Required land situations* — Flat land of plains (sea and river beds)
4. *Importance of crop* — Fruits (big), Plant vine type
5. *Fertilizer management* — N, P_2O_5 and K_2O @ 50, 30 and 30 kg/ha respectively
6. *Water management* — 1-2 irrigations, before and after flowering. It is raised in rainfed condition also.
7. *Duration of crop* — 4-5 months
8. *Parts used* — Fruit for fruits
9. *Where grown in India* — Where temperature is high, dry condition prevails
10. *Export possibility* — High, if properly packed
11. *Economic yields (q / ha)* — 200-300 (green)
12. *Economic value (Rs. / ha)* — 30,000 to 40,000
13. *By-product (If any)* — Foliage and vine
14. *Use of by-product* — Cattle feed / fuel / mulch / organic matter

84. *CROP* — **STRAW BERRY (Wild tomato) / Winter cherry**

Scientific Name — Physalis alkekengi L.

Family — Solanaceae

Herb / Shrub / Tree — Herb (vine / bush)

Annual / Biennial / Perennial — Annual

Used as — Fruits / medicine (used as urinary and skin disease)

1. *Climatic requirements* — Dry low temperature between 5°C and 16°C
2. *Soil requirements* — Clayey loam / sandy loam / loamy soils
3. *Required land situations* — High / medium plains / foot hills
4. *Importance of crop* — Fruits are diuretic, containing Vitamin C
5. *Fertilizer management* — 100 kg N, 50 kg P_2O_5 and 50 kg K_2O/ha
6. *Water management* — 2 or 3 irrigations, 6-7 cm depth of water
7. *Duration of crop* — 3-4 months
8. *Parts used* — Fruits
9. *Where grown in India* — Commercial cultivation in Maharashtra, Karnataka etc.
10. *Export possibility* — High, if proper processed
11. *Economic yields (q / ha)* — 60-70 (green)
12. *Economic value (Rs. / ha)* — 50,000 to 65,000
13. *By-product (If any)* — Haulm / vine
14. *Use of by-product* — For cattle feed / mulch material / organic matter

85. *CROP* — **CHERIMOYA / Hanumanphal / *Lakshman phal***

Scientific Name — Annona cherimolia Mill

Family — Annonaceae

Herb / Shrub / Tree — Small tree

Annual / Biennial / Perennial — Perennial

Used as — Fruits

1. *Climatic requirements* — Tropical climate (high temperature low rainfall, dry condition)
2. *Soil requirements* — All types of soils except heavy clay
3. *Required land situations* — High lands in plains / foot hills
4. *Importance of crop* — Fruits are edible
5. *Fertilizer management* — 80 kg N, 40 kg P_2O_5 and 40 kg K_2O/ha
6. *Water management* — Under natural rainfed condition

7. *Duration of crop* — Perennial, and hence 4 to 5 years
8. *Parts used* — Fruits for consumption, branches for fencing or fuel
9. *Where grown in India* — Mostly in Maharashtra, Tamil Nadu, Karnataka etc.
10. *Export possibility* — Less
11. *Economic yields (q / ha)* — 70-80 (green fruits)
12. *Economic value (Rs. / ha)* — 7,000 to 10,000
13. *By-product (If any)* — Branches and leaves
14. *Use of by-product* — Fencing / fuel

13. FLOWERS / SUCCULENTS / ORNAMENTALS

86. *CROP* — **MARI GOLD /** ***Gada*** **/ Geda /Big marigold / French marigold**

Scientific Name — Tagetes patula L. / ***T. erecta*** L.

Family — Compositae

Herb / Shrub / Tree — Herb / 30-40 cm height

Annual / Biennial / Perennial — Annual

Used as — Flower / poultry feed colouration

1. *Climatic requirements* — Dry winter with temperature 5-16°C
2. *Soil requirements* — Loamy / alluvial soils
3. *Required land situations* — High / medium
4. *Importance of crop* — Harvesting flowers for mixing in poultry feeds
5. *Fertilizer management* — 120kg N, 50 kg P_2O_5 and 50 kg K_2O/ha
6. *Water management* — 2 or 3 irrigations
7. *Duration of crop* — 3-4 months
8. *Parts used* — Flower – heads as aesthetic value / functions
9. *Where grown in India* — In all the states of India where temperature is low
10. *Export possibility* — High, if packed properly
11. *Economic yields (q / ha)* — 100-200 (green)
12. *Economic value (Rs. / ha)* — 20,000 to 60,000
13. *By-product (If any)* — Root / shoot / seeds
14. *Use of by-product* — Root, shoot and seeds used as purgative / plant as organic matter / fuel.

87. *CROP* — **CHERRY LAUREL / PRUNUS**

Scientific Name — Prunus laurocerasus L.

Family — Rosaceae

Herb / Shrub / Tree — Small tree

Annual / Biennial / Perennial — Perennial

Used as — Ornamental / medicine

1. *Climatic requirements* — Moderate temperature with moderate rainfall
2. *Soil requirements* — Loamy / sandy soils
3. *Required land situations* — High land
4. *Importance of crop* — Can be used as medicinal plant also
5. *Fertilizer management* — 60kg N, 30 kg P_2O_5 and 30 kg K_2O/ha
6. *Water management* — Normally rainfed; 1-2 Irrigations / annually if irrigated
7. *Duration of crop* — 2-3 years
8. *Parts used* — Leaves and plants
9. *Where grown in India* — In the hilly regions of India
10. *Export possibility* — High, if packed properly
11. *Economic yields (q / ha)* — For aesthetic value
12. *Economic value (Rs. / ha)* — Nil
13. *By-product (If any)* — Stem
14. *Use of by-product* — Timber / fuel / organic matter / fuel.

88. *CROP* — **ASTER / MICHAELMAS**

Scientific Name — Michaelmas daisies

Family — Compositae

Herb / Shrub / Tree — Herb 30-40 cm height

Annual / Biennial / Perennial — Annual

Used as — Aesthetic (flower) / ornamentals

1. *Climatic requirements* — Low temperature but dry conditions
2. *Soil requirements* — Loamy / alluvial soils
3. *Required land situations* — High land
4. *Importance of crop* — Aesthetic values increase
5. *Fertilizer management* — 60kg N, 60 kg P_2O_5 and 60 kg K_2O/ha
6. *Water management* — Frequent irrigations about 12 times
7. *Duration of crop* — 120 days (seed to seed)
8. *Parts used* — Flower

9. *Where grown in India* — Northern India where low temperature prevails
10. *Export possibility* — High, if packed properly
11. *Economic yields (q / ha)* — 120 (fresh flower)
12. *Economic value (Rs. / ha)* — 12,000/ha
13. *By-product (If any)* — Plant material / seeds
14. *Use of by-product* — Organic matter / seeds / fuel.

89. *CROP* — **GLADIOLUS / *Gladiolus***

Scientific Name* — *Gladiolus orientalis L.

Family — Iridaceae

Herb / Shrub / Tree — Herb, 50-80 cm height

Annual / Biennial / Perennial — Annual

Used as — Ornamental

1. *Climatic requirements* — Moderate temperature with moderate rainfall
2. *Soil requirements* — Loamy / sandy / alluvial soils
3. *Required land situations* — High land in plains
4. *Importance of crop* — Flower is used for decoration / function
5. *Fertilizer management* — 60kg N, 60 kg P_2O_5 and 60 kg K_2O/ha
6. *Water management* — Frequent irrigation nearly 12 times
7. *Duration of crop* — 120 days
8. *Parts used* — Flower and stalk
9. *Where grown in India* — All over India except southern states
10. *Export possibility* — Very high
11. *Economic yields (q / ha)* — 100-120 (fresh)
12. *Economic value (Rs. / ha)* — 1,20,000 to 1,50,000
13. *By-product (If any)* — Leaves and stalk
14. *Use of by-product* — Organic matter / fuel / mulch.

90. *CROP* — **PANSY / Heart's ease / *Pansi***

Scientific Name* — *Viola tricolor L.

Family — Violaceae

Herb / Shrub / Tree — Herb, Height 40 – 60 cm

Annual / Biennial / Perennial — Annual

Used as — Flower for beauty / ornamental / aesthetic value

1. *Climatic requirements* — Low or medium temperature with dry weather
2. *Soil requirements* — Loamy / alluvial soils
3. *Required land situations* — Flat lands of hills / foot hill / plains
4. *Importance of crop* — Flower (crop)spotted
5. *Fertilizer management* — N, P_2O_5 and K_2O @ 50, 50 & 30 kg/ha, respectively
6. *Water management* — 3-4 irrigations, at 20/25 days interval
7. *Duration of crop* — 4-5 months
8. *Parts used* — Flower for beauty / aesthetic sense
9. *Where grown in India* — All over the country where winter season prevails
10. *Export possibility* — High, for flower, if properly processed
11. *Economic yields (q / ha)* — 5-10
12. *Economic value (Rs. / ha)* — 25,000 to 50,000
13. *By-product (If any)* — Foliage
14. *Use of by-product* — Cattle feed / mulch / fuel.

91. *CROP* — **DAHLIA**

Scientific Name — Mexican genus

Family — Compositae

Herb / Shrub / Tree — Herb, 50-80 cm height

Annual / Biennial / Perennial — Annual

Used as — Flower (for aesthetic value)

1. *Climatic requirements* — Dry winter season
2. *Soil requirements* — Loamy / alluvial soils
3. *Required land situations* — High land
4. *Importance of crop* — Big bright flower
5. *Fertilizer management* — 60kg N, 60 kg P_2O_5 and 60 kg K_2O/ha
6. *Water management* — Frequent irrigation, 10-12 times
7. *Duration of crop* — 120 days
8. *Parts used* — Flower
9. *Where grown in India* — Northern parts of India
10. *Export possibility* — High, if packed properly under low temperature conditions
11. *Economic yields (q / ha)* — Aesthetic value
12. *Economic value (Rs. / ha)* — One lakh/ha

13. *By-product (If any)* — Plant materials
14. *Use of by-product* — Organic matter, fuel / mulch.

92. *CROP* — NASTERTIUM / Indian cress

Scientific Name* — *Tropaeolum majus L.

Family — Tropaeolaceae

Herb / Shrub / Tree — Herb

Annual / Biennial / Perennial — Annual

Used as — Flower / ornamental (flower like butterfly)

1. *Climatic requirements* — Cold climatic conditions
2. *Soil requirements* — Sandy / loamy / alluvial soils
3. *Required land situations* — High land
4. *Importance of crop* — Climbing in nature; small plant
5. *Fertilizer management* — 100kg N, 50 kg P_2O_5 and 50 kg K_2O/ha
6. *Water management* — Irrigation 6-7 times
7. *Duration of crop* — 120-150 days
8. *Parts used* — Flower
9. *Where grown in India* — All over the country where cold climate is prevails
10. *Export possibility* — High, if packed properly
11. *Economic yields (q / ha)* — 50-60
12. *Economic value (Rs. / ha)* — 30,000–50,000
13. *By-product (If any)* — Vines and leaves
14. *Use of by-product* — Cattle feed / fuel / organic matter

93. *CROP* — SANSEVIERA / Indian bowstring hemp / JENSIVERA / Jenseveram / Murahri / Mura / Ceylon bowstring hemp / Murva

Scientific Name — ***Sanseviera hyacinthoides*** (L) Willd / ***S. roxburghiana*** Schutt.f.

Family — Agavaceae

Herb / Shrub / Tree — Herb

Annual / Biennial / Perennial — Perennial

Used as — Plant / leaf beauty / ornamental

1. *Climatic requirements* — Dry tropical climate
2. *Soil requirements* — Sandy / loamy / sandy loam soils
3. *Required land situations* — High land
4. *Importance of crop* — Leaf long stripped, beautiful

5. *Fertilizer management* — 60kg N, 30 kg P_2O_5 and 20 kg K_2O/ha
6. *Water management* — Irrigation 10-12 times annually
7. *Duration of crop* — 3-4 years
8. *Parts used* — Plants / leaves
9. *Where grown in India* — All areas except too low or too high temperature regions
10. *Export possibility* — High
11. *Economic yields (q / ha)* — 150-200 (fresh)
12. *Economic value (Rs. / ha)* — 50,000 annually / ha
13. *By-product (If any)* — Nil
14. *Use of by-product* — Nil

94. *CROP* — **CROTON / Chuka / Patabahar / Bahari pata**

Scientific Name — Croton roxburghii Bal / ***C. oblongifolius*** Roxb.

Family — Euphorbiaceae

Herb / Shrub / Tree — Shrub

Annual / Biennial / Perennial — Perennial

Used as — Ornamental plant

1. *Climatic requirements* — Moderate warm climate
2. *Soil requirements* — Sandy / loam / alluvial soils
3. *Required land situations* — High land
4. *Importance of crop* — Variagated leaves / different shape and coloured
5. *Fertilizer management* — 100kg N, 50 kg P_2O_5 and 50 kg K_2O/ha
6. *Water management* — 7-8 times irrigations annually
7. *Duration of crop* — 4-5 days
8. *Parts used* — Aesthetic aspects
9. *Where grown in India* — All over India, except too hot/cold temperature area
10. *Export possibility* — Less
11. *Economic yields (q / ha)* — Nil
12. *Economic value (Rs. / ha)* — Nil
13. *By-product (If any)* — Branches
14. *Use of by-product* — Cattle feed / fuel

95. *CROP* — DRACONTIUM / Danab / Barakachu

Scientific Name* — *Dracontium polyphyllum L.

Family — Araceae

Herb / Shrub / Tree — Herb / shrub

Annual / Biennial / Perennial — Perennial

Used as — Ornamental plant

1. *Climatic requirements* — Moderate warm conditions
2. *Soil requirements* — Loamy / sandy / alluvium soils
3. *Required land situations* — High land
4. *Importance of crop* — Long petioled leaves
5. *Fertilizer management* — 100kg N, 50 kg P_2O_5 and 50 kg K_2O/ha annually
6. *Water management* — 6-7 time irrigations annually
7. *Duration of crop* — 2-4 years
8. *Parts used* — Leaves (for decoration)
9. *Where grown in India* — All over India except too hot or too cold areas
10. *Export possibility* — High
11. *Economic yields (q / ha)* — Aesthetic aspects
12. *Economic value (Rs. / ha)* — Nil
13. *By-product (If any)* — Branches
14. *Use of by-product* — Fuel/ organic matter

96. *CROP* — PARAPALM / Assai palm / Chhototal

Scientific Name* — *Euterpe edulis Mart

Family — Palmae

Herb / Shrub / Tree — Small tree

Annual / Biennial / Perennial — Perennial

Used as — Ornamentals

1. *Climatic requirements* — Moderately cold climate
2. *Soil requirements* — Alluvial / loamy soils
3. *Required land situations* — High land
4. *Importance of crop* — Decoration. Leaf palmy
5. *Fertilizer management* — 80kg N, 60 kg P_2O_5 and 40 kg K_2O/ha annually
6. *Water management* — 2-3 times annually
7. *Duration of crop* — 2-3 years

8. *Parts used* — Leaves for decoration
9. *Where grown in India* — All over India except too low/too high temperature regions
10. *Export possibility* — High
11. *Economic yields (q / ha)* — Aesthetic value
12. *Economic value (Rs. / ha)* — Nil
13. *By-product (If any)* — Branches
14. *Use of by-product* — Fuel / organic matter / hand fan

97. *CROP* — GARDEN CHRYSANTHEMUM / Crown daisy / Guldaudi / Chrysanthemum

Scientific Name — Chrysanthemum / carinatum Schousb

Family — Compositae

Herb / Shrub / Tree — Herb, 50-70 cm height

Annual / Biennial / Perennial — Annual

Used as — Flower / ornamental

1. *Climatic requirements* — Dry winter
2. *Soil requirements* — Loamy soil or alluvial soils, saline soil
3. *Required land situations* — High / medium land
4. *Importance of crop* — For beautification / for flower ornamental
5. *Fertilizer management* — 60 kg N, 60 kg P_2O_5 and 40 kg K_2O/ha
6. *Water management* — Irrigation 5-6 times
7. *Duration of crop* — 120 days
8. *Parts used* — Flower
9. *Where grown in India* — All over India in cold climate
10. *Export possibility* — High if properly packed
11. *Economic yields (q / ha)* — 15 (flower)
12. *Economic value (Rs. / ha)* — 80,000
13. *By-product (If any)* — Plant materials and leaves
14. *Use of by-product* — Organic matter / fuel / mulch.

98. *CROP* — KOCHIA / Bui / Kauraro / Badum

Scientific Name — Kochia indica Wight

Family — Chenopodiaceae

Herb / Shrub / Tree — Herb

Annual / Biennial / Perennial — Annual

Used as — Ornamental plant / fodder

1. *Climatic requirements* — High warm climate
2. *Soil requirements* — Loamy / sandy / alluvial soils
3. *Required land situations* — High land
4. *Importance of crop* — A diffuse, much-branched bush
5. *Fertilizer management* — 100 kg N, 50 kg P_2O_5 and 50 kg K_2O/ha
6. *Water management* — 4-5 irrigations
7. *Duration of crop* — 120-150 days
8. *Parts used* — Showy, green doom shaped small plant
9. *Where grown in India* — All over India except snowy areas
10. *Export possibility* — High
11. *Economic yields (q / ha)* — Aesthetic value (for green fodder 150-200)
12. *Economic value (Rs. / ha)* — 30,000 to 50,000
13. *By-product (If any)* — Entire plant is used as green fodder
14. *Use of by-product* — Nil

99. *CROP* — TRADESCANTIA

Scientific Name — Tradescantia discolor L 'Herit

Family — Commelinaceae

Herb / Shrub / Tree — Herb

Annual / Biennial / Perennial — Perennial

Used as — Ornamental (red colour)

1. *Climatic requirements* — Moderate rainfall with moderate temperature
2. *Soil requirements* — Loamy / sandy loam / clayey soils
3. *Required land situations* — High
4. *Importance of crop* — Foliage show / ornamental
5. *Fertilizer management* — 100 kg N, 50 kg P_2O_5 and 50 kg K_2O/ha annually
6. *Water management* — 6-7 irrigations annually
7. *Duration of crop* — 2-3 years
8. *Parts used* — Foliage as ornamental plants
9. *Where grown in India* — All over in India in gardens, except very hot / snowy areas
10. *Export possibility* — High
11. *Economic yields (q / ha)* — Aesthetic value
12. *Economic value (Rs. / ha)* — 40,000

13. *By-product (If any)* — Nil
14. *Use of by-product* — Nil

100. *CROP* — YUCCA / *UKKA* / Spanish dagger plant

***Scientific Name* — *Yucca gloriosa* L.**

Family — Agavaceae

Herb / Shrub / Tree — Shrub

Annual / Biennial / Perennial — Perennial

Used as — Ornamental

1. *Climatic requirements* — Dry climatic conditions
2. *Soil requirements* — Sandy / loam / alluvial soils
3. *Required land situations* — High dry land
4. *Importance of crop* — Spiny leaves
5. *Fertilizer management* — 50 kg N, 30 kg P_2O_5 and 30 kg K_2O/ha annually
6. *Water management* — Rainfed
7. *Duration of crop* — 9-10 years
8. *Parts used* — Showy leaves
9. *Where grown in India* — All over dry areas
10. *Export possibility* — High
11. *Economic yields (q / ha)* — Aesthetic value
12. *Economic value (Rs. / ha)* — 40,000
13. *By-product (If any)* — Stem
14. *Use of by-product* — Fuel / organic matter

14. BEVERAGE AND NARCOTICS

101. *CROP* — TOBACCO / VILAITI TAMBAKU / *TAMAK* / Tamaku

***Scientific Name* — *Nicotiana rustica / tabacum* L.**

Family — Solanaceae

Herb / Shrub / Tree — Herb, 30-40 cm / 70-80 cm height

Annual / Biennial / Perennial — Annual

Used as — Narcotics

1. *Climatic requirements* — Dry low temperature
2. *Soil requirements* — Loamy / sandy / entisol / alluvial soils
3. *Required land situations* — High land
4. *Importance of crop* — Nicotine in the plant / *jarda* is produced

5. *Fertilizer management* — 50 kg N, 50 kg P_2O_5 and 50 kg K_2O/ha
6. *Water management* — 3-4 irrigations
7. *Duration of crop* — 110-130 days
8. *Parts used* — Leaves and stem
9. *Where grown in India* — West Bengal, Bihar, Uttar Pradesh, Himachal Pradesh, A.P., Rajasthan, Gujarat
10. *Export possibility* — Very high
11. *Economic yields (q / ha)* — 5-10 (Leaves)
12. *Economic value (Rs. / ha)* — 50,000 to 80,000
13. *By-product (If any)* — Stem
14. *Use of by-product* — Used for narcotics / decoction / *Khaini* / Nassi

102. *CROP* — OPIUM POPPY / POSTO / *AFIM*

Scientific Name — Papaver somniferum L.

Family — Papaveraceae

Herb / Shrub / Tree — Herb, 70-90 cm height

Annual / Biennial / Perennial — Annual

Used as — Narcotics / flower / stem / seed (*posto*)

1. *Climatic requirements* —Low temperature in dry areas
2. *Soil requirements* — Loamy / sandy / stony / alluvial soils
3. *Required land situations* — High land
4. *Importance of crop* — Source of opium (latex from immature stem)
5. *Fertilizer management* — 50 kg N, 50 kg P_2O_5 and 50 kg K_2O/ha
6. *Water management* — 3-4 irrigations
7. *Duration of crop* — 110-130 days
8. *Parts used* — Seed for opium / flower stem for *afim*
9. *Where grown in India* — H.P., Punjab, Rajasthan, Gujarat, Bihar, West Bengal, Assam, etc.
10. *Export possibility* — Very high
11. *Economic yields (q / ha)* — 50-80 kg/ha (*afim*); 5 q/ha seed
12. *Economic value (Rs. / ha)* — 50,000 to 1,00,000
13. *By-product (If any)* — Stalk
14. *Use of by-product* — Cattle feed / fuel / organic matter.

Chapter - 2

SUMMER (PRE-*KHARIF*) SEASON

INTRODUCTION

Chapter 2 deals with different crops in cereals, pulses, oil seeds, fibres, tubers / rhizomes, sugar crops, vegetables, fodders / forage, green manuring crops, medicinal plants, spices, fruits, flowers (including succulents and ornamentals), beverage and narcotics, in summer (pre-kharif) season.

In India, summer season varies from place to place. In the northern, central and western parts, summer is severe (temperature ranges from 33 to 46°C), whereas it is mild in the eastern and southern parts of the country. No crop could be raised without irrigation during this period, except the eastern part where early monsoon is available. Besides, in the eastern part frequent but small amount of rainfall is obtained.

So crop cultivation is almost impossible in all the regions, without irrigation except the eastern part. In eastern part due to frequent (though small amount) rainfall, some crops like jute *aus* paddy, sesame, *dhaincha* etc. could be grown without irrigation, but in other parts, it is not possible.

1. CEREALS

103. *CROP* — RICE / *dhan* / Paddy / Chauval / *Dhanya* / Chal / *Bhat*

Scientific Name* — *Oryza sativa L.

Family — Gramineae

Herb / Shrub / Tree — Herb, Height 60-80 cm.

Annual / Biennial / Perennial — Annual

Used as — Seed as food, plant (straw) as cattle feed

1. *Climatic requirements* — High temperature, humidity and high rainfall
2. *Soil requirements* — Clay / alluvial soils
3. *Required land situations* — Flat lands of plains / hills

4. *Importance of crop* — Food crop (Carbohydrate). Water stagnation needed.
5. *Fertilizer management* — N, P_2O_5 and K_2O @ 100, 50 and 50 kg/ha respectively
6. *Water management* — 3-4 irrigations in dry areas but no irrigation in high rainfall areas
7. *Duration of crop* — 4-5 months
8. *Parts used* — Seeds as food and straw as cattle feed
9. *Where grown in India* — All wet areas of the country where rainfall, humidity and temperature are high, with clayey soils
10. *Export possibility* — Very high, if processed, particularly Basmati scented rice
11. *Economic yields (q / ha)* — 25-30
12. *Economic value (Rs. / ha)* — 20,000 to 50,000 (If exported 80,000 – 1,50,000)
13. *By-product (If any)* — Straw
14. *Use of by-product* — Cattle feed / fuel / mulch / organic matter / roofing

104. *CROP* — MAIZE / CORN / *BHUTTA / Makai*

Scientific Name — Zea mays L.

Family — Gramineae

Herb / Shrub / Tree — Herb, Height 3-4 m.

Annual / Biennial / Perennial — Annual

Used as — Food / Processed food

1. *Climatic requirements* — Medium temperature with dry/wet weather.
2. *Soil requirements* — Loamy / alluvial soils
3. *Required land situations* — Flat lands of plains and foothill
4. *Importance of crop* — Cereal (Carbohydrate). Silk and tussel present
5. *Fertilizer management* — N, P_2O_5 and K_2O @ 120-60, 60-50 kg/ha
6. *Water management* — 3-4 irrigations at 25 days intervals
7. *Duration of crop* — 4-6 months
8. *Parts used* — Cobs / grains for food, stalk (with leaves) as cattle feed
9. *Where grown in India* — All over the country where medium temperature with irrigation facilities are available
10. *Export possibility* — Very high, with proper processing

11. *Economic yields (q / ha)* — 25-30
12. *Economic value (Rs. / ha)* — 20,000 to 60,000
13. *By-product (If any)* — Straw / Stalk
14. *Use of by-product* — Cattle feed / fuel / base for trailing crops / organic matter / pegs.

105. *CROP* — SORGHUM / JOWAR / JOAAR / Milo

Scientific Name* — *Sorghum vulgare Pers

Family — Gramineae

Herb / Shrub / Tree — Herb, Height 2-3 m

Annual / Biennial / Perennial — Annual

Used as — Food (Cereals) / cattle feed

1. *Climatic requirements* — High temperature with less humidity and rainfall
2. *Soil requirements* — Loamy / alluvial / sandy loam soils
3. *Required land situations* — Flat lands in plains / foothills
4. *Importance of crop* — Cereal crop. Poor man's food
5. *Fertilizer management* — N, P_2O_5 and K_2O @ 60, 30 and 30 kg/ha respectively
6. *Water management* — 1-2 irrigations, 25 days interval
7. *Duration of crop* — 4-5 months
8. *Parts used* — Grain/stalk (with leaves)
9. *Where grown in India* — Bihar, Punjab, Rajasthan, Gujarat, M.P., Maharashtra, A.P., Orissa, U.P., etc.
10. *Export possibility* — High
11. *Economic yields (q / ha)* — 15-20
12. *Economic value (Rs. / ha)* — 15,000 to 20,000
13. *By-product (If any)* — Stalks and leaves
14. *Use of by-product* — Cattle feed / fuel / organic matter / mulch.

106. *CROP* — PEARL MILLET / *BAJRA*

Scientific Name* — *Pennisetum typoides Stapf

Family — Gramineae

Herb / Shrub / Tree — Herb / Height 1-2m.

Annual / Biennial / Perennial — Annual

Used as — Cereal food / cattle feed

1. *Climatic requirements* — High temperature with low humidity and rainfall.

2. *Soil requirements* — Loamy / sandy loam / alluvial soils
3. *Required land situations* — Flat lands of plains and foothills.
4. *Importance of crop* — Cereal crops, grown in dry areas
5. *Fertilizer management* — N, P_2O_5 and K_2O @ 60, 30 and 30 kg/ha respectively
6. *Water management* — 1-2 irrigations before and after flowering
7. *Duration of crop* — 4-5 months.
8. *Parts used* — Seeds for food, stalk for cattle feed
9. *Where grown in India* — All over the country where temperature is high and humidity and rainfall are low
10. *Export possibility* — High, if properly processed
11. *Economic yields (q / ha)* — 10-12.
12. *Economic value (Rs. / ha)* — 15,000 to 25,000
13. *By-product (If any)* — Stalk
14. *Use of by-product* — Cattle feed / fuel / organic matter

107. *CROP* — LANATUM GRASS

Scientific Name — Pennisetum lanatum Klotzsch

Family — Gramineae

Herb / Shrub / Tree — Herb, 1-2 m. height

Annual / Biennial / Perennial — Annual

Used as — Food / fodder / soil binding

1. *Climatic requirements* — Cold climate
2. *Soil requirements* — Stony / loamy / sandy soil
3. *Required land situations* — High dry land
4. *Importance of crop* — Used for small grain, fodder and soil-binding
5. *Fertilizer management* — 100 kg N, 50 kg P_2O_5 and 50 kg K_2O/ha
6. *Water management* — 4-5 times
7. *Duration of crop* — 120-150 days
8. *Parts used* — Grains (small)
9. *Where grown in India* — In hilly areas of Kashmir, H.P., U.P. etc.
10. *Export possibility* — High
11. *Economic yields (q / ha)* — 15-20
12. *Economic value (Rs. / ha)* — 15,000 to 20,000
13. *By-product (If any)* — Straw
14. *Use of by-product* — Cattle feed / fuel / mulch / organic matter

108. *CROP* — FINGER MILLET / MARUA / *Ragi / Marudhan*

Scientific Name — Eleusine coracana L. Gaertn

Family — Gramineae

Herb / Shrub / Tree — Herb, Height 60-70 cm

Annual / Biennial / Perennial — Annual

Used as — Food (cereal)

1. *Climatic requirements* — Very hot climate
2. *Soil requirements* — Loamy / alluvial soils
3. *Required land situations* — Flat land in plains / foot hill
4. *Importance of crop* — Cereal (small grains)
5. *Fertilizer management* — N, P_2O_5 and K_2O @ 100, 60, 30 and 30 kg/ha respectively
6. *Water management* — 3-4 irrigations, 25 days interval
7. *Duration of crop* — 4-5 months.
8. *Parts used* — Grain for food, straw for cattle feed
9. *Where grown in India* — All over the country except too low / too high temperature areas
10. *Export possibility* — High, if properly processed
11. *Economic yields (q / ha)* — 15-20
12. *Economic value (Rs. / ha)* — 20,000 to 40,000
13. *By-product (If any)* — Straw
14. *Use of by-product* — Cattle feed / fuel / organic matter / mulch

109. *CROP* — ITALIAN MILLET / KANGI / *KAON / CATS TAIL*

Scientific Name — Setaria italica (L.) Beauv

Family — Gramineae

Herb / Shrub / Tree — Herb, Height 60-70 cm.

Annual / Biennial / Perennial — Annual

Used as — Cereal food

1. *Climatic requirements* — High temperature with humidity or dry condition
2. *Soil requirements* — Loamy / alluvial / sandy loam soils
3. *Required land situations* — Flat lands in plains
4. *Importance of crop* — Cereal crop. Inflorescence like cat's tail
5. *Fertilizer management* — N, P_2O_5 and K_2O @ 50, 30 and 30 kg/ha respectively
6. *Water management* — 1-2 irrigations, before and after flowering

7. *Duration of crop* — 4-5 months.
8. *Parts used* — Grain for food, straw for cattle feed
9. *Where grown in India* — Andhra Pradesh, Maharashtra, Karnataka, Bengal, Assam etc. in summer
10. *Export possibility* — High, if properly processed
11. *Economic yields (q / ha)* — 10-15
12. *Economic value (Rs. / ha)* — 20,000 to 30,000
13. *By-product (If any)* — Straw
14. *Use of by-product* — Cattle feed / fuel / mulch / organic mater.

110. *CROP* — BARNYARD MILLET / *SHYAMA / SAMA*

Scientific Name* — *Echinochloa colona (L.) / ***E. crusgalli*** (L.)

Family — Gramineae

Herb / Shrub / Tree — Herb, Height 60-90 cm

Annual / Biennial / Perennial — Annual

Used as — Food (cereal, millet) / fodder

1. *Climatic requirements* — High temperature with less humidity and rainfall
2. *Soil requirements* — Loamy / alluvial / sandy loam soils
3. *Required land situations* — Flat land of plains
4. *Importance of crop* — Cereal / fodder crop. Weed in rice
5. *Fertilizer management* — N, P_2O_5 and K_2O @ 60, 30 and 30 kg/ha respectively
6. *Water management* — 1-2 irrigations, before and after flowering
7. *Duration of crop* — 4-5 months.
8. *Parts used* — Grain as millet / straw as fodder
9. *Where grown in India* — Dry areas with less rainfall in the country
10. *Export possibility* — High, with proper processing
11. *Economic yields (q / ha)* — 10-20
12. *Economic value (Rs. / ha)* — 20,000 to 40,000
13. *By-product (If any)* — Straw
14. *Use of by-product* — Cattle feed / fuel / mulch / organic matter

111. *CROP* — KODO MILLET / *KODO*

Scientific Name* — *Paspalum scrobiculatum L.

Family — Gramineae

Herb / Shrub / Tree — Herb, Height 60-80 cm

Annual / Biennial / Perennial — Annual

Used as — Food / fodder

1. *Climatic requirements* — Low / medium / high temperature, dry condition with medium humidity
2. *Soil requirements* — Sandy / loamy / alluvial soils
3. *Required land situations* — Flat lands of plains and hills
4. *Importance of crop* — Cereal / fodder crop
5. *Fertilizer management* — N, P_2O_5 and K_2O @ 60, 30 and 30 kg/ha respectively
6. *Water management* — 1-2 irrigations, before and after flowering
7. *Duration of crop* — 4-5 months.
8. *Parts used* — Grain for food, straw for cattle feed
9. *Where grown in India* — All over the country except too high or too low temperature areas
10. *Export possibility* — High, if properly processed
11. *Economic yields (q / ha)* — 10-15
12. *Economic value (Rs. / ha)* — 20,000 to 30,000
13. *By-product (If any)* — Straw
14. *Use of by-product* — Cattle feed / fuel / mulch / organic matter.

112. *CROP* — COMMON MILLET / CHIN / MORHA / ANU / HOG MILLET / *China dhan*

Scientific Name* — *Panicum miliaceum L.

Family — Gramineae

Herb / Shrub / Tree — Herb, 50-80 cm height

Annual / Biennial / Perennial — Annual

Used as — Food / cattle feed

1. *Climatic requirements* — Medium temperature in rainy season
2. *Soil requirements* — Loamy to sandy soils
3. *Required land situations* — High land
4. *Importance of crop* — Grown in poor soils
5. *Fertilizer management* — 60 kg N, 25 kg P_2O_5 and 25 kg K_2O/ha
6. *Water management* — Grown under rainfed condition
7. *Duration of crop* — 100-120 months.
8. *Parts used* — Grain for food and straw for cattle
9. *Where grown in India* — Mostly in U.P. and Bihar, A.P., Karnataka, etc.

10. *Export possibility* — High
11. *Economic yields (q / ha)* — 10-12
12. *Economic value (Rs. / ha)* — 10,000 to 12,000
13. *By-product (If any)* — Straw
14. *Use of by-product* — Cattle feed / fuel / mulch / organic matter

113. *CROP* — LITTLE MILLET / Shavan / Kutki / Kungu

Scientific Name* — *Panicum sumatrense Roth. Ex Roem & Schult

Family — Gramineae

Herb / Shrub / Tree — Herb, 60-80 cm height

Annual / Biennial / Perennial — Annual

Used as — Grains for food / straw for cattle feed

1. *Climatic requirements* — Rainy condition with higher temperature
2. *Soil requirements* — Loamy / sandy soil
3. *Required land situations* — High land
4. *Importance of crop* — Grains for human; straw for animal consumption
5. *Fertilizer management* — 60 kg N, 30 kg P_2O_5 and 30 kg K_2O/ha
6. *Water management* — Grown under rainfed condition in rainy season
7. *Duration of crop* — 100-120 days
8. *Parts used* — Grain for human beings and straw for cattle
9. *Where grown in India* — U.P., M.P., A.P., Orissa, etc.
10. *Export possibility* — Less
11. *Economic yields (q / ha)* — 10-12
12. *Economic value (Rs. / ha)* — 10,000 to 12,000
13. *By-product (If any)* — Straw
14. *Use of by-product* — Cattle feed / fuel / organic matter / mulch

2. PULSES

114. *CROP* — GREEN GRAM / Mungbean / *Mug / Mugdal / Moong*

Scientific Name* — *Vigna radiata (L.) Wilczek

Family — Papilionaceae

Herb / Shrub / Tree — Herb, 40-50 cm / Vine

Annual / Biennial / Perennial — Annual

Used as — Grain as *dal* / green plant as fodder

1. *Climatic requirements* — Pre-*kharif* (summer), under rainfall
2. *Soil requirements* — Loamy / clayey / sandy soils
3. *Required land situations* — High land
4. *Importance of crop* — Grain and green plant. Nodule bacteria present
5. *Fertilizer management* — 40 kg N, 60 kg P_2O_5 and 40 kg K_2O
6. *Water management* — Mainly rainfed / If irrigated 1-2 irrigations
7. *Duration of crop* — 85-95 days
8. *Parts used* — Grain as *dal* (Pulse) / stover as fodder / manure
9. *Where grown in India* — All over India except too low or too hot temperature areas
10. *Export possibility* — High, if properly packed
11. *Economic yields (q / ha)* — 10-12
12. *Economic value (Rs. / ha)* — 20,000 to 30,000
13. *By-product (If any)* — Stalk
14. *Use of by-product* — Cattle feed / fuel / green manure / mulch.

115. *CROP* — COWPEA / Lobia / Chowli / *Barbati* / *Lalsa*

Scientific Name* — *Vigna unguiculata (L.) Walp

Family — Papilionaceae

Herb / Shrub / Tree — Herb, Vine – 1-2m, but 50 cm height

Annual / Biennial / Perennial — Annual

Used as — Pod as vegetable / Plant as fodder / green manure crop / seed as pulse

1. *Climatic requirements* — Rainy condition with higher temperature
2. *Soil requirements* — Loamy / sandy soils
3. *Required land situations* — High land
4. *Importance of crop* — Vegetable or fodder / pulse / green manure
5. *Fertilizer management* — 40 kg N, 60 kg P_2O_5 and 30 kg K_2O/ha
6. *Water Management*— Normally rainfed, if irrigated 1-2 irrigations
7. *Duration of crop* — 90-120 days
8. *Parts used* — Pod as vegetable, whole plant as fodder or as green manure crop
9. *Where grown in India* — All over India except too low or too high temperature areas.
10. *Export possibility* — High (for seed/pod)
11. *Economic yields (q / ha)* — 10-12

12. *Economic value (Rs. / ha)* — 20,000 to 30,000
13. *By-product (If any)* — Stalk
14. *Use of by-product* — Used as cattle feed / fuel / organic matter / mulch.

116. *CROP* — HORSE GRAM / Kulthi / *Kultikalai*

Scientific Name* — *Dolichos uniflorus Lam.

Family — Papilionaceae

Herb / Shrub / Tree — Trailing herb

Annual / Biennial / Perennial — Annual

Used as — Seed used as pulse or frying / Vine for cattle food

1. *Climatic requirements* — Rainy season with high temperature
2. *Soil requirements* — Loam / laterite / sandy soils
3. *Required land situations* — High land
4. *Importance of crop* — Fodder or seeds / green manure
5. *Fertilizer management* — 40 kg N, 60 kg P_2O_5 and 30 kg K_2O/ha
6. *Water management* — Normally rainfed, irrigations 1-2 times, if irrigated
7. *Duration of crop* — 90-120 days
8. *Parts used* — Whole plant or seed and stalk
9. *Where grown in India* — A.P., Karnataka, Tamil Nadu and Bihar for fodder / pulse.
10. *Export possibility* — High
11. *Economic yields (q / ha)* — 8-10
12. *Economic value (Rs. / ha)* — 8,000 to 10,000
13. *By-product (If any)* — Stalk
14. *Use of by-product* — Cattle feed / fuel / organic matter / mulch / green maure.

117. *CROP* — MOTH BEAN / KULATRA KALAI / MOTH / *Bhringga*

Scientific Name* — *Vigna aconitifolia Marechal

Family — Papilionanceae

Herb / Shrub / Tree — Herb, trailing

Annual / Biennial / Perennial — Annual

Used as — Pulse / fodder / medicine

1. *Climatic requirements* — High hot climate with rains
2. *Soil requirements* — Loamy / alluvial / sandy soils
3. *Required land situations* — High land

4. *Importance of crop* — Can be used as cattle feed. Fine vines
5. *Fertilizer management* — 40 kg N, 60 kg P_2O_5 and 60 kg K_2O/ha
6. *Water management* — Normally rainfed, 1-2 irrigations, if irrigated
7. *Duration of crop* — 10-140 days.
8. *Parts used* — Grains/vines
9. *Where grown in India* — West Bengal, Bihar, A.P., Orissa, Karnataka, U.P., Maharashta etc.
10. *Export possibility* — High
11. *Economic yields (q / ha)* — 10-15
12. *Economic value (Rs. / ha)* — 20,000 to 25,000
13. *By-product (If any)* — Stalk
14. *Use of by-product* — Cattle feed / organic matter / fuel / mulch.

118. *CROP* — PIGEONPEA / RED GRAM / Arhar / Tur / *Arahar*

Scientific Name — Cajanus cajan Mill sp

Family — Papilionaceae

Herb / Shrub / Tree — Shrub

Annual / Biennial / Perennial — Annual

Used as — Pulse (*dal*) with seeds / stem for fuel

1. *Climatic requirements* — Rainy season with high temperature, (some varieties may be grown in winter also)
2. *Soil requirements* — Loamy or sandy / alluvial soils
3. *Required land situations* — High land
4. *Importance of crop* — Seeds as pulse (*dal*), big plant, yellow flower
5. *Fertilizer management* — 40kg N, 60 kg P_2O_5 and 30 kg K_2O/ha
6. *Water management* — Rainfed crop, 1-2 times, if irrigated
7. *Duration of crop* — 100-120 days.
8. *Parts used* — Seeds and stalk
9. *Where grown in India* — Warm climate with rain
10. *Export possibility* — High
11. *Economic yields (q / ha)* — 10-12
12. *Economic value (Rs. / ha)* — 20,000 to 30,000
13. *By-product (If any)* — Stalk
14. *Use of by-product* — Stalk used as fuel or stick in trailing vegetables.

3. OIL SEEDS

119. *CROP* — GROUNDNUT / PEANUT / Mungphalli / China *Badam* / *Badam*

Scientific Name* — *Arachis hypogaea L.

Family — Papilionaceae

Herb / Shrub / Tree — Herb, erect / trailing diffuse herb

Annual / Biennial / Perennial — Annual (30-80 cm height)

Used as — Seeds for oil seeds / Vine for fodder / manure

1. *Climatic requirements* — Dry or rainy season with high temperature
2. *Soil requirements* — Loamy or Sandy soil
3. *Required land situations* — High land
4. *Importance of crop* — Seeds are used for oil/butter
5. *Fertilizer management* — 40 kg N, 60 kg P_2O_5 and 40 kg K_2O ha
6. *Water management* — Rainfed / irrigated, 1-2 irrigations
7. *Duration of crop* — 100-120 days
8. *Parts used* — Seeds / Haulm
9. *Where grown in India* — Gujarat, Maharashtra, Tamil Nadu, W.B., Orissa etc.
10. *Export possibility* — Very high
11. *Economic yields (q / ha)* — 15-20
12. *Economic value (Rs. / ha)* — 30,000 to 60,000
13. *By-product (If any)* — Haulm
14. *Use of by-product* — Fuel / green manure / fodder / organic manure

Note : In Bengal, this groundnut (short duration variety 1 CGV 86124) may be grown in pre-*kharif* season and haulm may be utilized for fodder/green manuring the next *Kharif* (*aman*) rice.

120. *CROP* — SESAME / GINGELLY / Til / *Kalotil*

Scientific Name* — *Sesamum indicum L.

Family — Pedaliaceae

Herb / Shrub / Tree — Herb, 40-80 cm height

Annual / Biennial / Perennial — Annual

Used as — Oils / confectionary / cosmetics

1. *Climatic requirements* — Rains with high temperature also in mild winter
2. *Soil requirements* — Loamy / sandy soils

3. *Required land situations* — High / medium land
4. *Importance of crop* — Seeds used for oil / confectionery, white flower
5. *Fertilizer management* — 60 kg N, 30 kg P_2O_5 and 30 kg K_2O ha
6. *Water management* — Mainly rainfed, 1-2 times irrigation, if irrigated
7. *Duration of crop* — 100-120 days
8. *Parts used* — Seed
9. *Where grown in India* — Rajasthan, U.P., M.P., Orissa, Bihar, W.B., Gujarat, A.P., T.N., Maharashtra etc.
10. *Export possibility* — High
11. *Economic yields (q / ha)* — 5-10
12. *Economic value (Rs. / ha)* — 10,000 to 30,000
13. *By-product (If any)* — Stalk
14. *Use of by-product* — Fuel / organic matter / pegs

121. *CROP* — SUNFLOWER / Surajmukhi / *Suryamukhi*

Scientific Name* — *Helianthus annuus L.

Family — Compositae

Herb / Shrub / Tree — Herb, 70-90 cm height

Annual / Biennial / Perennial — Annual

Used as — Oil seed / fuel / margarine preparation

1. *Climatic requirements* — In high / low / medium temperature in dry conditions
2. *Soil requirements* — Loam / sandy soil
3. *Required land situations* — High
4. *Importance of crop* — Seeds used for oil / margarine preparation; white flower
5. *Fertilizer management* — 60 kg N, 30 kg P_2O_5 and 30 kg K_2O/ha
6. *Water management* — Normally rainfed; 1-2 times irrigations, if irrigated
7. *Duration of crop* — 100-120 days
8. *Parts used* — Seeds / flower (beautification)
9. *Where grown in India* — All over India except too low / too high temperature areas
10. *Export possibility* — High
11. *Economic yields (q / ha)* — 8-10
12. *Economic value (Rs. / ha)* — 24,000 to 35,000

13. *By-product (If any)* — Stalk
14. *Use of by-product* — Fuel / organic matter / mulch.

122. ***CROP*** **— CASTOR /** ***Arandi*** **/** ***Reri*** **/ CASTOR-BEAN**

Scientific Name **—** ***Ricinus communis*** L.

Family — Euphorbiaceae

Herb / Shrub / Tree — Shrub, 80 cm to 2 m height

Annual / Biennial / Perennial — Annual

Used as — Lubricant from seed / seed purgative / soap

1. *Climatic requirements* — Medium / high temperature with rainfall
2. *Soil requirements* — Loamy / sandy soil
3. *Required land situations* — Medium / high land
4. *Importance of crop* — May be used as bio-fuel
5. *Fertilizer management* — 60 kg N, 30 kg P_2O_5 and 30 kg K_2O/ha
6. *Water management* — Normally rainfed; 1-2 times irrigation, if irrigated
7. *Duration of crop* — 120-140 days
8. *Parts used* — Seeds used for confectionery / perfume / medicine
9. *Where grown in India* — A.P., Maharashtra, Karnataka and Orissa, W.B., etc.
10. *Export possibility* — Very high
11. *Economic yields (q / ha)* — 5-10
12. *Economic value (Rs. / ha)* — 25,000 to 1,00,000
13. *By-product (If any)* — Stalk
14. *Use of by-product* — Fuel / paper pulp /fencing / pegs

4. FIBRES

123. ***CROP*** **— JUTE / Pat / Titapat / Mitha pat / Jhute /** ***Nalita***

Scientific Name — Corchorus capsularis / C. olitorius L.

Family — Tiliaceae

Herb / Shrub / Tree — Shrub, 2-3 m. height

Annual / Biennial / Perennial — Annual

Used as — Fibre (bast) / Vegetable

1. *Climatic requirements* — High rainfall with high temperature and high humidity
2. *Soil requirements* — Alluvial / loamy / clayey loam soil

3. *Required land situations* — Low land for *Capsularis* and highland for *Olitorius* jute in plains
4. *Importance of crop* — Long pod for *Olitorius* and round pod for *Capsularis*
5. *Fertilizer management* — 80 kg N, 40 kg P_2O_5 and 80 kg K_2O/ha
6. *Water management* — Normally rainfed; 2-3 irrigations, if irrigated in high land
7. *Duration of crop* — 120-140 days
8. *Parts used* — Fibre (bast) / leaf / stick
9. *Where grown in India* — West Bengal, Assam, Tripura, Eastern part of Bihar and Orissa and U.P.
10. *Export possibility* — Very high
11. *Economic yields (q / ha)* — 15-30
12. *Economic value (Rs. / ha)* — 30,000 to 60,000
13. *By-product (If any)* — Stick / *pakati*
14. *Use of by-product* — Fencing / stick in *boraj* / fuel / paper pulp

124. *CROP* — RAMIE / Kankhura / *Patta*

Scientific Name* — *Boehmeria nivea (L.) Goud.

Family — Urticaceae

Herb / Shrub / Tree — Shrub

Annual / Biennial / Perennial — Perennial

Used as — Fibre (bast)

1. *Climatic requirements* — High temperature with high rainfall and high humidity
2. *Soil requirements* — Alluvial / loamy / sandy loam soils
3. *Required land situations* — High land
4. *Importance of crop* — High quality fibre for notes (currency)
5. *Fertilizer management* — 100 kg N, 50 kg P_2O_5 and 50 kg K_2O/ha annually
6. *Water management* — In dry season 3-4 irrigations
7. *Duration of crop* — More than 2 years
8. *Parts used* — Bark of stem, for fibre
9. *Where grown in India* — Assam, West Bengal and other states of North East regions
10. *Export possibility* — Very high
11. *Economic yields (q / ha)* — 5-10
12. *Economic value (Rs. / ha)* — 20,000 to 40,000

13. *By-product (If any)* — Stick and leaves
14. *Use of by-product* — Paper pulp (for boards) / fuel / mulch / fodder

125. *CROP* — SISAL / AGAVE / *Agave*

Scientific Name — Agave sisalana Perr

Family — Agavaceae

Herb / Shrub / Tree — Shrub, 0.5-1.0m height

Annual / Biennial / Perennial — Perennial (xerophytic)

Used as — Ornamental

1. *Climatic requirements* — High temperature with slight rainfall, dry weather
2. *Soil requirements* — Loamy, sandy loam or sandy soils
3. *Required land situations* — High land sandy in nature
4. *Importance of crop* — Fibre for ropes / twines / cordage, xerophytic plant
5. *Fertilizer management* — 60 kg N, 40 kg P_2O_5 and 30 kg K_2O/ha
6. *Water management* — Normally rainfed crop; but 1-2 irrigation, if irrigated
7. *Duration of crop* — 2-3 years
8. *Parts used* — Leaf
9. *Where grown in India* — Assam, Bihar, Maharashtra, W. Bengal and States of South India
10. *Export possibility* — High
11. *Economic yields (q / ha)* — 5-10
12. *Economic value (Rs. / ha)* — 15,000 to 30,000
13. *By-product (If any)* — Stalk (short and thick)
14. *Use of by-product* — Fuel / organic matter.

126. *CROP* — MESTA / KENAF / ROSELLE / *Patwa / Lalambari*

Scientific Name — Hibiscus sabdariffa / H. cannabinus L.

Family — Malvaceae

Herb / Shrub / Tree — Shrub, 2-3 m. height

Annual / Biennial / Perennial — Annual

Used as — Fibre (bast)

1. *Climatic requirements* — High temperature with high rainfall along with high humidity
2. *Soil requirements* — Alluvial / loamy / sandy loam soils
3. *Required land situations* — Low, medium / high lands

4. *Importance of crop* — Fibre (for ropes / cordage) / fleshy fruit with red colour; rich in acids; curry (meat) is coloured with just set-fruits
5. *Fertilizer management* — 60 kg N, 40 kg P_2O_5 and 40 kg K_2O/ha
6. *Water management* — Normally rainfed, 1-2 irrigations, if irrigated
7. *Duration of crop* — 100-130 days
8. *Parts used* — Bark of the stem for fibre / stem or stick
9. *Where grown in India* — Andhra Pradesh, Bihar, Punjab, Assam Tamil Nadu, West Bengal and U.P.
10. *Export possibility* — High
11. *Economic yields (q / ha)* — 10-20
12. *Economic value (Rs. / ha)* — 20,000 to 40,000
13. *By-product (If any)* — Long stick
14. *Use of by-product* — Indoor fencing / paper pulp / fuel / mulch

127. ***CROP*** **— SUNNHEMP / Shan / San / Sunn.**

Scientific Name — Crotalaria juncea L.

Family — Papilionaceae

Herb / Shrub / Tree — Shrub, (tall), 2-3 m. height

Annual / Biennial / Perennial — Annual

Used as — Fibre / green manure crop

1. *Climatic requirements* — High temperature with high rainfall and high humidity
2. *Soil requirements* — Alluvial / loamy / clayey soils
3. *Required land situations* — Low / medium / high land
4. *Importance of crop* — Fibre for cigarette paper / tissue paper
5. *Fertilizer management* — 60 kg N, 40 kg P_2O_5 and 40 kg K_2O/ha
6. *Water management* — Normally rainfed, 1-2 irrigations, if irrigated
7. *Duration of crop* — 100-130 days
8. *Parts used* — Stem for fibre / leaves for vegetables and stick for paper pulp
9. *Where grown in India* — West Bengal, Assam, Bihar, Uttar Pradesh etc.
10. *Export possibility* — High
11. *Economic yields (q / ha)* — 10-20
12. *Economic value (Rs. / ha)* — 20,000 to 40,000

13. *By-product (If any)* — Stick (long)
14. *Use of by-product* — Fuel / for vegetable / staking / paper pulp

5. TUBERS / RHYZOMES

128. ***CROP*** **— ELEPHANT FOOT YAM / TELUGU POTATO / Surankand / Ol /** ***Zamin-kand***

Scientific Name **—** ***Amorphophallus campanulatus*** Blume ex Dcne.

Family — Araceae

Herb / Shrub / Tree — Herb, 50 cm to 1.5 m height

Annual / Biennial / Perennial — Perennial

Used as — Vegetable (tubers)

1. *Climatic requirements* — High temperature with rainfall and high humidity
2. *Soil requirements* — All types of soils except clay
3. *Required land situations* — High lands
4. *Importance of crop* — Vegetables (big corm tuber) / stem soft & long
5. *Fertilizer management* — 100-120 kg N, 50 kg P_2O_5 and 50 kg K_2O/ha
6. *Water management* — Irrigations 3-4 times needed under dry condition
7. *Duration of crop* — 120-150 days
8. *Parts used* — Tuber (or corm) / soft stem
9. *Where grown in India* — All over India except too low / too high temperature areas
10. *Export possibility* — High
11. *Economic yields (q / ha)* — 150-200 (green)
12. *Economic value (Rs. / ha)* — 25,000 to 60,000
13. *By-product (If any)* — Stalks
14. *Use of by-product* — Fuel / organic matter / vegetable

129. ***CROP*** **— ARUM / TARO / Arvi / Kachalu / Ghuiyan /** ***Kachu***

Scientific Name **—** ***Colocasia esculenta*** (L.) Schott.

Family — Araceae

Herb / Shrub / Tree — Herb, (rhizomatous)

Annual / Biennial / Perennial — Parennial, 0.50-1.5 m height

Used as — Vegetable

1. *Climatic requirements* — High temperature with high rainfall and humidity
2. *Soil requirements* — All types of soils except sandy
3. *Required land situations* — Low or medium land
4. *Importance of crop* — Rhizomes / runners used to get 'seeds'.
5. *Fertilizer management* — 100-120 kg N, 40 kg P_2O_5 and 40 kg K_2O/ha
6. *Water management* — Grown in rainfed, marshy land and 3-4 times irrigations, if grown under transplanted dry conditions in high lands
7. *Duration of crop* — 2-3 years
8. *Parts used* — Rhizomes / peteol / runners used as vegetables
9. *Where grown in India* — All over India in marshy / dry land except too low / too high temperature areas.
10. *Export possibility* — High
11. *Economic yields (q / ha)* — 150-200
12. *Economic value (Rs. / ha)* — 25,000 to 30,000
13. *By-product (If any)* — Leaves (leaves of some varieties eaten)
14. *Use of by-product* — Organic matter / fuel / mulch / plate

130. *CROP* — YAM BEAN / *Sakalu*

Scientific Name* — *Pachyrhizus erosus L.

Family — Papilionaceae

Herb / Shrub / Tree — Trailing herb (Vine 5-10 m. length)

Annual / Biennial / Perennial — Perennial

Used as — Vegetables (tuberous root) / fodder

1. *Climatic requirements* — High temperature with high rainfall and high humidity
2. *Soil requirements* — Alluvial / loamy / sandy and clayey loam soils
3. *Required land situations* — High lands
4. *Importance of crop* — Tuberous roots and young pods / vines trailed on trees / supports
5. *Fertilizer management* — 50-60 kg N, 60 kg P_2O_5 and 60 kg K_2O/ha
6. *Water management* — Normally rainfed crop; 3-4 irrigations throughout the year, if irrigated
7. *Duration of crop* — 2-3 years.

8. *Parts used* — Tuberous roots / vine
9. *Where grown in India* — West Bengal, Assam, Orissa, A.P., Bihar, Orissa etc.
10. *Export possibility* — High
11. *Economic yields (q / ha)* — 25-30
12. *Economic value (Rs. / ha)* — 25,000 to 30,000
13. *By-product (If any)* — Vine or stock or leaves
14. *Use of by-product* — Organic mater / cattle feeds / fuel / mulch.

Note : Arum has different varieties which may be grown in water / marshy land / high land soil, for rhizomes/petiole / runners / leaves.

6. SUGARS

131. *CROP* — SUGARCANE / Sweet cane / *Akh* / *Ikshu* / Noble cane / Paunda / *Ganna*

Scientific Name* — *Saccharum officinarum L.

***Family* —** Gramineae

***Herb / Shrub / Tree* —** Herb (tall), 2-3m height

***Annual / Biennial / Perennial* —** Perennial grass

***Used as* —** Basic material for sugar or molasses / bagasse for fuel

1. *Climatic requirements* — Hot high temperature with humidity
2. *Soil requirements* — Clayey loam / loam / black cotton / loamy soils
3. *Required land situations* — High or medium land
4. *Importance of crop* — Production of sugar / rum / alcohol / molasses etc.
5. *Fertilizer management* — 120 kg N, 60 kg P_2O_5 and 60 kg K_2O/ha
6. *Water management* — 7-10 irrigations needed according to places
7. *Duration of crop* — 365-540 days (*Eksali* / *Adhsali* crop)
8. *Parts used* — Stem (for juice-for-sugar) / leaf for cattle feed
9. *Where grown in India* — U.P., Karnataka, Bihar, Maharashtra, Tamil Nadu, etc.
10. *Export possibility* — Very high (sugar)
11. *Economic yields (q / ha)* — 600-1000 q/ha (*eksali* crop) in North India but 800-1300 q/ha (*adsali* in South India)
12. *Economic value (Rs. / ha)* — 45,000 to 65,000
13. *By-product (If any)* — Bagasse and leaves
14. *Use of by-product* — Bag paper, wrapping, writing and printing paper, cardboard and as fuel also for soil ameliorant / cattle feed.

7. VEGETABLES

132. ***CROP* — RED GOURD / Sitaphal / Misti *kumra* / *Red pumpkin***

Scientific Name* — *Cucurbita maxima Duch

Family — Cucurbitaceae

Herb / Shrub / Tree — Trailing shrub, 1-10 m length

Annual / Biennial / Perennial — Annual

Used as — Vegetable (fruits and vines)

1. *Climatic requirements* — Dry cold climate with low moderate temperature
2. *Soil requirements* — Loamy / sandy loam soil
3. *Required land situations* — High land
4. *Importance of crop* — Fruits and vines used as vegetables
5. *Fertilizer management* — 100-120 kg N, 60 kg P_2O_5 and 60 kg K_2O/ha.
6. *Water management* — Frequent irrigations 6-7 times
7. *Duration of crop* — 140-150 days
8. *Parts used* — Fruits / stems and leaves (as vegetable)
9. *Where grown in India* — All over India except too low / too high temperature areas
10. *Export possibility* — High
11. *Economic yields (q / ha)* — 150-200
12. *Economic value (Rs. / ha)* — 25,000 to 30,000
13. *By-product (If any)* — Haulm and leaves
14. *Use of by-product* — Cattle feed / fuel / vegetable / mulch

133. ***CROP* — POINTED GOURD / Parwal / *Patol* / *Patal***

Scientific Name* — *Trichosanthes dioica Rodb

Family — Cucurbitaceae

Herb / Shrub / Tree — Trailing herb

Annual / Biennial / Perennial — Annual

Used as — Vegetable

1. *Climatic requirements* — High temperature, high rainfall and high humidity
2. *Soil requirements* — Alluvial / loamy / sandy loam soils
3. *Required land situations* — High land
4. *Importance of crop* — Trailing type, 0.5 to 2 m length

5. *Fertilizer management* — 100-120 kg N, 50 kg P_2O_5 and 50 kg K_2O/ha.
6. *Water management* — 3-4 times irrigation
7. *Duration of crop* — 140-150 days
8. *Parts used* — Fruits and leaves
9. *Where grown in India* — U.P., West Bengal, Bihar, Assam, Orissa etc.
10. *Export possibility* — High
11. *Economic yields (q / ha)* — 250-300
12. *Economic value (Rs. / ha)* — 25,000 to 30,000
13. *By-product (If any)* — Vines and leaves
14. *Use of by-product* — Cattle feed / fuel / mulch / organic matter / vegetable

134. *CROP* — LEMON / LIME / Nimbu / Kaghzi Nimbu / *Lebu*

Scientific Name* — *Citrus aurantifolia Swingle

Family — Rutaceae

Herb / Shrub / Tree — Shrub, (small tree)

Annual / Biennial / Perennial — Perennial

Used as — Flavour, taste and confectionery. It helps in digestion

1. *Climatic requirements* — Moderate temperature moderate rainfall and moderate humidity
2. *Soil requirements* — All types of soils except clay and sand
3. *Required land situations* — High lands
4. *Importance of crop* — It gives good flavour in food / good smell; plant thorny
5. *Fertilizer management* — 120 kg N, 60 kg P_2O_5 and 60 kg K_2O/ha
6. *Water management* — 5-6 irrigations in dry period
7. *Duration of crop* — 2-3 years
8. *Parts used* — Fruits
9. *Where grown in India* — All over India except too low / too high temperature and waterlogged soil areas
10. *Export possibility* — Very high
11. *Economic yields (q / ha)* — 30-40
12. *Economic value (Rs. / ha)* — 60,000 to 80,000
13. *By-product (If any)* — Branches and leaves
14. *Use of by-product* — Fuel and organic matter / for trailing vine crops / indoor decoration

135. ***CROP* — GREEN BANANA / Kela / Kela / *Kachkala***

***Scientific Name* — *Musa paradisiaca* L.**

Family — Musaceae

Herb / Shrub / Tree — Shrub, with pseudo stem, Height 3-4 m

Annual / Biennial / Perennial — Annual / Perennial

Used as — Fruits / Vegetables

1. *Climatic requirements* — High temperature, high rainfall with high humidity
2. *Soil requirements* — Clayey / loamy / sandy loam soils
3. *Required land situations* — High
4. *Importance of crop* — Fruits / vegetables. It is propagated by suckers
5. *Fertilizer management* — 120 kg N, 60 kg P_2O_5 and 60 kg K_2O/ha.
6. *Water management* — 5-6 irrigations in dry period
7. *Duration of crop* — 300-360 days
8. *Parts used* — Fruits / leaves stem (pseudo)
9. *Where grown in India* — All wet tropical / sub-tropical areas of India
10. *Export possibility* — High (for starch)
11. *Economic yields (q / ha)* — 150-250
12. *Economic value (Rs. / ha)* — 25,000 to 60,000
13. *By-product (If any)* — Pseudo stem and leaves
14. *Use of by-product* — Vegetable / cattle feed / plates / fuel / packing material for pan.

136. ***CROP* — SQUASH / Mitha kaddu / *Mistikoash***

Scientific Name* — *Cucurbita moschata (Duch) Poir

Family — Cucurbitaceae

Herb / Shrub / Tree — Herb (vine), height 1-3 m

Annual / Biennial / Perennial — Annual

Used as — Fruits

1. *Climatic requirements* — High rainfall, high temperature and high humidity
2. *Soil requirements* — Alluvial / loamy / sandy loam soils
3. *Required land situations* — High
4. *Importance of crop* — Vegetable. Vine crop
5. *Fertilizer management* —100 kg N, 50 kg P_2O_5 and 50 kg K_2O/ha.

6. *Water management* — 3-4 irrigation
7. *Duration of crop* — 100-140 days
8. *Parts used* — Fruits / stem
9. *Where grown in India* — In humid and semi humid region of India
10. *Export possibility* — High
11. *Economic yields (q / ha)* — 150-200 (green)
12. *Economic value (Rs. / ha)* — 30,000 to 50,000
13. *By-product (If any)* — Vines / leaves
14. *Use of by-product* — Cattle feed / vegetable / fuel / mulch.

137. *CROP* — PAPAYA / PAPAW / Papita / *Papua / Pepe*

Scientific Name* — *Carica papaya L.

Family — Caricaceae

Herb / Shrub / Tree — Shrub

Annual / Biennial / Perennial — Perennial

Used as — Vegetable / papain / fruits

1. *Climatic requirements* — High rainfall, high temperature and high humidity
2. *Soil requirements* — Alluvial / loamy / sandy loam soils
3. *Required land situations* — High land
4. *Importance of crop* — Vegetables and Papain (latex)
5. *Fertilizer management* — 100 kg N, 60 kg P_2O_5 and 60 kg K_2O/ha annually
6. *Water management* — 6-7 irrigation in dry season, annually
7. *Duration of crop* — 2-3 years
8. *Parts used* — Fruits (green fruits / ripe fruits)
9. *Where grown in India* — In humid region of India
10. *Export possibility* — Very high
11. *Economic yields (q / ha)* — 150-200 (green)
12. *Economic value (Rs. / ha)* — 40,000 to 50,000
13. *By-product (If any)* — Stem / leaves
14. *Use of by-product* — Organic matter / fuel / mulch

138. *CROP* — AMARANTH/Chaulai / Nate / data / Lalsag / TAMPALA

Scientific Name* — *Amaranthus tricolor L.

Family — Amaranthaceae

Herb / Shrub / Tree — Herb

***Annual / Biennial / Perennial* —** Annual

***Used as* —** Vegetables

1. *Climatic requirements* — Moderate rainfall, temperature and humidity
2. *Soil requirements* — All types of soils
3. *Required land situations* — High lands
4. *Importance of crop* — Vegetables (Leaves and stems)
5. *Fertilizer management* — 120 kg N, 60 kg P_2O_5 and 60 kg K_2O/ha.
6. *Water management* — 3-4 irrigation
7. *Duration of crop* — 100-120 days
8. *Parts used* — Stems / leaves / seed
9. *Where grown in India* — All over India having normal humidity, rainfall and temperature
10. *Export possibility* — High
11. *Economic yields (q / ha)* — 100-200
12. *Economic value (Rs. / ha)* — 25,000 to 30,000
13. *By-product (If any)* — Nil
14. *Use of by-product* — Nil.

139. *CROP* — INDIAN SPINACH / Poi / *Pui / Puisag*

Scientific Name* — *Basella alba L.

***Family* —** Basellaceae

***Herb / Shrub / Tree* —** Trailing shrub (1-3m length)

***Annual / Biennial / Perennial* —** Perennial

***Used as* —** Vegetable (stem and leaves)

1. *Climatic requirements* — Rainfall high, temperature high with high humidity
2. *Soil requirements* — Alluvial / loamy / sandy loam / clayey soils.
3. *Required land situations* — High / medium land
4. *Importance of crop* — Both succulent stem and leaves used as vegetable
5. *Fertilizer management* —100 kg N, 50 kg P_2O_5 and 50 kg K_2O/ha.
6. *Water management* — 5-6 irrigations
7. *Duration of crop* — 2-3 years
8. *Parts used* —Entire plant
9. *Where grown in India* — Humid areas in India (West Bengal, Assam, Bihar, Orissa, Tripura etc.

10. *Export possibility* — Less
11. *Economic yields (q / ha)* — 100-200 (Green)
12. *Economic value (Rs. / ha)* — 25,000 to 30,000
13. *By-product (If any)* — Nil
14. *Use of by-product* — Nil

140. *CROP* — ARUM / TARO (Arvi / Ghuiyan / Kachalu / *Kachu*

Scientific Name* — *Colocasia esculenta L.

Family — Araceae

Herb / Shrub / Tree — Herb, 0.5 to 1.5 m height

Annual / Biennial / Perennial — Perennial

Used as — Vegetables

1. *Climatic requirements* — Moderate to high rainfall, high temperature, moderate humidity
2. *Soil requirements* — Clayey / loamy / sandy loam soils
3. *Required land situations* — High / medium / low land (according to varieties)
4. *Importance of crop* — Vegetable
5. *Fertilizer management* —100 kg N, 50 kg P_2O_5 and 50 kg K_2O/ha.
6. *Water management* — 3-4 times irrigations when grown in high land or under transplanted conditions
7. *Duration of crop* — 150-160 days
8. *Parts used* — Rhizomes / runners / leaves / petiole
9. *Where grown in India* — All over India (mostly humid and semi-humid areas)
10. *Export possibility* — High
11. *Economic yields (q / ha)* — 500-600 (Green all parts)
12. *Economic value (Rs. / ha)* — 50,000 to 60,000
13. *By-product (If any)* — Leaves
14. *Use of by-product* — Plates / fuel / organic matter / mulch / wrapping *pan*

141. *CROP* — ELEPHANT-FOOT-YAM/TELUGU POTATO/ Surankand / Ol

Scientific Name* — *Amorphophallus campanulatus Roxb.

Family — Araceae

Herb / Shrub / Tree — Herb, 0.5-1.5 m. height

Annual / Biennial / Perennial — Annual (stem less)

Used as — Vegetable

1. *Climatic requirements* — High temperature with high rainfall and high humidity
2. *Soil requirements* — Loamy / clayey / sandy loam / loamy soils
3. *Required land situations* — High land
4. *Importance of crop* — Stem less, rhizomes
5. *Fertilizer management* —120 kg N, 60 kg P_2O_5 and 60 kg K_2O/ha.
6. *Water management* — 3-4 irrigations, if irrigated
7. *Duration of crop* — 140-180 days
8. *Parts used* — Rhizomes / corm / stem
9. *Where grown in India* — Humid and semi humid areas
10. *Export possibility* — High
11. *Economic yields (q / ha)* — 200-400 (raw)
12. *Economic value (Rs. / ha)* — 20,000 to 40,000
13. *By-product (If any)* — Soft stem and leaves
14. *Use of by-product* — Vegetable / organic matter / fuel / mulch.

142. *CROP* — KOVAIFRUIT / Kundru / *Kundli* / Ivy gourd / Kanduri

Scientific Name — Coccinia grandis /C. indica (L) Voigt

Family — Cucurbitaceae

Herb / Shrub / Tree — Spreading shrub

Annual / Biennial / Perennial — Perennial

Used as — Vegetable (green fruits) / medicine

1. *Climatic requirements* — High rainfall with high temperature and high humidity
2. *Soil requirements* — Clayey / loamy / alluvial / sandy loam soils
3. *Required land situations* — Low, medium or high land
4. *Importance of crop* — Twining or spreading type of plant
5. *Fertilizer management* —100 kg N, 50 kg P_2O_5 and 50 kg K_2O/ha.
6. *Water management* — 4-6 irrigation, if irrigated
7. *Duration of crop* — 2-3 years
8. *Parts used* — Fruits (as vegetable), juice (from leaves or roots) as medicine in diabetes
9. *Where grown in India* — Humid and sub-humid areas
10. *Export possibility* — High
11. *Economic yields (q / ha)* — 100-150
12. *Economic value (Rs. / ha)* — 25,000 to 30,000

13. *By-product (If any)* — Stem and leaves
14. *Use of by-product* — Organic matter / vegetable / fuel / mulch.

143. *CROP* — BITTER GOURD (Kerela / *Karala* or Uchhe)

Scientific Name* — *Momordica charantia L.

Family — Cucurbitaceae

Herb / Shrub / Tree — Spreading / trailing herb, 1-3m length

Annual / Biennial / Perennial — Annual

Used as — Vegetable

1. *Climatic requirements* — High rainfall and high temperature
2. *Soil requirements* — All-types of soil except sandy soil
3. *Required land situations* — High land
4. *Importance of crop* — Bitterness in the fruit
5. *Fertilizer management* —120 kg N, 60 kg P_2O_5 and 60 kg K_2O/ha.
6. *Water management* — 3-4 irrigation, if irrigated
7. *Duration of crop* — 140-150 days
8. *Parts used* — Fruits (as vegetable)
9. *Where grown in India* — All areas except dry areas
10. *Export possibility* — High
11. *Economic yields (q / ha)* — 150-200 (green)
12. *Economic value (Rs. / ha)* — 25,000 to 40,000
13. *By-product (If any)* — Stems and leaves
14. *Use of by-product* — Organic matter / fuel / mulch / fodder.

144. *CROP* — BRINJAL / EGG PLANT / Baigun / *Begun* / Aubergine

Scientific Name* — *Solanum melongena L.

Family — Solanaceae

Herb / Shrub / Tree — Shrub, 0.5-1.5m height

Annual / Biennial / Perennial — Annual (sometimes perennial)

Used as — Vegetable (fruits)

1. *Climatic requirements* — Moderate temperature in dry or wet condition
2. *Soil requirements* — Loamy / sandy / sandy loam / alluvial soils
3. *Required land situations* — High land
4. *Importance of crop* — Big fruits coloured (velvet / green)
5. *Fertilizer management* — 100 kg N, 50 kg P_2O_5 and 50 kg K_2O/ha
6. *Water management* — 4-5 irrigations

7. *Duration of crop* — 120-140 days
8. *Parts used* — Fruits
9. *Where grown in India* — Mostly all over India except dry regions
10. *Export possibility* — High
11. *Economic yields (q / ha)* — 150-200 (Green)
12. *Economic value (Rs. / ha)* — 50,000 to 60,000
13. *By-product (If any)* — Plants and leaves
14. *Use of by-product* — Fuel or organic matter / for trailing cowpea on it (plant)

(N.B. Trailing types of plants like cowpea etc. could be trailed over the standing crop of brinjal after removing all fruits).

145. *CROP* — CHILLIES / RED PEPPER / Lalmircha / Lanka / dhani lanka / Phapalanka / Marich

Scientific Name — Capsicum annuum L.

Family — Solanaceae

Herb / Shrub / Tree — Herb / Shrub, Height 0.25-1.5 m

Annual / Biennial / Perennial — Annual / Perennial

Used as — Spice / condiment / vegetable / medicine

1. *Climatic requirements* — Warm climate / weather (also cold climate)
2. *Soil requirements* — Alluvial / sandy / loamy / clayey soils
3. *Required land situations* — High land
4. *Importance of crop* — Pungent (bitter) smell
5. *Fertilizer management* — 100 kg N, 50 kg P_2O_5 and 50 kg K_2O/ha
6. *Water management* — 4-6 irrigations annually
7. *Duration of crop* — 120-150 days / 2-3 years
8. *Parts used* — Fruits
9. *Where grown in India* — All over India mainly in A.P., Maharashtra, Assam, Bihar, Orissa, T.N., M.P., Rajasthan, Kerala, W.B., Karnataka etc.
10. *Export possibility* — High
11. *Economic yields (q / ha)* — 150-200 (green)
12. *Economic value (Rs. / ha)* — 30,000 to 40,000
13. *By-product (If any)* — Plants / leaves
14. *Use of by-product* — Fuel / organic matter / trailing the vine crops on it.

146. *CROP* — CUCUMBER / Khira / *Shasa*/ *Sasa*

***Scientific Name* — *Cucumis sativa* L.**

Family — Cucurbitaceae

Herb / Shrub / Tree — Trailing shrub / herb, 0.5-1.5 m length

Annual / Biennial / Perennial — Annual

Used as — Vegetable (salad / fruits)

1. *Climatic requirements* — High / medium rainfall with high / medium temperature with moderate humidity
2. *Soil requirements* — Loamy / sandy / alluvial soils
3. *Required land situations* — Land without water stagnation
4. *Importance of crop* — Salad
5. *Fertilizer management* — 60 kg N, 30 kg P_2O_5 and 30 kg K_2O/ha
6. *Water management* — 3-4 irrigations, if irrigated
7. *Duration of crop* — 140-150 days
8. *Parts used* — Fruits
9. *Where grown in India* — All humid or semi humid areas of India
10. *Export possibility* — High
11. *Economic yields (q / ha)* — 150-200 (green)
12. *Economic value (Rs. / ha)* — 25,000 to 30,000
13. *By-product (If any)* — Vines and leaves
14. *Use of by-product* — Cattle feed / organic matter / fuel / mulch.

147. CROP — MUSKMELON / Kharbuza / Kharmuz

***Scientific Name* — *Cucumis melo* L.**

Family — Cucurbitaceae

Herb / Shrub / Tree — Creeping shrub, 0.5-1.5 length, vine

Annual / Biennial / Perennial — Annual

Used as — Fruits as vegetable / fruit

1. *Climatic requirements* — High temperature with hot and dry condition in summer
2. *Soil requirements* — Sandy soil (riverbed) / alluvial / loamy soils
3. *Required land situations* — Land, with no water stagnation
4. *Importance of crop* — Vine with different flowers (male and female)
5. *Fertilizer management* — 60 kg N, 30 kg P_2O_5 and 30 kg K_2O/ha
6. *Water management* — 1-2 irrigations, if irrigated
7. *Duration of crop* — 110-130 days

8. *Parts used* — Fruits
9. *Where grown in India* — All over India in summer, except too hot/too cool areas.
10. *Export possibility* — High
11. *Economic yields (q / ha)* — 160-170 (green)
12. *Economic value (Rs. / ha)* — 30,000 to 70,000
13. *By-product (If any)* — Vine and leaves
14. *Use of by-product* — Cattle feed / organic matter / fuel / mulch.

148. *CROP* — WATER MELON / Torbooz / *Tarmuj*

Scientific Name* — *Citrullus lanatus (Thunb) Mansf

Family — Cucurbitaceae

Herb / Shrub / Tree — Creeping / streading herb

Annual / Biennial / Perennial — Annual, 0.5-1.5 m long, vine

Used as — Fruits (red inside) / medicine

1. *Climatic requirements* — Dry condition with high temperature
2. *Soil requirements* — Sandy / loamy / sandy loam soils
3. *Required land situations* — Nearer to river / sea beds
4. *Importance of crop* — Fruits
5. *Fertilizer management* — 100 kg N, 50 kg P_2O_5 and 50 kg K_2O/ha
6. *Water management* — 1-2 irrigations, if irrigated
7. *Duration of crop* — 110-130 days
8. *Parts used* — Fruits
9. *Where grown in India* — All river / seabeds of India (summer)
10. *Export possibility* — High
11. *Economic yields (q / ha)* — 200-250 (green)
12. *Economic value (Rs. / ha)* — 40,000 to 50,000
13. *By-product (If any)* — Vine and leaf
14. *Use of by-product* — Cattle feed / organic matter / fuel / mulch.

149. *CROP* — BOTTLE GOURD / PUMPKIN / Vilaiti Kaddu / *Lau*

Scientific Name* — *Cucurbita pepo L., ***C. siceraria*** Molina

Family — Cucurbitaceae

Herb / Shrub / Tree — Spreading shrub 0.5-1.5 m long, vine

Annual / Biennial / Perennial — Annual

Used as — Vegetable

1. *Climatic requirements* — Moderate rainfall, temperature and humidity
2. *Soil requirements* — Loamy / alluvial / sandy loam / sandy soils
3. *Required land situations* — High land
4. *Importance of crop* — Vegetable; plant trailing type
5. *Fertilizer management* — 60 kg N, 30 kg P_2O_5 and 30 kg K_2O/ha
6. *Water management* — 3-4 irrigations, if irrigated
7. *Duration of crop* — 110-130 days
8. *Parts used* — Fruits and vine
9. *Where grown in India* — Assam, West Bengal, Bihar, Orissa A.P., Tripura, U.P. (East) etc.
10. *Export possibility* — High
11. *Economic yields (q / ha)* — 150-200 (green)
12. *Economic value (Rs. / ha)* — 60,000 to 80,000
13. *By-product (If any)* — Vine and leaves used as *sag*
14. *Use of by-product* — Cattle feed / organic matter / fuel / mulch / vegetable.

150. *CROP* — SPONGE OR RIDGED GOURD / LOOFAH / Luffa / *Jinga*

Scientific Name — Luffa acutangula (L.) Roxb.

Family — Cucurbitaceae

Herb / Shrub / Tree — Spreading / climbing shrub, 0.5-1.5 m long, vine

Annual / Biennial / Perennial — Annual

Used as — Vegetable

1. *Climatic requirements* — High rainfall with high temperature and humidity
2. *Soil requirements* — Loamy / clayey / sandy / alluvial soils
3. *Required land situations* — High land
4. *Importance of crop* — Long spongy fruits with coloured patches / ridges
5. *Fertilizer management* — 100 kg N, 50 kg P_2O_5 and 50 kg K_2O/ha
6. *Water management* — 1-2 irrigations, if irrigated
7. *Duration of crop* — 110-130 days
8. *Parts used* — Long stripped ridged fruits
9. *Where grown in India* — Humid and sub humid regions
10. *Export possibility* — High
11. *Economic yields (q / ha)* — 100-150 (green)

12. *Economic value (Rs. / ha)* — 20,000 to 30,000
13. *By-product (If any)* — Vine and leaves
14. *Use of by-product* — Cattle feed / organic matter / fuel / mulch.

151. *CROP* — ROUND GOURD / *Tinda*

Scientific Name* — *Citrullus lanatus (Thunb) Mansf. var. ***fislulosus*** Mansf.

Family — Cucurbitaceae

Herb / Shrub / Tree — Spreading / climbing herb, 0.5-1.5m long vine

Annual / Biennial / Perennial — Annual

Used as — Vegetable

1. *Climatic requirements* — Moderate dry condition with moderate temperature
2. *Soil requirements* — Loamy / sandy soils
3. *Required land situations* — High land
4. *Importance of crop* — Fistulous stem and petioles from main axis
5. *Fertilizer management* — 60 kg N, 30 kg P_2O_5 and 30 kg K_2O/ha
6. *Water management* — 1-2 irrigation, if irrigated
7. *Duration of crop* — 110-140 days
8. *Parts used* — Fruits
9. *Where grown in India* — U.P., Punjab, Haryana, Rajasthan, Bihar, W. Bengal (West), Orissa etc.
10. *Export possibility* — High
11. *Economic yields (q / ha)* — 100-200 (green)
12. *Economic value (Rs. / ha)* — 25,000 to 30,000
13. *By-product (If any)* — Vine and leaves
14. *Use of by-product* — Cattle feed / organic matter / fuel / mulch

152. *CROP* — SNAKE GOURD / *Chichinga* / SERPENT / *Chachinga*

Scientific Name* — *Trichosanthes anguina L.

Family — Cucurbitaceae

Herb / Shrub / Tree — Climbing / trailing herb, 0.5-1.5 m, long, vine

Annual / Biennial / Perennial — Annual

Used as — Vegetable

1. *Climatic requirements* — Moderate temperature and high rainfall
2. *Soil requirements* — Loamy / sandy / alluvial soils
3. *Required land situations* — High / medium land

4. *Importance of crop* — Black and white striped of long fruit, like snake
5. *Fertilizer management* — 60 kg N, 30 kg P_2O_5 and 30 kg K_2O/ha
6. *Water management* — 3-4 irrigations, if irrigated
7. *Duration of crop* — 110-140 days
8. *Parts used* — Long fruit, like snakes
9. *Where grown in India* — All over India, mostly in humid zone
10. *Export possibility* — Less
11. *Economic yields (q / ha)* — 100-200 (green)
12. *Economic value (Rs. / ha)* — 25,000 to 30,000
13. *By-product (If any)* — Vine and leaves
14. *Use of by-product* — Cattle feed / organic matter / fuel / mulch

153. CROP — OKRA / LADY'S FINGER / Bhendi / Dheras / Bhindi

Scientific Name — Abelmoschus esculentus (L.) Moench

Family — Malvaceae

Herb / Shrub / Tree — Erect herb / shrub, 0.5-1.5 m height

Annual / Biennial / Perennial — Annual

Used as — Vegetable (Tender fruits)

1. *Climatic requirements* — High rainfall, high temperature and high humidity
2. *Soil requirements* — Loamy, clayey and sandy soils
3. *Required land situations* — Medium / high land
4. *Importance of crop* — Mucilage in stem and tender fruits
5. *Fertilizer management* — 100 kg N, 50 kg P_2O_5 and 50 kg K_2O/ha
6. *Water management* — 3-4 irrigations, if irrigated
7. *Duration of crop* — 110-140 days
8. *Parts used* — Tender fruits
9. *Where grown in India* — All over India under humid conditions
10. *Export possibility* — High
11. *Economic yields (q / ha)* — 50-60 (green)
12. *Economic value (Rs. / ha)* — 50,000 to 60,000
13. *By-product (If any)* — Whole plant
14. *Use of by-product* — Fibres and fuel; cowpea etc. could be trailed on it.

154. *CROP* — LETTUCE / Salad / *Letus*

***Scientific Name* — *Lactuca sativa* L. var. *sativa* Clarke**

Family — Compositae

Herb / Shrub / Tree — Leafy herb, 0.25-0.6 m height

Annual / Biennial / Perennial — Annual

Used as — Salad (vegetable)

1. *Climatic requirements* — Low temperature with low rainfall
2. *Soil requirements* — Loamy / sandy / alluvial soils
3. *Required land situations* — High land
4. *Importance of crop* — Leaves are used as salad in vegetable foods
5. *Fertilizer management* — 120 kg N, 60 kg P_2O_5 and 60 kg K_2O/ha
6. *Water management* — 4-5 irrigations
7. *Duration of crop* — 110-140 days
8. *Parts used* — Leaves (with small stem)
9. *Where grown in India* — All over India during winter, except too low / too high temperature areas
10. *Export possibility* — High
11. *Economic yields (q / ha)* — 50-60 (green)
12. *Economic value (Rs. / ha)* — 50,000 to 60,000
13. *By-product (If any)* — Nil
14. *Use of by-product* — Nil.

155. *CROP* — SWEET POTATO / Shakarkand / Mithalu / *Mistialu*

***Scientific Name* — *Ipomoea batatas* (L.) Lamk**

Family — Convolvulaceae

Herb / Shrub / Tree — Diffusely spreading herb, 0.5-1.5m vine on soil surface

Annual / Biennial / Perennial — Annual

Used as — Tuberous root as vegetable / food

1. *Climatic requirements* — Dry climate with low rainfall
2. *Soil requirements* — Sandy soils / loamy / alluvial soils
3. *Required land situations* — High / medium land
4. *Importance of crop* — Root tuberous plant (rooting at the nodes)
5. *Fertilizer management* — 60 kg N, 30 kg P_2O_5 and 30 kg K_2O/ha
6. *Water Management*—Normally rainfed; 1-2 irrigations, if irrigated
7. *Duration of crop* — 140-160 days

8. *Parts used* — Root tubers as vegetable / foods / manufacturing starch / alcohol / glucose
9. *Where grown in India* — All over India, mostly in dry areas except too low / too high temperataure areas
10. *Export possibility* — Very high
11. *Economic yields (q / ha)* — 150-160 (green)
12. *Economic value (Rs. / ha)* — 50,000 to 60,000
13. *By-product (If any)* — Vines and leaves
14. *Use of by-product* — Cattle feed / medicine / vegetables / fuel / mulch.

156. *CROP* — CELERIAC / GARDEN CELERY / Salari / Ajmud / Karas

Scientific Name* — *Apium graveolens var. ***dulce*** (Mill) DC

Family — Umbelliferae

Herb / Shrub / Tree — Herb

Annual / Biennial / Perennial — Annual

Used as — Vegetables (roots / long leaf stalk) / medicine

1. *Climatic requirements* — Low temperature with low rainfall in hilly region
2. *Soil requirements* — Loamy / stony soils
3. *Required land situations* — High land situation
4. *Importance of crop* — Long leaf-stalk used as salad / medicine
5. *Fertilizer management* — 60 kg N, 30 kg P_2O_5 and 30 kg K_2O/ha
6. *Water Management*—Normally rainfed; 1-2 irrigations, if irrigated
7. *Duration of crop* — 120-140 days
8. *Parts used* — Root / leaf-stalk
9. *Where grown in India* — In hilly regions of U.P. / South India, H.P. etc.
10. *Export possibility* — High
11. *Economic yields (q / ha)* — 30-40 (green)
12. *Economic value (Rs. / ha)* — 15,000 to 20,000
13. *By-product (If any)* — Fruits / seeds
14. *Use of by-product* — Spice / perfumes (if seed / fruits crushed)

157. *CROP* — PARSLEY

Scientific Name* — *Petroselinum crispum (Mill) Nym

Family — Umbelliferae

Herb / Shrub / Tree — Herb

***Annual / Biennial / Perennial* —** Annual

***Used as* —** Vegetable (as salad) / medicine

1. *Climatic requirements* — Low temperature with low rainfall
2. *Soil requirements* — Loamy / stony / sandy soils
3. *Required land situations* — High
4. *Importance of crop* — Leaf is eaten / juice from leaf for medicine (urine troubles)
5. *Fertilizer management* — 60 kg N, 30 kg P_2O_5 and 30 kg K_2O/ha
6. *Water management* — Normally rainfed, 1-2 irrigation, if irrigated
7. *Duration of crop* — 130-140 days, some times more
8. *Parts used* — Leaf / stem / fruits
9. *Where grown in India* — All the hilly regions of India
10. *Export possibility* — High
11. *Economic yields (q / ha)* — 30-40
12. *Economic value (Rs. / ha)* — 30,000 to 40,000
13. *By-product (If any)* — Stalks
14. *Use of by-product* — Cattle feed / crushed oil used for flavour of feeds and food products.

158. *CROP* — Yambean / *Sakalu*

Scientific Name* — *Pachyrhizus erosus (L.) Urb.

Family — Papilionaceae

Herb / Shrub / Tree — Herb (trailing)

***Annual / Biennial / Perennial* —** Perennial

***Used as* —** Vegetable (tuberous roots and young pods)

1. *Climatic requirements* — High temperature with moderate rainfall and humidity
2. *Soil requirements* — Alluvial / loamy soils
3. *Required land situations* — High lands
4. *Importance of crop* — Tuberous sweet roots (in soils)
5. *Fertilizer management* — 60 kg N, 30 kg P_2O_5 and 30 kg K_2O/ha
6. *Water management* — Normally rainfed, 1-2 times irrigations, if irrigated
7. *Duration of crop* — 2-3 years
8. *Parts used* — Tuberous sweet roots / tender pods (vegetable)
9. *Where grown in India* — West Bengal, Assam, Bihar, Orissa and A.P.

10. *Export possibility* — High
11. *Economic yields (q / ha)* — 100-200 (green)
12. *Economic value (Rs. / ha)* — 25,000 to 30,000
13. *By-product (If any)* — Vines and leaves
14. *Use of by-product* — Cattle feed / organic manure / fuel / mulch

159. *CROP* — TAPIOCA / MANIOC / CASSAVA / Sakarkanda / Maravuli / Satamuli

Scientific Name* — *Manihot esculenta Crantz.

Family — Euphorbiaceae

Herb / Shrub / Tree — Shrub, 0.5-1.5 m height

Annual / Biennial / Perennial — Annual

Used as — Food / vegetable

1. *Climatic requirements* — High temperature with low rainfall
2. *Soil requirements* — Sandy / stony / Loamy / alluvial soils
3. *Required land situations* — High land
4. *Importance of crop* — Tuberous roots (swollen)
5. *Fertilizer management* — 60 kg N, 30 kg P_2O_5 and 30 kg K_2O/ha
6. *Water Management*—Normally rainfed; 1-2 irrigations, if irrigated
7. *Duration of crop* — 110-140 days
8. *Parts Used*—Tubers (used to obtain starch / sago / semolina / flour)
9. *Where grown in India* — Kerala, A.P., T. Nadu, Karnataka, etc.
10. *Export possibility* — High
11. *Economic yields (q / ha)* — 50-60
12. *Economic value (Rs. / ha)* — 25,000 to 30,000
13. *By-product (If any)* — Stems, branches and leaves
14. *Use of by-product* — Fuel / organic matter / mulch / stick

8. FODDER / FORAGE (GRASSY)

160. *CROP* — NAPIER GRASS / ELEPHANT GRASS (Napier)

Scientific Name* — *Pennisetum purpureum Schum

Family — Gramineae

Herb / Shrub / Tree — Herb

Annual / Biennial / Perennial — Annual

Used as — Fodder / Hay

1. *Climatic requirements* — High temperature with medium rainfall

2. *Soil requirements* — Alluvial / loamy soils
3. *Required land situations* — High / medium land
4. *Importance of crop* — Long stem with big leaves
5. *Fertilizer management* — 100 kg N, 50 kg P_2O_5 and 50 kg K_2O/ha
6. *Water management* — 3-4 irrigations
7. *Duration of crop* — 110-140 days
8. *Parts used* — Whole plant (Stems + leaves) cut at flowering
9. *Where grown in India* — Assam, West Bengal, Bihar, U.P., Punjab, etc.
10. *Export possibility* — High
11. *Economic yields (q / ha)* — 200-250 (green)
12. *Economic value (Rs. / ha)* — 50,000 to 75,000
13. *By-product (If any)* — Nil
14. *Use of by-product* — Nil.

161. CROP — THIN NAPIER / Dinanath grass / Chhotonapier

Scientific Name — Pennisetum pedicellatum Trin.

Family — Gramineae

Herb / Shrub / Tree — Herb, 1.0-2.5 m height

Annual / Biennial / Perennial — Annual

Used as — Fodder / decoration

1. *Climatic requirements* — High rainfall with high temperature
2. *Soil requirements* — Alluvial loamy / sandy loam soils
3. *Required land situations* — High / medium land
4. *Importance of crop* — Good pinkish inflorescence
5. *Fertilizer management* — 100 kg N, 50 kg P_2O_5 and 50 kg K_2O/ha
6. *Water management* — 3-4 irrigations
7. *Duration of crop* — 110-140 days
8. *Parts used* — Entire plant (Stem + leaves and inflorescence) cut at flowering
9. *Where grown in India* — Assam, W. Bengal, Bihar, U.P. and Punjab, etc.
10. *Export possibility* — High
11. *Economic yields (q / ha)* — 200-250 (Green)
12. *Economic value (Rs. / ha)* — 25,000 to 75,000
13. *By-product (If any)* — Nil
14. *Use of by-product* — Nil.

162. *CROP* — TEOSINTE / Tiosenti / Nakalbhutta

Scientific Name* — *Euchlaena mexicana Schrad

Family — Gramineae

Herb / Shrub / Tree — Shrub (tall), 1.0 to 3.0 m height

Annual / Biennial / Perennial — Annual

Used as — Fodder

1. *Climatic requirements* — Moderate temperature with moderate rainfall
2. *Soil requirements* — Loamy / stony / alluvial soils
3. *Required land situations* — High land
4. *Importance of crop* — Sparing cob-like structure, like maize
5. *Fertilizer management* — 100 kg N, 50 kg P_2O_5 and 50 kg K_2O/ha
6. *Water management* — 4-5 irrigations
7. *Duration of crop* — 110-140 days
8. *Parts used* — Entire plant for fodder, cut at flowering
9. *Where grown in India* — U.P., Assam, Maharashtra, Punjab Bengal etc.
10. *Export possibility* — High
11. *Economic yields (q / ha)* — 80-260 (green)
12. *Economic value (Rs. / ha)* — 40,000 to 80,000
13. *By-product (If any)* — Nil
14. *Use of by-product* — Nil.

163. *CROP* — MAIZE / CORN / INDIAN CORN / Makai / *Bhutta*

Scientific Name* — *Zea mays L.

Family — Gramineae

Herb / Shrub / Tree — Herb (tall), 1.0-3.0 m height

Annual / Biennial / Perennial — Annual

Used as — Fodder / Food

1. *Climatic requirements* — Moderate temperature with moderate rainfall
2. *Soil requirements* — Alluvial / loamy / sandy loam soil
3. *Required land situations* — High land
4. *Importance of crop* — Cob with silk and tussel
5. *Fertilizer management* — 120 kg N, 60 kg P_2O_5 and 60 kg K_2O/ha
6. *Water management* — 4-5 irrigations

7. *Duration of crop* — 110-140 days
8. *Parts used* — Entire green plant as fodder, cut at flowering
9. *Where grown in India* — U.P., Punjab, M.P., Bihar, West Bengal, A.P. and J. & K.
10. *Export possibility* — High
11. *Economic yields (q / ha)* — 180-260 (green)
12. *Economic value (Rs. / ha)* — 40,000 to 80,000
13. *By-product (If any)* — Nil
14. *Use of by-product* — Nil.

164. *CROP* — PEARL MILLET / *Bajra*

Scientific Name* — *Penisetum typhoides Stapf and C.E

Family — Gramineae

Herb / Shrub / Tree — Herb

Annual / Biennial / Perennial — Annual

Used as — Fodder and also for food grains

1. *Climatic requirements* — Less rainfall and high temperature
2. *Soil requirements* — Sandy loam / loamy / sandy soils
3. *Required land situations* — High land
4. *Importance of crop* — Blunt compact inflorescence
5. *Fertilizer management* — 60 kg N, 30 kg P_2O_5 and 30 kg K_2O/ha
6. *Water management* — Normally rainfed; 1-2 times irrigation, if irrigated
7. *Duration of crop* — 110-130 days
8. *Parts used* — Entire plant (stem + leaf + inflorescence) cut at flowering
9. *Where grown in India* — In drier tracts
10. *Export possibility* — High
11. *Economic yields (q / ha)* — 175-250 (green)
12. *Economic value (Rs. / ha)* — 25,000 to 50,000
13. *By-product (If any)* — Nil
14. *Use of by-product* — Nil.

165. *CROP* — SORGHUM / Joar / Jowar

Scientific Name* — *Sorghum vulgare Pers.

Family — Gramineae

Herb / Shrub / Tree — Herb, 1.0-3.0 m height

Annual / Biennial / Perennial — Annual

Used as — Fodder and also for food grains

1. *Climatic requirements* — Low rainfall with high temperature
2. *Soil requirements* — Loamy / sandy loam / sandy soils
3. *Required land situations* — High
4. *Importance of crop* — Growth both for grains and fodder
5. *Fertilizer management* — 60 kg N, 30 kg P_2O_5 and 30 kg K_2O/ha
6. *Water management* — Normally rainfed; 1-2 irrigation, if irrigated
7. *Duration of crop* — 110-140 days
8. *Parts used* — Entire plant (stem + leaves + inflorescence) cut at flowering
9. *Where grown in India* — U.P., Punjab, Madhya Pradesh, A.P., Karnataka, T.N., Maharashtra and Rajasthan
10. *Export possibility* — High
11. *Economic yields (q / ha)* — 170-200 (green)
12. *Economic value (Rs. / ha)* — 30,000 to 50,000
13. *By-product (If any)* — Nil
14. *Use of by-product* — Nil.

166. *CROP* — PEARL MILLET / Bajra

Scientific Name — Pennisetum typoides Stapf and C.E.

Family — Gramineae

Herb / Shrub / Tree — Herb, 0.5-1.5 m height

Annual / Biennial / Perennial — Annual

Used as — Fodder / food

1. *Climatic requirements* — Warm climate
2. *Soil requirements* — Sandy / loamy / alluvial soils
3. *Required land situations* — High / medium land
4. *Importance of crop* — Sparing cob-like structure
5. *Fertilizer management* — 60 kg N, 30 kg P_2O_5 and 30 kg K_2O/ha
6. *Water management* — 3-4 irrigations
7. *Duration of crop* — 120-130 Days
8. *Parts used* — Whole plant
9. *Where grown in India* — In the drier tracts of India
10. *Export possibility* — High
11. *Economic yields (q / ha)* — 150-200 (green)

12. *Economic value (Rs. / ha)* — 15,000 to 20,000
13. *By-product (If any)* — Nil
14. *Use of by-product* — Nil

167. CROP — ANJAN GRASS / Cenchrus ghash / Anjan

Scientific Name — Cenchrus setigerus Vahl

Family — Gramineae

Herb / Shrub / Tree — Herb, Height 30-80 cm

Annual / Biennial / Perennial — Annual

Used as — Fodder / soil binding material

1. *Climatic requirements* — Dry climate with high temperature and less water
2. *Soil requirements* — Loamy / sandy soils
3. *Required land situations* — High land
4. *Importance of crop* — Used in soil conservation due to more rooting and growing in dry condition
5. *Fertilizer management* — 100 kg N, 30 kg P_2O_5 and 30 kg K_2O/ha
6. *Water management* — Normally rainfed crops; 1-2 irrigations, irrigated
7. *Duration of crop* — 110-140 days
8. *Parts used* — Entire plant (stem + inflorescence + leaves) cut at flowering
9. *Where grown in India* — All over India, except too hot or too low temperature areas
10. *Export possibility* — High
11. *Economic yields (q / ha)* — 160-180 (green)
12. *Economic value (Rs. / ha)* — 30,000 to 40,000
13. *By-product (If any)* — Nil
14. *Use of by-product* — Nil

168. CROP — PARAGRASS /Para /Dalghash / Mauritius grass

Scientific Name — Brachiaria mutica Stapf

Family — Gramineae

Herb / Shrub / Tree — Herb, 0.5-1.5 long length

Annual / Biennial / Perennial — Perennial

Used as — Fodder

1. *Climatic requirements* — High rainfall and high temperature

2. *Soil requirements* — Clayey / loamy / alluvial soils
3. *Required land situations* — Low / medium / marshy land
4. *Importance of crop* — It can be grown in waterlogged low land and in saline soils
5. *Fertilizer management* — 60 kg N, 30 kg P_2O_5 and 30 kg K_2O/ha
6. *Water management* — Normally grown in low marshy land; 3-4 irrigations, if in medium land
7. *Duration of crop* — 2-3 years
8. *Parts used* — Entire plant cut at flowering or before
9. *Where grown in India* — In Eastern parts of India (West Bengal, Assam, Bihar, Orissa, Tripura and Manipur)
10. *Export possibility* — High
11. *Economic yields (q / ha)* — 100-200
12. *Economic value (Rs. / ha)* — 50,000 to 60,000
13. *By-product (If any)* — Nil
14. *Use of by-product* — Nil

169. *CROP* — RHODES GRASS / *Rode ghash*

Scientific Name* — *Chloris gayana Kunth

Family — Gramineae

Herb / Shrub / Tree — Herb, 0.5-1.0 m height

Annual / Biennial / Perennial — Perennial

Used as — Fodder

1. *Climatic requirements* — Moderate rainfall and moderate temperature
2. *Soil requirements* — Loamy / sandy soils
3. *Required land situations* — High / medium land
4. *Importance of crop* — Excellent 'hay' is produced from it
5. *Fertilizer management* — 60 kg N, 30 kg P_2O_5 and 30 kg K_2O/ha
6. *Water management* — 3-4 irrigations
7. *Duration of crop* — 110-140 days
8. *Parts used* — Whole plant cut at flowering
9. *Where grown in India* — U.P., Punjab, West Bengal, Bihar, Orissa, Maharashtra etc.
10. *Export possibility* — High
11. *Economic yields (q / ha)* — 70-80 (Hay)
12. *Economic value (Rs. / ha)* — 35,000 to 40,000

13. *By-product (If any)* — Nil
14. *Use of by-product* — Nil

170. *CROP* — SUDAN GRASS / *Sudan ghash*

Scientific Name* — *Sorghum sudanense (Piper) Stapf

Family — Gramineae

Herb / Shrub / Tree — Herb, 1.0-3.0 m height

Annual / Biennial / Perennial — Annual

Used as — Fodder

1. *Climatic requirements* — Medium rainfall and high temperature
2. *Soil requirements* — Alluvial / loamy / sandy loam soils
3. *Required land situations* — High
4. *Importance of crop* — Tall plants
5. *Fertilizer management* — 60 kg N, 30 kg P_2O_5 and 30 kg K_2O/ha
6. *Water management* — 3-4 irrigations
7. *Duration of crop* — 110-140 Days
8. *Parts used* — Entire plant cut at flowering
9. *Where grown in India* — Maharashtra, Assam, Punjab, W.B., T.N. etc.
10. *Export possibility* — High
11. *Economic yields (q / ha)* — 180-200 (green)
12. *Economic value (Rs. / ha)* — 40,000 to 50,000
13. *By-product (If any)* — Nil
14. *Use of by-product* — Nil

NON–GRASS

171. CROP — BLACK GRAM / Urd / Mash kalai / Biuri dal

Scientific Name* — *Vigna mungo Hepper

Family — Papilionaceae

Herb / Shrub / Tree — Herb, 25-50 cm long length

Annual / Biennial / Perennial — Annual

Used as — Fodder / grain / green manure / cover crop

1. *Climatic requirements* — High / Medium rainfall with high temperature
2. *Soil requirements* — Loam / alluvial / sandy loam soils
3. *Required land situations* — High land

4. *Importance of crop* — This can be used as grain (pulse), fodder and green manuring crop even as *paira* crop
5. *Fertilizer management* — 40 kg N, 60 kg P_2O_5 and 40 kg K_2O/ha
6. *Water management* — Normally rainfed; 1-2 irrigations, if irrigated
7. *Duration of crop* — 100-130 days
8. *Parts used* — Whole plant for fodder, cut at flowering
9. *Where grown in India* — All over India, mostly in high rainfall areas
10. *Export possibility* — High
11. *Economic yields (q / ha)* — 100-200
12. *Economic value (Rs. / ha)* — 25,000 to 50,000
13. *By-product (If any)* — Nil
14. *Use of by-product* — Nil

172. *CROP* — RICE BEAN / Sutri / *Gaimug*

Scientific Name* — *Vigna umbellata (Thunb) Ohwi and Ohashi

Family — Papilionaceae

Herb / Shrub / Tree — Trailing herb, 25-90 cm long length

Annual / Biennial / Perennial — Annual

Used as — Fodder (also as pulse or green manure crop)

1. *Climatic requirements* — Medium rainfall with high temperature
2. *Soil requirements* — Alluvial / loamy / sandy loam soils
3. *Required land situations* — High land
4. *Importance of crop* — This crop can be used as fodder, pulse; pods (vegetable) and as green manure crop and also to control weeds
5. *Fertilizer management* — 50 kg N, 60 kg P_2O_5 and 50 kg K_2O/ha
6. *Water Management*—Normally rainfed; 1-2 irrigations, if irrigated
7. *Duration of crop* — 110-140 days
8. *Parts used* — Whole plant as fodder, cut at flowering
9. *Where grown in India* — All over in India except too low / too high temperature areas
10. *Export possibility* — High
11. *Economic yields (q / ha)* — 100-180
12. *Economic value (Rs. / ha)* — 25,000 to 45,000
13. *By-product (If any)* — Nil
14. *Use of by-product* — Nil

173. *CROP* — MUNG BEAN / Mung / *Mug* / Pessara / Green gram

Scientific Name* — *Vigna radiata (L.) Wilczek

Family — Papilionaceae

Herb / Shrub / Tree — Herb, Height 30-50 cm

Annual / Biennial / Perennial — Annual

Used as — Fodder (as pulse or as green manure crop)

1. *Climatic requirements* — High/low rainfall with high temperature
2. *Soil requirements* — Alluvial / loamy / sandy loam soils
3. *Required land situations* — High land
4. *Importance of crop* — This crop can be used as a pulse crop or green manure crop
5. *Fertilizer management* — 40 kg N, 60 kg P_2O_5 and 40 kg K_2O/ha
6. *Water management* — 1-2 irrigations
7. *Duration of crop* — 100-120 days
8. *Parts used* — Whole plant for fodder, cut at flowering
9. *Where grown in India* — All over India except too low / too high temperature areas.
10. *Export possibility* — High
11. *Economic yields (q / ha)* — 100-150
12. *Economic value (Rs. / ha)* — 25,000 to 40,000
13. *By-product (If any)* — Nil
14. *Use of by-product* — Nil

174. *CROP* — COWPEA / Lobia / Chowli / *Barbati / Lalsa*

Scientific Name* — *Vigna unguiculata (L.) Walp

Family — Papilionaceae

Herb / Shrub / Tree — Spreading herb, 1.0-3.0 m long length

Annual / Biennial / Perennial — Annual

Used as — Fodder [used as vegetable (pod) or pulse]

1. *Climatic requirements* — Medium rainfall with high temperature
2. *Soil requirements* — Alluvial / loamy / sandy loam soils
3. *Required land situations* — High land
4. *Importance of crop* — This crop is used as vegetable / pulse crop also
5. *Fertilizer management* — 40 kg N, 60 kg P_2O_5 and 50 kg K_2O/ha
6. *Water management* — Normally rainfed; 1-2 irrigations, if irrigated

7. *Duration of crop* — 110-130 days
8. *Parts used* — Entire plant, cut at flowering
9. *Where grown in India* — All over India except too low / too high temperature areas
10. *Export possibility* — High
11. *Economic yields (q / ha)* — 100-150
12. *Economic value (Rs. / ha)* — 25,000 to 75,000
13. *By-product (If any)* — Nil
14. *Use of by-product* — Nil

175. *CROP* — CLUSTER BEAN / Guar / Gawar / Jatabin

Scientific Name* — *Cyamopsis tetragonoloba (L.) Taub

Family — Papilionaceae

Herb / Shrub / Tree — Herb, 50-90 cm height

Annual / Biennial / Perennial — Annual

Used as — Fodder / green manure crop / vegetable

1. *Climatic requirements* — Moderate temperature with moderate rainfall
2. *Soil requirements* — Alluvial / loamy / sandy loam soils
3. *Required land situations* — High land
4. *Importance of crop* — Bushy crop with clusterbean used as vegetable; green manure crop also
5. *Fertilizer management* — 40 kg N, 60 kg P_2O_5 and 40 kg K_2O/ha
6. *Water management* — Normally rainfed; 1-2 irrigations if irrigated
7. *Duration of crop* — 100-130 days
8. *Parts used* — Entire plant for fodder cut at flowering
9. *Where grown in India* — In dry tracts except too low / too high temperature areas.
10. *Export possibility* — High
11. *Economic yields (q / ha)* — 100-160
12. *Economic value (Rs. / ha)* — 25,000 to 40,000
13. *By-product (If any)* — Nil
14. *Use of by-product* — Nil

176. *CROP* — MOTH BEAN / MAT BEAN / Moth / Bhringga / *Kultikalai*

Scientific Name* — *Vigna aconitifolia Marechal.

Family — Papilionaceae

***Herb / Shrub / Tree* —** Herb, (trailing) 1.0-2.5 m length

***Annual / Biennial / Perennial* —** Annual

***Used as* —** Fodder / vegetable / pulse

1. *Climatic requirements* — Moderate rainfall with high temperature
2. *Soil requirements* — Loamy / sandy loam soils
3. *Required land situations* — High land
4. *Importance of crop* — Can be grown as pulse, green manure also
5. *Fertilizer management* — 40 kg N, 60 kg P_2O_5 and 40 kg K_2O/ha
6. *Water management* — Normally rainfed; 1-2 irrigations, if irrigated
7. *Duration of crop* — 110-130 days
8. *Parts used* — Whole plant cut at flowering
9. *Where grown in India* — All over India except too low / too high temperature areas
10. *Export possibility* — High
11. *Economic yields (q / ha)* — 100-150
12. *Economic value (Rs. / ha)* — 20,000 to 40,000
13. *By-product (If any)* — Nil
14. *Use of by-product* — Nil

177. *CROP* — FIELD PEA / GARDEN PEA / Matar / *Karaisuti*

Scientific Name* — *Pisum sativum L. var. ***arvense*** Poir.

***Family* —** Papilionaceae

***Herb / Shrub / Tree* —** Herb, 0.5-1.0 m length

***Annual / Biennial / Perennial* —** Annual

***Used as* —** Fodder / vegetable / pulse (green seed used as vegetable).

1. *Climatic requirements* — Low temperature with low rainfall
2. *Soil requirements* — Loamy / sandy / alluvial soils
3. *Required land situations* — High land
4. *Importance of crop* — Some are big or small seeds, big seeds (pulse)
5. *Fertilizer management* — 40 kg N, 60 kg P_2O_5 and 50 kg K_2O/ha
6. *Water management* — 3-4 irrigations
7. *Duration of crop* — 110-130 days
8. *Parts used* — Whole plant for fodder, cut at flowering
9. *Where grown in India* — U.P., Punjab, Haryana, Rajasthan, Madhya Pradesh, Bihar, Karnataka H.P., West Bengal, etc.

10. *Export possibility* — High
11. *Economic yields (q / ha)* — 100-150
12. *Economic value (Rs. / ha)* — 25,000 to 40,000
13. *By-product (If any)* — Nil
14. *Use of by-product* — Nil

178. *CROP* — HORSE GRAM / Kulthi / *Kulatrakalai*

Scientific Name* — *Dolichos uniflorus Lam.

Family — Papilionaceae

Herb / Shrub / Tree — Herb, (Trailing), 0.5-1.0 m length

Annual / Biennial / Perennial — Annual

Used as — Fodder / pulse / medicine (seeds)

1. *Climatic requirements* — High / medium temperature with low rainfall
2. *Soil requirements* — Loamy / sandy loam soils
3. *Required land situations* — High land
4. *Importance of crop* — It is used as creeper in fence also
5. *Fertilizer management* — 40 kg N, 60 kg P_2O_5 and 50 kg K_2O/ha
6. *Water management* — Normally rainfed; 1-2 irrigations, if irrigated
7. *Duration of crop* — 110-130 days
8. *Parts used* — Entire plant for fodder, cut at flowering
9. *Where grown in India* — A.P., Karnataka, T.N. and Bihar etc.
10. *Export possibility* — High
11. *Economic yields (q / ha)* — 80-160 (green fodder), seed 5-6
12. *Economic value (Rs. / ha)* — 20,000 to 80,000
13. *By-product (If any)* — Nil
14. *Use of by-product* — Nil

179. *CROP* — TURNIP / Shaljam / Salgam

Scientific Name* — *Brassica rapa L.

Family — Cruciferae

Herb / Shrub / Tree — Herb, 25-50 cm height

Annual / Biennial / Perennial — Annual

Used as — Fodder (branched) / vegetable (swollen roots)

1. *Climatic requirements* — Low temperature with dry weather
2. *Soil requirements* — Loamy / alluvial / sandy loam soils

3. *Required land situations* — High land
4. *Importance of crop* — Swollen roots used as vegetable also
5. *Fertilizer management* — 100 kg N, 50 kg P_2O_5 and 50 kg K_2O/ha
6. *Water management* — 2-3 irrigations
7. *Duration of crop* — 110-130 days
8. *Parts used* — Whole plant, cut at flowering
9. *Where grown in India* — Punjab, U.P., Bihar, Maharashtra etc.
10. *Export possibility* — Less
11. *Economic yields (q / ha)* — 180-260
12. *Economic value (Rs. / ha)* — 20,000 to 40,000
13. *By-product (If any)* — Nil
14. *Use of by-product* — Nil

180. *CROP* — BUTTERFLY PEA / Chouko matar

Scientific Name* — *Centrosema pubescens Bength

Family — Papilionaceae

Herb / Shrub / Tree — Herb, (Trailing / spreading)

Annual / Biennial / Perennial — Perennial

Used as — Fodder / for checking soil erosion

1. *Climatic requirements* — Moderate rainfall and moderate temperature
2. *Soil requirements* — Loamy / sandy loam soils
3. *Required land situations* — High land
4. *Importance of crop* — Cover crop for coffee / prevents soil erosion
5. *Fertilizer management* — 40 kg N, 60 kg P_2O_5 and 50 kg K_2O/ha
6. *Water management* — 3-4 irrigations
7. *Duration of crop* — 2-3 years
8. *Parts used* — Entire plant cut at flowering
9. *Where grown in India* — Grown in the South, in coffee plantation
10. *Export possibility* — High
11. *Economic yields (q / ha)* — 80-160 (green)
12. *Economic value (Rs. / ha)* — 20,000 to 40,000
13. *By-product (If any)* — Nil
14. *Use of by-product* — Nil

9. GREEN MANURE CROPS

LEGUMES

181. ***CROP* — DHAINCHA / Dhencha / Baghadhaincha / *Jayanti***

Scientific Name* — *Sesbania aculeata Pers.

Family — Papilionaceae

Herb / Shrub / Tree — Herb, tall-1.0-3.0 m height)

Annual / Biennial / Perennial — Annual

Used as — Green manure crop / used for reclaiming saline or alkaline soils

1. *Climatic requirements* — High rainfall with high temperature
2. *Soil requirements* — Alluvial / clayey / loamy / sandy loam soils
3. *Required land situations* — Low land / medium land
4. *Importance of crop* — This crop can be used to reclaim saline
5. *Fertilizer management* — 40 kg N, 60 kg P_2O_5 and 50 kg K_2O/ha
6. *Water management* — Normally rainfed; 1-2 irrigation, if irrigated
7. *Duration of crop* — 110-130 Days
8. *Parts used* — Entire plant ploughed-under (at tender age)
9. *Where grown in India* — In the humid areas of India
10. *Export possibility* — Less
11. *Economic yields (q / ha)* — 150-200 (biomass)
12. *Economic value (Rs. / ha)* — Incorporated into soils, hence no value
13. *By-product (If any)* — Nil
14. *Use of by-product* — Nil

182. ***CROP* — SUNNHEMP / San / Sunn / *Shan***

Scientific Name* — *Crotalaria juncea L.

Family — Papilionaceae

Herb / Shrub / Tree — Shrub, tall-1.0 to 3.0 m height

Annual / Biennial / Perennial — Annual

Used as — Green manure crop (at tender age) / Fibre

1. *Climatic requirements* — High rainfall with high temperature
2. *Soil requirements* — Loamy / alluvial / sandy loam soils
3. *Required land situations* — High / medium land

4. *Importance of crop* — This can be grown as fibre crop. For seed it can be grown as *paira* crop also in winter after *Kharif* rice.
5. *Fertilizer management* — 40 kg N, 60 kg P_2O_5 and 60 kg K_2O/ha
6. *Water management* — Normally rainfed; 1-2 irrigations, if irrigated
7. *Duration of crop* — 110-130 days
8. *Parts used* — Entire plant ploughed-under (at tender age)
9. *Where grown in India* — In humid / semi humid areas in India
10. *Export possibility* — Less
11. *Economic yields (q / ha)* — 150-200 (biomass)
12. *Economic value (Rs. / ha)* — Incorporated into soils, hence no value
13. *By-product (If any)* — Nil
14. *Use of by-product* — Nil

183. *CROP* — GIANT DHAINCHA / Bagha dhaincha / Sada dhaincha

Scientific Name — Sesbania rostrata

Family — Papilionaceae

Herb / Shrub / Tree — Shrub, (tall) 2.0 to 3.5 m height

Annual / Biennial / Perennial — Annual

Used as — Green manure crop / paper pulp

1. *Climatic requirements* — Moderate rainfall with moderate temperature
2. *Soil requirements* — Alluvial / loamy / sandy loam soils
3. *Required land situations* — High land
4. *Importance of crop* — This stem nodulating *dhaincha*, can be used in the forestry for N-economy and for paper-pulp (white wood)
5. *Fertilizer management* — 40 kg N, 60 kg P_2O_5 and 50 kg K_2O/ha
6. *Water management* — Normally rainfed but 1-2 irrigations, if irrigated
7. *Duration of crop* — 110-140 days
8. *Parts used* — Whole plant ploughed-under (at tender age)
9. *Where grown in India* — In humid / sub-humid areas of India
10. *Export possibility* — Less
11. *Economic yields (q / ha)* — 150-200 (biomass)
12. *Economic value (Rs. / ha)* — Incorporated into soils
13. *By-product (If any)* — Nil
14. *Use of by-product* — Nil

184. ***CROP*** **— BLACK GRAM / Urd. / Mash kalai / Kalai / Biuridal**

Scientific Name — Vigna mungo L. Hepper

Family — Papilionaceae

Herb / Shrub / Tree — Herb, 30-60 cm height

Annual / Biennial / Perennial — Annual

Used as — Green manure crop / pulse / cover crop / fodder

1. *Climatic requirements* — Medium rainfall with high temperature
2. *Soil requirements* — Alluvial / loamy / sandy loam soils
3. *Required land situations* — High land
4. *Importance of crop* — This is a good pulse crop. South Indian foods (*dosa*) are prepared from it mostly
5. *Fertilizer management* — 40 kg N, 60 kg P_2O_5 and 50 kg K_2O/ha
6. *Water management* — Normally rainfed; 1-2 irrigations, if irrigated
7. *Duration of crop* — 110-130 days
8. *Parts used* — Entire plant ploughed under (at tender age)
9. *Where grown in India* — In humid / semi humid areas of India
10. *Export possibility* — High
11. *Economic yields (q / ha)* — 150-200 (biomass)
12. *Economic value (Rs. / ha)* — Incorporated into soils
13. *By-product (If any)* — Nil
14. *Use of by-product* — Nil

185. ***CROP*** **— GREEN GRAM / Mung / Mug / Mungbean / Mong**

Scientific Name — Vigna radiata (L.) Wilczek.

Family — Papilionaceae

Herb / Shrub / Tree — Herb, Height 0.5-0.7 m

Annual / Biennial / Perennial — Annual

Used as — Green manure crop / Pulse (seed)

1. *Climatic Requirements*— Medium rainfall with high temperature
2. *Soil requirements* — Loamy / alluvial / sandy loam soils
3. *Required land situations* — High land
4. *Importance of crop* — This is used as pulse crop in rainfed cropping system
5. *Fertilizer management* — 40 kg N, 60 kg P_2O_5 and 50 kg K_2O/ha
6. *Water management* — 1-2 irrigations, if irrigated; normally rainfed

7. *Duration of crop* — 90-110 days
8. *Parts used* — Whole plant ploughed under for green manure
9. *Where grown in India* — All over India except desert / snowy areas
10. *Export possibility* — High
11. *Economic yields (q / ha)* — 150-200 (biomass)
12. *Economic value (Rs. / ha)* — Incorporated into soils
13. *By-product (If any)* — Nil
14. *Use of by-product* — Nil

186. *CROP* — COW PEA / Lobia / Chowli / *Barbati* / *Lalsa*

Scientific Name* — *Vigna unguiculata (L.) Walp

Family — Papilionaceae

Herb / Shrub / Tree — Spreading / trailing herb, (1.0 to 3.0 m length)

Annual / Biennial / Perennial — Annual

Used as — Green manure crop / vegetable / fodder / cover crop

1. *Climatic Requirements*— Moderate rainfall with high temperature
2. *Soil requirements* — Alluvial / loamy / sandy loam soils
3. *Required land situations* — High land
4. *Importance of crop* — Improves the soil organic matter and used as vegetable (pods)
5. *Fertilizer management* — 40 kg N, 60 kg P_2O_5 and 50 kg K_2O/ha
6. *Water Management*—Normally rainfed; 1-2 irrigations, if irrigated
7. *Duration of crop* — 110-130 days
8. *Parts used* — Entire plant ploughed-under (at tender age)
9. *Where grown in India* — All over India except too low / too high temperature
10. *Export possibility* — High (for long pods) if fredging temperature in container is available
11. *Economic yields (q / ha)* — 150-200 (green biomass)
12. *Economic value (Rs. / ha)* — Incorporated into soils
13. *By-product (If any)* — Nil
14. *Use of by-product* — Nil

187. *CROP* — CLUSTER BEAN / Guar / Gawar / Jatabin

Scientific Name* — *Cyamopsis tetragonoloba (L.) Taub

Family — Papilionaceae

Herb / Shrub / Tree — Bushy Herb, 50-80 cm height

***Annual / Biennial / Perennial* —** Annual

***Used as* —** Green manure crop / vegetable / fodder

1. *Climatic requirements* — Moderate rainfall with high temperature
2. *Soil requirements* — Loamy / alluvial / sandy loam soils
3. *Required land situations* — High land
4. *Importance of crop* — This can be used as vegetable (pod) crop also
5. *Fertilizer management* — 40 kg N, 60 kg P_2O_5 and 50 kg K_2O/ha
6. *Water management* — Normally rainfed; 1-2 irrigations, if irrigated
7. *Duration of crop* — 90-110 days
8. *Parts used* — Whole plant ploughed-under (before flowering stage)
9. *Where grown in India* — In the drier tract of India
10. *Export possibility* — Less
11. *Economic yields (q / ha)* — 150-200 (biomass)
12. *Economic value (Rs. / ha)* — Incorporated into soils for increasing soil fertility
13. *By-product (If any)* — Nil
14. *Use of by-product* — Nil

188. *CROP* — RICE BEAN / Sutri / *Gaimug*

Scientific Name* — *Vigna umbellata (Thunb.) Ohwi and Ohashi

***Family* —** Papilionaceae

***Herb / Shrub / Tree* —** Trailing / spreading Herb, 0.5-1.0 m length

***Annual / Biennial / Perennial* —** Annual

***Used as* —** Green manure crop / fodder / cover crop

1. *Climatic requirements* — Moderate rainfall with high temperature
2. *Soil requirements* — Alluvial / loamy / sandy loam soils
3. *Required land situations* — High land
4. *Importance of crop* — This may be grown as pulse crop also
5. *Fertilizer management* — 40 kg N, 60 kg P_2O_5 and 50 kg K_2O/ha
6. *Water Management*—Normally rainfed; 1-2 irrigations, if irrigated
7. *Duration of crop* — 110-130 days
8. *Parts used* — Entire plant (luxuriant growth) ploughed-under
9. *Where grown in India* — In humid / semi humid areas
10. *Export possibility* — Less
11. *Economic yields (q / ha)* — 200-250 (biomass)

12. *Economic value (Rs. / ha)* — Incorporated into soils
13. *By-product (If any)* — Nil
14. *Use of by-product* — Nil

189. *CROP* — GLIRICIDIA / MADRE TREE / Spotted gliricidia

Scientific Name* — *Gliricidia maculata Stend or ***G. sepium*** Walp

Family — Papilionaceae

Herb / Shrub / Tree — Tree (small)

Annual / Biennial / Perennial — Perennial

Used as — Green leaf manure / ornamental plant

1. *Climatic requirements* — Moderate temperature with moderate rainfall
2. *Soil requirements* — Alluvial / loam / sandy loam soils
3. *Required land situations* — High
4. *Importance of crop* — Good ornamental flowers (for beautification)
5. *Fertilizer management* — 40 kg N, 60 kg P_2O_5 and 60 kg K_2O/ha
6. *Water management* — Normally rainfed
7. *Duration of crop* — 2-4 years
8. *Parts used* — Leaves and twigs (for green-leaf-manure)
9. *Where grown in India* — T.N., A.P., Karnataka, Kerala, Maharashtra, West Bengal
10. *Export possibility* — High
11. *Economic yields (q / ha)* — 200-250 (green leaf and twig)
12. *Economic value (Rs. / ha)* — Incorporated into soils (rice fields)
13. *By-product (If any)* — Branches and trunk
14. *Use of by-product* — Fuel / organic mater / timber / pegs / fuel.

190. *CROP* — Maize / Corn / *Bhutta* / *Makai* / Indian corn

Scientific Name* — *Zea mays L.

Family — Gramineae

Herb / Shrub / Tree — Tall herb, 2.0 to 3.5 m height

Annual / Biennial / Perennial — Annual

Used as — Green manure crop / food crop / fodder

1. *Climatic requirements* — Moderate rainfall with moderate temperature
2. *Soil requirements* — Alluvial / loamy / sandy loam soils

3. *Required land situations* — High land
4. *Importance of crop* — Silks and tussles are present and cob is a speciality
5. *Fertilizer management* — 120kg N, 60 kg P_2O_5 and 60 kg K_2O/ha
6. *Water management* — 4-5 irrigations
7. *Duration of crop* — 110-130 days
8. *Parts used* — Ploughed-under entire plant (before flowering)
9. *Where grown in India* — U.P., Punjab, M.P., Bihar, A.P., W.B. and J. & K.
10. *Export possibility* — High
11. *Economic yields (q / ha)* — 200-400 (biomass)
12. *Economic value (Rs. / ha)* — Incorporated into soils
13. *By-product (If any)* — Nil
14. *Use of by-product* — Nil

191. *CROP* — NAPIER GRASS / ELEPHANT GRASS

Scientific Name* — *Pennisetum purpureum Schum

Family — Gramineae

Herb / Shrub / Tree — Tall herb, (Height 1.5-2.5 m)

Annual / Biennial / Perennial — Annual

Used as — Green manure crop

1. *Climatic requirements* — High rainfall with high temperature
2. *Soil requirements* — Alluvial / loamy / sand loam soils
3. *Required land situations* — High land
4. *Importance of crop* — Stem like sugarcane
5. *Fertilizer management* — 100kg N, 50 kg P_2O_5 and 50 kg K_2O/ha
6. *Water management* — 4-5 irrigations
7. *Duration of crop* — 110-130 days
8. *Parts used* — Entire plant ploughed under (before flowering)
9. *Where grown in India* — Assam, Bengal, Bihar, U.P., Punjab
10. *Export possibility* — High
11. *Economic yields (q / ha)* — 200-250 (biomass)
12. *Economic value (Rs. / ha)* — Incorporated into soils
13. *By-product (If any)* — Nil
14. *Use of by-product* — Nil

192. *CROP* — MESTA / ROSELLE / *Mestapat*

***Scientific Name* — *Hibiscus sabdariffa* L. / *H. cannabinus* L.**

Family — Malvaceae

Herb / Shrub / Tree — Tall shrub (1.5-2.5 m height)

Annual / Biennial / Perennial — Annual

Used as — Green manure crop / fibre crop

1. *Climatic Requirements*—High rainfall and high temperature areas
2. *Soil requirements* — Alluvial / loamy / sandy soils
3. *Required land situations* — Low / medium land
4. *Importance of crop* — A bast fibre crop, having big flowers
5. *Fertilizer management* — 60 kg N, 30 kg P_2O_5 and 30 kg K_2O/ha
6. *Water management* — Normally rainfed, 1-2 irrigations, if irrigated
7. *Duration of crop* — 110-140 days
8. *Parts used* — Entire plant ploughed under (before flowering)
9. *Where grown in India* — W.B., Assam, Bihar, Orissa, A.P. etc.
10. *Export possibility* — Less
11. *Economic yields (q / ha)* — 200-250 (biomass)
12. *Economic value (Rs. / ha)* — Entire plant is used as green manure
13. *By-product (If any)* — Nil
14. *Use of by-product* — Nil

193. *CROP* — JUTE / WHITE JUTE / Pat / *Titapat* / *Mithapat* / *Pattam* / Jhute

***Scientific Name* — *Corchorus capsularis* / *olitorius* L.**

Family — Tiliaceae

Herb / Shrub / Tree — A tall shrub (1.5-2.5 m height)

Annual / Biennial / Perennial — Annual

Used as — Green manure crop (for making unavailable P2O5 available)

1. *Climatic requirements* — High rainfall, high temperature and high humidity
2. *Soil requirements* — Alluvial / loamy / sandy loam / clayey soils
3. *Required land situations* — Low / medium lands
4. *Importance of crop* — Also a bast fibre crop
5. *Fertilizer management* — 80-100kg N, 50 kg P_2O_5 and 50 kg K_2O/ha

6. *Water Management*—Normally rainfed; 1-2 irrigations, if irrigated
7. *Duration of crop* — 110-140 days
8. *Parts used* — Entire plant ploughed-under (at tender age)
9. *Where grown in India* — West Bengal, Assam, Bihar, Orissa, U.P. and Tripura
10. *Export possibility* — High
11. *Economic yields (q / ha)* — 150-200 (biomass)
12. *Economic value (Rs. / ha)* — Incorporated into soils at tender age
13. *By-product (If any)* — Nil
14. *Use of by-product* — Nil

10. MEDICINAL PLANTS (ANNUAL)

194. *CROP* — RAUVOLFIA ROOT / SERPENTINE / Chandra bhaga / Chota-chand / Sarpa gandha

Scientific Name* — *Rauvolfia serpentina Bength ex Kurtz.

Family — Apocynaceae

Herb / Shrub / Tree — Trailing / spreading herb, 1.5-2.0 m long length

Annual / Biennial / Perennial — Perennial

Used as — Medicinal plans against snake, other diseases

1. *Climatic requirements* — Low or moderate temperature in hills / plains
2. *Soil requirements* — Stony soil with organic matter / alluvial / loamy soils
3. *Required land situations* — High land
4. *Importance of crop* — Widely distributed in hills roots and leaves are valuable against snakes.
5. *Fertilizer management* — Soils full of organic matter containing nitrogen, phosphate and potash. N, P_2O_5 and K_2O @ 80, 50 and 30 kg/ha respectively.
6. *Water management* — Rainfed; 1-2 irrigations, if irrigated
7. *Duration of crop* — 2-3 years
8. *Parts used* — Roots for medicine (air dried roots for antihypertension and as sedatives, snake repelling)
9. *Where grown in India* — Hilly regions of Himalayas, Western ghats
10. *Export possibility* — Very high
11. *Economic yields (q / ha)* — 50-90 (dry roots)

12. *Economic value (Rs. / ha)* — 1,00,000–2,00,000
13. *By-product (If any)* — Stem and leaves
14. *Use of by-product* — Fuel, cattle feed / medicine / organic matter / mulch.

195. *CROP* — GINGER / Adrak / Ada

Scientific Name— ***Zingiber officinale*** Rosc

Family — Zingiberaceae

Herb / Shrub / Tree — Rhizomatous herb

Annual / Biennial / Perennial — Annual

Used as — Medicine (as carminative and digestive stimulant) / spice and condiment / medicine

1. *Climatic requirements* — Low / medium temperature with high rainfall
2. *Soil requirements* — Soils with full of organic matter; loamy or clayey soils
3. *Required land situations* — High / medium land
4. *Importance of crop* — Rhizome is used as spice
5. *Fertilizer management* — 120kg N, 60 kg P_2O_5 and 60 kg K_2O/ha
6. *Water management* — 3-4 irrigations
7. *Duration of crop* — 110-140 days
8. *Parts used* — Rhizomes
9. *Where grown in India* — Kerala, Maharashtra, Uttar Pradesh, West Bengal, H.P., A.P., Bihar, Orissa, Assam etc.
10. *Export possibility* — Very high
11. *Economic yields (q / ha)* — 100-200 (green)
12. *Economic value (Rs. / ha)* — 1,00,000 to 2,00,000
13. *By-product (If any)* — Stem and leaves
14. *Use of by-product* — Fuel, cattle feed / organic matter / vegetable / spice.

196. CROP — TURMERIC / Haldi / Halud

Scientific Name — Curcuma domestica Vahl

Family — Zingiberaceae

Herb / Shrub / Tree — Herb (Height 50-80 cm height)

Annual / Biennial / Perennial — Perennial

Used as — Medicine as, stomachic tonic, blood purifier, antiseptic to wound and skin / spice

1. *Climatic requirements* — Moderate rainfall with moderate temperature
2. *Soil requirements* — Loamy / sandy / alluvial soils
3. *Required land situations* — High / medium land
4. *Importance of crop* — This rhizome is used as spice also
5. *Fertilizer management* — 60kg N, 30 kg P_2O_5 and 30 kg K_2O/ha
6. *Water management* — Normally rainfed; 1-2 irrigations, if irrigated
7. *Duration of crop* — 2-3 years
8. *Parts used* — Rhizomes
9. *Where grown in India* — Tamil Nadu, A.P., Maharashtra, Bihar, West Bengal, Kerala and Orissa
10. *Export possibility* — Very High
11. *Economic yields (q / ha)* — 100-200 (green)
12. *Economic value (Rs. / ha)* — 1,00,000 to 2,00,000
13. *By-product (If any)* — Stem and leaves
14. *Use of by-product* — Fuel / cattle feed / organic matter / spice / mulch

197. *CROP* — INDIAN SENNA / Sena

Scientific Name — Cassia angustifolia Vahl

Family — Caesalpiniaceae

Herb / Shrub / Tree — Shrub

Annual / Biennial / Perennial — Perennial

Used as — Medicine (for purgative / laxative)

1. *Climatic requirements* — Moderate temperature with moderate rainfall
2. *Soil requirements* — Loamy / sandy loam and alluvial soils
3. *Required land situations* — High land
4. *Importance of crop* — Leaves and fruits are used for laxative, cathartic, vermifuge and purgative
5. *Fertilizer management* — 50kg N, 30 kg P_2O_5 and 30 kg K_2O/ha annually
6. *Water management* — Mostly rainfed, 1-2 irrigations, if irrigated
7. *Duration of crop* — 2-3 years
8. *Parts used* — Leaves and fruits
9. *Where grown in India* — Mostly in Tamil Nadu, Karnataka and A.P.

10. *Export possibility* — High
11. *Economic yields (q / ha)* — Leaves-10 q/ha + Fruits 15 q/ha
12. *Economic value (Rs. / ha)* — 50,000 to 70,000
13. *By-product (If any)* — Stem
14. *Use of by-product* — Fuel / pegs / mulch / timber

198. *CROP* — WHITE YAM / GREATER YAM / ASIATIC YAM / *Kkhamalu / Chuprialu / Gechhoalu*

Scientific Name — Dioscorea alata L. / ***D. bulbifera*** L.

Family — Dioscoreaceae

Herb / Shrub / Tree — Climbing herb, 3-5 m long length

Annual / Biennial / Perennial — Perennial

Used as — Bulbils / tubers as medicine (impotency)

1. *Climatic requirements* — High rainfall and high temperature
2. *Soil requirements* — Alluvial / loamy / sandy soils
3. *Required land situations* — High land
4. *Importance of crop* — Bulbils / tubers are used as vegetable also
5. *Fertilizer management* — Soils with organic matter (no cultivation)
6. *Water management* — Rainfed
7. *Duration of crop* — 2-3 years
8. *Parts used* — tubers / corm / bulbils
9. *Where grown in India* — Assam, Tamil Nadu, W.B. and M.P.
10. *Export possibility* — High (Rhizomes for checking population), hence demand will be very high
11. *Economic yields (q / ha)* — 100-200 (green)
12. *Economic value (Rs. / ha)* — 50,000 to 60,000
13. *By-product (If any)* — Vine and leaves
14. *Use of by-product* — Fuel / cattle feed / organic matter / mulch

199. *CROP* — KING-OF-BITTER / Kiryat / *Kalmegh*

Scientific Name — Andrographis paniculata (Burm) Wall ex Nees

Family — Acanthaceae

Herb / Shrub / Tree — Herb, branched but small, 20-40 cm. height

Annual / Biennial / Perennial — Annual / Perennial

Used as — Medicine (used as tonic, fever, worms, dysentery, liver, digestion)

1. *Climatic requirements* — Moderate rainfall and temperature

2. *Soil requirements* — Alluvial / loamy / sandy soils
3. *Required land situations* — High land
4. *Importance of crop* — Very bitter in taste, if leaves taken on tongue
5. *Fertilizer management* — 60 kg N, 30 kg P_2O_5 and 30 kg K_2O/ha
6. *Water management* — Normally rainfed, 1-2 irrigations, if irrigated
7. *Duration of crop* — 110-140 days
8. *Parts used* — Stem, leaves, roots, inflorescence
9. *Where grown in India* — All over India, except too low / high temperature areas
10. *Export possibility* — Now high, it will be higher because of medicinal value
11. *Economic yields (q / ha)* — 160-180 (total plants)
12. *Economic value (Rs. / ha)* — 60,000 to 80,000
13. *By-product (If any)* — Nil
14. *Use of by-product* — Nil

200. *CROP* — NAYANTARA / Madagascar periwinkle / *Sada sawagan*

Scientific Name* — *Catharanthus roseus (L) G.Don

Family — Apocynaceae

Herb / Shrub / Tree — Shrub / Herb, 40-80 cm height

Annual / Biennial / Perennial — Parennial

Used as — Medicine (in cancer therapy) / flower

1. *Climatic requirements* — Moderate rainfall and temperature
2. *Soil requirements* — Sandy / loamy / alluvial soils
3. *Required land situations* — High and medium land
4. *Importance of crop* — This is used as flower also for beautification
5. *Fertilizer management* — 50 kg N, 30 kg P_2O_5 and 30 kg K_2O/ha
6. *Water management* — 3-4 irrigations annually
7. *Duration of crop* — 2-3 years
8. *Parts used* — Leaves, root and flowers
9. *Where grown in India* — Except extreme hot and cold, all over India
10. *Export possibility* — High
11. *Economic yields (q / ha)* — 15-20 (dry) (leaves)
12. *Economic value (Rs. / ha)* — 40,000 to 60,000

13. *By-product (If any)* — Stalk
14. *Use of by-product* — Fuel / mulch / organic matter

201. *CROP—* GRITAKUMARI / GRITKUMARI / Barbados aloe / *Ghikanvar*

***Scientific Name* — *Aloe barbadensis* Mill**

***Family* —** Liliaceae

***Herb / Shrub / Tree* —** Shrub

***Annual / Biennial / Perennial* —** Perennial

***Used as* —** Medicine (Purgative from leaves)

1. *Climatic requirements* — Moderate climatic conditions
2. *Soil requirements* — Sandy / loamy / alluvial soils
3. *Required land situations* — High land
4. *Importance of crop* — Succulent small body but salt resistant
5. *Fertilizer management* — **50 kg N, 30 kg P_2O_5 and 30 kg K_2O/ha** annually
6. *Water management* — 4-5 irrigations annually
7. *Duration of crop* — 2-3 years
8. *Parts used* — Whole succulent body
9. *Where grown in India* — All over India except extreme hot and low temperature (mostly in sea shore)
10. *Export possibility* — High
11. *Economic yields (q / ha)* — 150-180 (biomass-green)
12. *Economic value (Rs. / ha)* — Aesthetic value
13. *By-product (If any)* — Nil
14. *Use of by-product* — Nil

11. SPICES

202. *CROP* — CHILLIES / RED PEPPER / Lal mircha / *Lanka/ Marich*

***Scientific Name* — *Capsicum annuum* L.**

***Family* —** Solanaceae

***Herb / Shrub / Tree* —** Herb

***Annual / Biennial / Perennial* —** Annual / Perennial

***Used as* —** Spice / condiment / vegetable / medicine

1. *Climatic requirements* — Moderate to high temperature with moderate to high rainfall

2. *Soil requirements* — Loamy / sandy loam / alluvial soils
3. *Required land situations* — High land
4. *Importance of crop* — Pungency / bitterness in fruits
5. *Fertilizer management* — 100 kg N, 50 kg P_2O_5 and 50 kg K_2O/ha
6. *Water management* — 3-4 irrigations, if irrigated
7. *Duration of crop* — 120 days / 2-3 years
8. *Parts used* — Fruits (green and ripe)
9. *Where grown in India* — All over India except too low / too high temperature areas
10. *Export possibility* — High
11. *Economic yields (q / ha)* — 50-60 annually (green)
12. *Economic value (Rs. / ha)* — 50,000 to 90,000
13. *By-product (If any)* — Stem and leaves
14. *Use of by-product* — Fuel / for climbing cowpea / organic matter / mulch

203. *CROP* — GINGER / Adrak / *Ada*

Scientific Name* — *Zingiber officinale Rose

Family — Zingiberaceae

Herb / Shrub / Tree — Herb

Annual / Biennial / Perennial — Perennial

Used as — Medicine / spice / condiments

1. *Climatic requirements* — Warm / cool climate with humidity
2. *Soil requirements* — Sandy / loamy soils with organic matter
3. *Required land situations* — High land (with running water)
4. *Importance of crop* — Rhizomes in the under ground with smell
5. *Fertilizer management* — 100 kg N, 50 kg P_2O_5 and 50 kg K_2O/ha
6. *Water management* — 8-10 irrigations (or rain water flowing)
7. *Duration of crop* — 120-150 days
8. *Parts used* — Rhizomes
9. *Where grown in India* — Kerala, West Bengal, Maharashtra, Assam, H.P. and H.P.
10. *Export possibility* — High
11. *Economic yields (q / ha)* — 80-160 (green)
12. *Economic value (Rs. / ha)* — 40,000 to 80,000
13. *By-product (If any)* — Stalk

14. *Use of by-product* — Spice / organic matter / other medicines / cattle feeds / fuel

204. *CROP* — TURMERIC / Haldi / Halud

Scientific Name* — *Curcuma domestica Val

Family — Zingiberaceae

Herb / Shrub / Tree — Herb

Annual / Biennial / Perennial — Perennial

Used as — Spice / medicine / condiments

1. *Climatic requirements* — Warm climate with rainfall
2. *Soil requirements* — Sandy / loamy / alluvial soils
3. *Required land situations* — High and medium lands
4. *Importance of crop* — Yellow colour is due to pigment of *curcumin*; it can be grown in shade. Medicine for cancer
5. *Fertilizer management* — 100 kg N, 50 kg P_2O_5 and 50 kg K_2O/ha annually
6. *Water management* — 4-5 irrigations annually, if irrigated
7. *Duration of crop* — 2-3 years
8. *Parts used* — Rhizomes under the soil level
9. *Where grown in India* — T.N., A.P., Maharashtra, Bihar, West Bengal, Uttar Pradesh, Kerala, Karnataka, Orissa, etc.
10. *Export possibility* — Very high
11. *Economic yields (q / ha)* — 20-25 (dry)
12. *Economic value (Rs. / ha)* — 40,000 to 60,000
13. *By-product (If any)* — Stalks
14. *Use of by-product* — Spice / organic matter / cattle feeds / fuel / mulch

205. *CROP* — CURRY LEAF TREE / Mitha neem / Gandhela / *Karipata* / Kadhi neem

Scientific Name* — *Murraya koenigii (L.) Spreng

Family — Rutaceae

Herb / Shrub / Tree — Shrub, (large) / small tree

Annual / Biennial / Perennial — Perennial

Used as — Spice (leaves are used for flavouring curry)

1. *Climatic requirements* — Moderate temperature with moderate rainfall
2. *Soil requirements* — Loamy / sandy / alluvial / sandy loam soil
3. *Required land situations* — High land

4. *Importance of crop* — Flavouring smell in leaves
5. *Fertilizer management* — 100 kg N, 50 kg P_2O_5 and 50 kg K_2O/ha annually
6. *Water management* — Rainfed; 1-2 times (annually) irrigations annually
7. *Duration of crop* — 2-3 years
8. *Parts used* — Leaves
9. *Where grown in India* — Tamil Nadu, Maharashtra, Karnataka, Kerala, West Bengal etc.
10. *Export possibility* — High
11. *Economic yields (q / ha)* — 50-60 (green leaves)
12. *Economic value (Rs. / ha)* — 50,000 to 60,000
13. *By-product (If any)* — Stems
14. *Use of by-product* — Fuel / pegging

206. *CROP* — FIELD MINT / *Pudina*

Scientific Name* — *Mentha arvensis L.

Family — Labiatae

Herb / Shrub / Tree — Herb

Annual / Biennial / Perennial — Perennial

Used as — Medicine (Carminative / refrigerant / stimulant) / spice

1. *Climatic requirements* — High rainfall and low / moderate temperature
2. *Soil requirements* — Loamy / sandy loam / alluvial soils
3. *Required land situations* — High land
4. *Importance of crop* — A good smell due to essential oils
5. *Fertilizer management* — 100 kg N, 50 kg P2O5 and 50 kg K2O/ha annually
6. *Water management* — 4-5 irrigations annually
7. *Duration of crop* — 2-4 Years
8. *Parts used* — Leaves
9. *Where grown in India* — Mostly all hilly stations in India
10. *Export possibility* — High
11. *Economic yields (q / ha)* — 50-60 (green)
12. *Economic value (Rs. / ha)* — 50,000 to 60,000
13. *By-product (If any)* — Stem
14. *Use of by-product* — Fuel / organic matter / further medicine.

207. *CROP* — CORIANDER / DHANIA / *Dhane*

***Scientific Name* — *Coriandrum sativum* L.**

Family — Umbelliferae

Herb / Shrub / Tree — Herb, aromatic

Annual / Biennial / Perennial — Annual

Used as — Spice / medicine / vegetable / condiment

1. *Climatic requirements* — Low temperature in dry / wet places
2. *Soil requirements* — Sandy / loamy / alluvial / sandy loam soils
3. *Required land situations* — High / medium or low lands
4. *Importance of crop* — It is used for flavouring curry and soups
5. *Fertilizer management* — 60 kg N, 30 kg P_2O_5 and 30 kg K_2O/ha annually
6. *Water management* — 3-4 irrigations (can also be raised as rainfed *paira* crop)
7. *Duration of crop* — 100-120 days
8. *Parts used* — Leaves and seeds
9. *Where grown in India* — M.P., Maharashtra, Rajasthan, A.P., T.N., Karnataka, West Bengal, Bihar and Orissa
10. *Export possibility* — High
11. *Economic yields (q / ha)* — 10-15 (seeds) + leaves (50-100 green) as spice
12. *Economic value (Rs. / ha)* — 40,000 to 80,000
13. *By-product (If any)* — Stalk
14. *Use of by-product* — Cattle feed / fuel / organic matter.

12. FRUITS (ANNUAL AND PERENNIAL)

208. *CROP* — WATER MELON / Tarbooz / *Tarmuj*

Scientific Name* — *Citrullus lanatus (Thunb) Mansf

Family — Cucurbitaceae

Herb / Shrub / Tree — Climbing herb

Annual / Biennial / Perennial — Annual

Used as — Fruits

1. *Climatic requirements* — Warm climatic conditions
2. *Soil requirements* — Loamy / sandy / rocky alluvial soils
3. *Required land situations* — High / medium lands
4. *Importance of crop* — Trailing dioceous plant

5. *Fertilizer management* — 50 kg N, 30 kg P_2O_5 and 30 kg K_2O/ha
6. *Water management* — 3-4 irrigations
7. *Duration of crop* — 120-140 days
8. *Parts used* — Fruits
9. *Where grown in India* — All over India (sea shores / river beds)
10. *Export possibility* — High
11. *Economic yields (q / ha)* — 200-250 (green)
12. *Economic value (Rs. / ha)* — 20,000 to 25,000
13. *By-product (If any)* — Vine
14. *Use of by-product* — Cattle feed / fuel / organic matter / mulch.

209. *CROP* — LITCHI / Leechee / Lichi / Lichu

Scientific Name* — *Litchi chinensis Sonn

Family — Sapindaceae

Herb / Shrub / Tree — Tree

Annual / Biennial / Perennial — Perennial

Used as — Fruits

1. *Climatic requirements* — Hot humid areas
2. *Soil requirements* — Loamy / sandy / alluvial / sandy loam soils
3. *Required land situations* — High lands
4. *Importance of crop* — Fleshy mass in small fruits (with deep red colour in the coat)
5. *Fertilizer management* — 100 kg N, 50 kg P_2O_5 and 50 kg K_2O/ha annually
6. *Water management* — 3-4 irrigations annually
7. *Duration of crop* — 8-10 years
8. *Parts used* — Fruits
9. *Where grown in India* — Bihar (Muzaffapur and Darbhangha), West Bengal (Nadia), Uttaranchal (Dehradun), U.P. (Muzaffar-nagar), Pathankot of H.P. etc.
10. *Export possibility* — High
11. *Economic yields (q / ha)* — 50-100 (green)
12. *Economic value (Rs. / ha)* — 25,000 to 50,000
13. *By-product (If any)* — Trunk, branches and leaves
14. *Use of by-product* — Fuel / cattle feed / timber / organic matter / mulch / peg.

210. *CROP* — CUCUMBER / Khira / *Sasa* / *Khirai*

Scientific Name* — *Cucumis sativus L.

Family — Cucurbitaceae

Herb / Shrub / Tree — Trailing herb

Annual / Biennial / Perennial — Annual

Used as — Fruits

1. *Climatic requirements* — Warm season with rains
2. *Soil requirements* — Loamy / sandy / alluvial soils
3. *Required land situations* — High / medium land
4. *Importance of crop* — Climbing diocious plants
5. *Fertilizer management* — 50 kg N, 50 kg P_2O_5 and 50 kg K_2O/ha
6. *Water management* — 3-4 irrigations
7. *Duration of crop* — 120-140 days
8. *Parts used* — Fruits / vegetable / fodder / organic matter
9. *Where grown in India* — Grown all over India except too cold or too hot areas, in warm season
10. *Export possibility* — High
11. *Economic yields (q / ha)* — 100-150 (green)
12. *Economic value (Rs. / ha)* — 25,000 to 35,000
13. *By-product (If any)* — Vine and leaves
14. *Use of by-product* — Cattle feed / mulching materials / fuel / organic matter.

211. *CROP* — BALE / WOOD-APPLE / Elephant-apple / *Bel* / *Bael*

Scientific Name* — *Limonia acidissima L. or ***Aegle marmelos*** (L.) Corr.

Family — Rutaceae

Herb / Shrub / Tree — Tree

Annual / Biennial / Perennial — Perennial

Used as — Fruits (seet aromatic pulp)

1. *Climatic requirements* — Moderate warm season
2. *Soil requirements* — Alluvial / loamy / sandy / sandy loam soils
3. *Required land situations* — High lands
4. *Importance of crop* — Hard cover of the big fruits; thorny deciduous plants
5. *Fertilizer management* — 100 kg N, 50 kg P_2O_5 and 50 kg K_2O/ha annually

6. *Water management* — 3-4 irrigations annually
7. *Duration of crop* — 8-10 years
8. *Parts used* — Fruits
9. *Where grown in India* — In entire India except too cold areas
10. *Export possibility* — High
11. *Economic yields (q / ha)* — 100-150 (green)
12. *Economic value (Rs. / ha)* — 25,000 to 35,000
13. *By-product (If any)* — Trunk, branches and leaves
14. *Use of by-product* — Cattle feed / fuel / timber / organic matter

212. *CROP* — MANGO / Am / Amra / Rasala

Scientific Name* — *Mangifera indica L.

Family — Anacardiaceae

Herb / Shrub / Tree — Tree

Annual / Biennial / Perennial — Perennial

Used as — Fruits (with juice)

1. *Climatic requirements* — Medium warm season
2. *Soil requirements* — Alluvial / loamy / sandy / clayey / sandy loam soils
3. *Required land situations* — High land
4. *Importance of crop* — Juicy fruits with good taste
5. *Fertilizer management* — 100 kg N, 50 kg P_2O_5 and 50 kg K_2O/ha annually
6. *Water management* — 3-4 irrigations annually
7. *Duration of crop* — 20-25 years
8. *Parts used* — Fruits
9. *Where grown in India* — U.P., Bihar, West Bengal, Karnataka, Uttaranchal, Punjab, Maharashtra, A.P., T.N. etc.
10. *Export possibility* — Very high
11. *Economic yields (q / ha)* — 200-400 (green)
12. *Economic value (Rs. / ha)* — 80,000 to 1,80,000
13. *By-product (If any)* — Trunk, branch and leaves
14. *Use of by-product* — Timber / fuel / cattle feed / organic matter

213. *CROP* — JACK TREE / JACK FRUIT / *Kathal / Kanthal / Panas*

Scientific Name* — *Artocarpus heterophyllus Lamk

Family — Moraceae

Herb / Shrub / Tree — Tree

Annual / Biennial / Perennial — Perennial

Used as — Fruits / Vegetable

1. *Climatic requirements* — Warm climate with high humidity
2. *Soil requirements* — Loamy / alluvial / sandy loam soil
3. *Required land situations* — High land
4. *Importance of crop* — Big fruits come from the trunk
5. *Fertilizer management* — 100 kg N, 50 kg P_2O_5 and 50 kg K_2O/ha annually
6. *Water management* — 3-4 irrigations annually
7. *Duration of crop* — 8-15 years
8. *Parts used* — Fruits / seeds
9. *Where grown in India* — All over India except too hot and too cold areas
10. *Export possibility* — High
11. *Economic yields (q / ha)* — 200-800 (green)
12. *Economic value (Rs. / ha)* — 40,000 to 80,000
13. *By-product (If any)* — Trunk, branches and leaves
14. *Use of by-product* — Timber / fuel / cattle feed / organic matter

214. *CROP* — JAMUN / *Jam* / Java plum / Jambolana / Jambu

Scientific Name* — *Syzygium cumini Skeels

Family — Myrtaceae

Herb / Shrub / Tree — Tree (big)

Annual / Biennial / Perennial — Perennial

Used as — Fruits / fodder (extract from seeds reduces blood-sugar in diabetic patients)

1. *Climatic requirements* — Warm climates with humidity
2. *Soil requirements* — Loamy / sandy / alluvial soils
3. *Required land situations* — High land
4. *Importance of crop* — Fleshy mass in berry type fruits (black out side red inside)
5. *Fertilizer management* — 100 kg N, 50 kg P_2O_5 and 50 kg K_2O/ha annually
6. *Water management* — 3-4 irrigations annually
7. *Duration of crop* — 10-20 years
8. *Parts used* — Fruits / seeds

9. *Where grown in India* — All over India except too cold / too high temperature areas
10. *Export possibility* — High
11. *Economic yields (q / ha)* — 100-150 (green)
12. *Economic value (Rs. / ha)* — 20,000 to 40,000
13. *By-product (If any)* — Trunk / branches / leaves
14. *Use of by-product* — Timber / fuel / cattle feed / organic matter

13. FLOWERS / SUCCULENTS / ORNAMENTALS

215. *CROP* — HOLIYHOCK / Gulkhera / Holihak

Scientific Name — Althaea rosea (L.) Cav.

Family — Malvaceae

Herb / Shrub / Tree — Herb

Annual / Biennial / Perennial — Annual (Perennial in hills)

Used as — Flower / ornamentals

1. *Climatic requirements* — Warm weathers
2. *Soil requirements* — Loamy / alluvial / sandy soils
3. *Required land situations* — High / medium land
4. *Importance of crop* — Fibres are obtained from stem
5. *Fertilizer management* — 100 kg N, 50 kg P_2O_5 and 50 kg K_2O/ha
6. *Water management* — 3-4 irrigations
7. *Duration of crop* — 120-130 days
8. *Parts used* — Flower
9. *Where grown in India* — All over India except extreme cold / hot areas
10. *Export possibility* — High
11. *Economic yields (q / ha)* — Aesthetic purposes
12. *Economic value (Rs. / ha)* — Aesthetic value
13. *By-product (If any)* — Plants with leaves
14. *Use of by-product* — Organic matter / fuel / mulch / cattle feed

216. *CROP* — ZINNIA/ *Ginia*

Scientific Name — Zinnia angustifolia H.B. and K.

Family — Compositae

Herb / Shrub / Tree — Herb

Annual / Biennial / Perennial — Annual

Used as — Flower

1. *Climatic requirements* — Cold or low cold climatic conditions
2. *Soil requirements* — Alluvial / loamy / sandy soils
3. *Required land situations* — High land
4. *Importance of crop* — Star like flower with different colours
5. *Fertilizer management* — 50 kg N, 30 kg P_2O_5 and 30 kg K_2O/ha
6. *Water management* — 4-5 irrigations
7. *Duration of crop* — 120-140 days
8. *Parts used* — Flower
9. *Where grown in India* — All over India except extreme cold or hot areas
10. *Export possibility* — High
11. *Economic yields (q / ha)* — 50 - 90
12. *Economic value (Rs. / ha)* — Aesthetic value
13. *By-product (If any)* — Plants with leaves
14. *Use of by-product* — Fuel / organic matter / mulch / cattle feeds

217. *CROP* — GARDEN BALSAM / Gul-mehndi / *Balsum*

Scientific Name — Impatiens balsamina L.

Family — Balsaminaceae

Herb / Shrub / Tree — Herb

Annual / Biennial / Perennial — Annual

Used as — Flower / medicine

1. *Climatic requirements* — Warm conditions
2. *Soil requirements* — Alluvial / loamy / sandy soil
3. *Required land situations* — High / medium lands
4. *Importance of crop* — Bushy plants with coloured flowers
5. *Fertilizer management* — 50 kg N, 30 kg P_2O_5 and 30 kg K_2O/ha
6. *Water management* — 4-5 irrigations
7. *Duration of crop* — 120-140 days
8. *Parts used* — Flower
9. *Where grown in India* — All over India except low temperature and hot areas
10. *Export possibility* — High
11. *Economic yields (q / ha)* — 70 - 100

12. *Economic value (Rs. / ha)* — Aesthetic value
13. *By-product (If any)* — Plants with leaves
14. *Use of by-product* — Fuel / mulch / organic mater / cattle feed

218. *CROP* — INDIAN RUBBER TREE / Assam rubber / Bor / *Rabar* / *Attahbar*

Scientific Name* — *Ficus elastica Roxb.

Family — Moraceae

Herb / Shrub / Tree — Tree (small)

Annual / Biennial / Perennial — Perennial

Used as — Ornamentals / Industrial use (rubber from latex)

1. *Climatic requirements* — Moderately warm climate
2. *Soil requirements* — Loamy / sandy loam / sandy soils
3. *Required land situations* — High land
4. *Importance of crop* — Good for indoor and garden decoration
5. *Fertilizer management* — 50 kg N, 30 kg P_2O_5 and 30 kg K_2O/ha annually
6. *Water management* — 10-12 times annually
7. *Duration of crop* — 3-4 years
8. *Parts used* — Leaf (beauty) and latex for rubber
9. *Where grown in India* — All over India except too low / too high temperature areas (mostly grows in high rainfall areas)
10. *Export possibility* — High
11. *Economic yields (q / ha)* — 7-20 (rubber)
12. *Economic value (Rs. / ha)* — Aesthetic value
13. *By-product (If any)* — Nil
14. *Use of by-product* — Nil

219. *CROP* — KOCHIA / Bui / *Kauraro*

Scientific Name* — *Kochia indica Wight

Family — Chenopodiaceae

Herb / Shrub / Tree — Herb

Annual / Biennial / Perennial — Annual

Used as — Ornamental

1. *Climatic requirements* — Warm season
2. *Soil requirements* — Loamy / sandy / alluvial soils
3. *Required land situations* — High medium land

4. *Importance of crop* — Much branched bushy plant like doom
5. *Fertilizer management* — 50 kg N, 30 kg P2O5 and 30 kg K2O/ha
6. *Water management* — 4-5 irrigations
7. *Duration of crop* — 120-140 Days
8. *Parts used* — Entire bushy plants
9. *Where grown in India* — All over India, except too low / high temperature areas
10. *Export possibility* — High
11. *Economic yields (q / ha)* — Nil
12. *Economic value (Rs. / ha)* — Aesthetic value
13. *By-product (If any)* — Plant with leaves (after removal)
14. *Use of by-product* — Fuel / organic manure / mulch / cattle feed

220. *CROP* — TRADESCANTIA

Scientific Name — Tradescantia discolor L.' Herit

Family — Commelinaceae

Herb / Shrub / Tree — Herb

Annual / Biennial / Perennial — Perennial

Used as — Ornamental

1. *Climatic requirements* — Moderate warm climate
2. *Soil requirements* — Loamy / sandy / alluvial soils
3. *Required land situations* — High medium lands
4. *Importance of crop* — Leaves are red in colour
5. *Fertilizer management* — 50 kg N, 30 kg P_2O_5 and 30 kg K_2O/ha annually
6. *Water management* — 4-5 irrigations annually
7. *Duration of crop* — 2 – 3 years
8. *Parts used* — Entire plant
9. *Where grown in India* — All over India except too cold areas
10. *Export possibility* — High
11. *Economic yields (q / ha)* — Nil
12. *Economic value (Rs. / ha)* — Aesthetic value
13. *By-product (If any)* — Plant with leaves
14. *Use of by-product*—Fuel/propagation/organic matter/mulch / cattle.

221. *CROP* — CROTON / *Patabahar* / Ornamental leaf-plant

Scientific Name — Codiaeum variegatum (L.) Blume

***Family* —** Euphorbiaceae

***Herb / Shrub / Tree* —** Shrub

***Annual / Biennial / Perennial* —** Perennial

***Used as* —** Ornamental / Hedge plants

1. *Climatic requirements* — Moderate warm climate
2. *Soil requirements* — Alluvial / loamy / sandy soils
3. *Required land situations* — High lands
4. *Importance of crop* — Leaves are of different sizes and colours
5. *Fertilizer management* — 100 kg N, 50 kg P_2O_5 and 50 kg K_2O/ha annually
6. *Water management* — 4-5 irrigations annually
7. *Duration of crop* — 2-5 years
8. *Parts used* — Leaves (for beauty)
9. *Where grown in India* — All over India except extreme low / too high temperature areas
10. *Export possibility* — High
11. *Economic yields (q / ha)* — Nil
12. *Economic value (Rs. / ha)* — Aesthetic value
13. *By-product (If any)* — Plants / branch / leaves
14. *Use of by-product* — Fuel / propagation / pegs / cattle fed.

222. CROP — WHITE CEDER / Jhaw / Aborvitae / Bagan thuja

Scientific Name* — *Thuja occidentalis L.

***Family* —** Cupressaceae

***Herb / Shrub / Tree* —** Tree (small)

***Annual / Biennial / Perennial* —** Perennial

***Used as* —** Ornamental (in garden)

1. *Climatic requirements* — Moderate warm climate
2. *Soil requirements* — Sandy / loamy / alluvial soils
3. *Required land situations* — High / medium / low lands
4. *Importance of crop* — Like doom shaped plant (good looking)
5. *Fertilizer management* — 100 kg N, 50 kg P_2O_5 and 50 kg K_2O/ha annually
6. *Water management* — 4-5 irrigations annually
7. *Duration of crop* — 5-10 years
8. *Parts used* — Entire plant for beautification

9. *Where grown in India* — All over India except too cold / too hot areas
10. *Export possibility* — High
11. *Economic yields (q / ha)* — Nil
12. *Economic value (Rs. / ha)* — Aesthetic value
13. *By-product (If any)* — Plants with leaves
14. *Use of by-product* — Decoration / fuel / organic matter / pegs / cattle feeds

223. *CROP* — ORIENTAL / GIANT-ARBOR-VITAE / Bara jhaw / Barojhau

Scientific Name* — *Thuja plicata D. Don / ***Thuja orientalis*** L.

Family — Cupressaceae

Herb / Shrub / Tree — Tree

Annual / Biennial / Perennial — Perennial

Used as — Ornamental (In park) / wind break (sea side)

1. *Climatic requirements* — Moderate warm conditions
2. *Soil requirements* — Sandy / loamy / alluvial soils
3. *Required land situations* — High / medium lands
4. *Importance of crop* — Horse tail leaves; plant fixes N
5. *Fertilizer management* — 100 kg N, 50 kg P_2O_5 and 50 kg K_2O/ha annually
6. *Water management* — 4-5 irrigations annually
7. *Duration of crop* — 5-10 years
8. *Parts used* — Entire plant
9. *Where grown in India* — H.P., Punjab, W. Bengal (for gardens)
10. *Export possibility* — High
11. *Economic yields (q / ha)* — Nil
12. *Economic value (Rs. / ha)* — Aesthetic value
13. *By-product (If any)* — Stems / branches / leaves
14. *Use of by-product* — Timbers / pegs / decoration / fuel / cattle feed.

224. *CROP* — AGAVE / Century plant / *Kantala* / Ageb / Aloe

Scientific Name* — *Agave americana L. / ***A. cantala*** Roxb.

Family — Agavaceae

Herb / Shrub / Tree — Shrub (short stemed, long leaves)

Annual / Biennial / Perennial — Perennial

Used as — Ornamentals / Industrial use (fibre from leaves)

1. *Climatic requirements* — Dry condition and warm weather
2. *Soil requirements* — Sandy / loamy / alluvial soils
3. *Required land situations* — High / medium lands
4. *Importance of crop* — Short stem pointed long sword like leaves with spines
5. *Fertilizer management* — 50 kg N, 30 kg P_2O_5 and 30 kg K_2O/ha annually
6. *Water management* — 1-2 irrigations annually, if irrigated
7. *Duration of crop* — 5-10 years
8. *Parts used* — Whole plant (for beautification) / leaves for fibre
9. *Where grown in India* — All over India in dry areas except too low / too high temperature areas
10. *Export possibility* — High
11. *Economic yields (q / ha)* — Nil / 10-20 (leaves)
12. *Economic value (Rs. / ha)* — Aesthetic value
13. *By-product (If any)* — Stem and leaves
14. *Use of by-product* — Fuel / organic matter / propagation.

14. NARCOTICS AND BEVERAGE

225. *CROP* — BHANG / HEMP / *GANJA* / Charas

Scientific Name* — *Canabis sativa L.

Family — Cannabinaceae

Herb / Shrub / Tree — Smelling herb (tall)

Annual / Biennial / Perennial — Annual

Used as — Narcotics (*bhang, ganja* and *charas* from leaves)

1. *Climatic requirements* — Warm season with humidity
2. *Soil requirements* — Alluvial / loamy / sandy loam soils
3. *Required land situations* — High / medium / low land
4. *Importance of crop* — Smells in the plant itself
5. *Fertilizer management* — 100 kg N, 50 kg P_2O_5 and 50 kg K_2O/ha
6. *Water management* — 4-5 irrigations (in W.B. no irrigation)
7. *Duration of crop* — 120-140 days
8. *Parts used* — Leaves
9. *Where grown in India* — Kerala, Assam, W.B., Bihar, T.N. etc.
10. *Export possibility* — Very high
11. *Economic yields (q / ha)* — 10-15

12. *Economic value (Rs. / ha)* — 40,000 to 80,000
13. *By-product (If any)* — Stalk
14. *Use of by-product* — Fuel / pegs / cattle feed / other narcotics use / mulch.

226. *CROP* — STRAMONIUM / *Dhatura* / DATURA / Thornapple / Jimson weed / *Dhutura*

Scientific Name* — *Datura stramonium L.

Family — Solanaceae

Herb / Shrub / Tree — Herb / Shrub

Annual / Biennial / Perennial — Annual

Used as — Narcotics (leaves and seeds) / medicine (asthma)

1. *Climatic requirements* — Warm season with high humidity
2. *Soil requirements* — Alluvial / sandy / loamy soils
3. *Required land situations* — High / medium lands
4. *Importance of crop* — Poisonous fruits and leaves
5. *Fertilizer management* — 50 kg N, 30 kg P_2O_5 and 30 kg K_2O/ha
6. *Water management* — 3-4 irrigations, if irrigated.
7. *Duration of crop* — 120-150 days
8. *Parts used* — Leaves and seeds
9. *Where grown in India* — All over India except too low / too high temperature areas.
10. *Export possibility* — Very High
11. *Economic yields (q / ha)* — 15-20 (leaves and seeds)
12. *Economic value (Rs. / ha)* — 40,000 to 80,000
13. *By-product (If any)* — Stalk and branches
14. *Use of by-product* — Fuel / fencing / mulch / organic matter.

227. *CROP* — BETELVINE / *Pan* / Betel pepper / *Paan*

Scientific Name* — *Piper betle L.

Family — Piperaceae

Herb / Shrub / Tree — Shrub, (Climbing)

Annual / Biennial / Perennial — Perennial

Used as — Beverage (leaves) with dough arecanut

1. *Climatic requirements* — Warm climatic conditions
2. *Soil requirements* — Alluvial / loamy soils
3. *Required land situations* — High lands

4. *Importance of crop* — Leaves like heart; raw (leaves) consumed
5. *Fertilizer management* — 100 kg N, 50 kg P_2O_5 and 50 kg K_2O/ha annually mostly from organic manures
6. *Water management* — 3-4 irrigations annually
7. *Duration of crop* — 10-15 years
8. *Parts used* — Leaves for raw consumption
9. *Where grown in India* — West Bengal, Assam, Orissa, Bihar, T.N., U.P. Karnataka, Kerala etc.
10. *Export possibility* — Very High
11. *Economic yields (q / ha)* — 140-150 (green leaves)
12. *Economic value (Rs. / ha)* — 80,000 to 100,000
13. *By-product (If any)* — Vines
14. *Use of by-product* — Cattle feeds / organic matter / fuel / propagation.

Chapter - 3

RAINY / WET (*KHARIF*) SEASON CROPS

INTRODUCTION

Chapter 3 deals with different crops in cereals, pulses, oil seeds, fibres, tubers/rhizomes, sugar crops, vegetables, fodders, forages, green manuring crops, medicinal plans, spices, fruits, flowers (including succulents and ornamentals) and beverage and narcotics, in rainy or wet (*kharif*) season.

Rainy season in India varies from place to place. In the eastern part, the rainfall is high (ranging from 1200 mm to 3,000 mm), whereas, in the central tract rainfall is medium (800 mm to 1400 mm). In the western tract, rainfall is very low (ranges in between 600 mm to 900 mm). In the southern tract, of course, rainfall is high (except some areas) ranging from 1,000 mm to 2,500 mm. During this period, temperature is also more or less bearable in most of the areas (33°C to 41°C). Due to excessive rainfall (water), low-lying areas in eastern parts are also in inundated and as a result these low-lying areas are suited for rice cultivation. Rainfall is high in the hilly areas of the north also. Normal crops are grown in any land, except in eastern parts where high lands are used for arable crops in this season.

1. CEREALS

228. *CROP*—RICE / CHAUVAL / DHAN / PADDY / Anna / Bhat / Tandul

Scientific Name — Oryza sativa L.

Family — Gramineae

Herb / Shrub / Tree — Herb (60-80 cm height)

Annual / Biennial / Perennial — Annual

Used as — Food (staple) / food products

1. *Climatic requirements* — High hot temperature with rainfall and humidity
2. *Soil requirements* — Clayey or loamy soils
3. *Required land situations* — Low or medium land
4. *Importance of crop* — Major carbohydrate food for nearly 50% of the world population
5. *Fertilizer management* — 100 kg N, 50 kg P_2O_5 and 50 kg K_2O/ha
6. *Water management* — 4-5 irrigations in *boro*, if irrigated (normally water stagnated)
7. *Duration of crop* — 120-150 days
8. *Parts used* — Grain and straw
9. *Where grown in India* — All low lying areas of West Bengal, Orissa, Bihar, A.P., T. Nadu, Karnataka, U.P., M.P., Kerala, Assam, Tripura etc.
10. *Export possibility* — Very high
11. *Economic yields (q / ha)* — 25-30
12. *Economic value (Rs. / ha)* — 20,000-50,000 (scented rice is costly)
13. *By-product (If any)* — Straw and husk
14. *Use of by-product* — Cattle feed / bedding for cattle and poultry / fuel (potash fertilizer with burnt husk).

229. *CROP* — MAIZE / CORN / INDIAN CORN / Makai / Bhutta

***Scientific Name* — *Zea mays* L.**

Family — Gramineae

Herb / Shrub / Tree — Herb (grass), tall

Annual / Biennial / Perennial — Annual (1.5-3.0 m height)

Used as — Food / food-products

1. *Climatic requirements* — Moderate temperature, moderate rainfall
2. *Soil requirements* — Loamy or sandy loam soils
3. *Required land situations* — Flat lands of plains and foot hill
4. *Importance of crop* — Cereal food, having silk and tussels
5. *Fertilizer Management*—N, P_2O_5 and K_2O @ 120, 60 and 60 kg/ha, respectively
6. *Water management* — 5-6 irrigations
7. *Duration of crop* — 120-140 days
8. *Parts used* — Cobs (grains) and stalk
9. *Where grown in India* — U.P., Punjab, M.P., Bihar, A.P., J & K., W.B.

10. *Export possibility* — Very high
11. *Economic yields (q / ha)* — 25-50
12. *Economic value (Rs. / ha)* — 50,000 to 80,000
13. *By-product (If any)* — Stalk with leaves
14. *Use of by-product* — Fuel / trailing other crops / organic manure / pegs./ cattle feed

230. *CROP* — SORGHUM / Jowar / Joar

Scientific Name* — *Sorghum vulgare Pers

Family — Gramineae

Herb / Shrub / Tree — Herb (grass), tall (1.5-2.5 m height)

Annual / Biennial / Perennial — Annual

Used as — Foods grain / fodder

1. *Climatic requirements* — High hot climate with low rainfall
2. *Soil requirements* — Loamy or sandy loam soils
3. *Required land situations* — High land
4. *Importance of crop* — Cluster inflorescence
5. *Fertilizer management* — 60 kg N, 30 kg P_2O_5 and 30 kg K_2O/ha
6. *Water management* — Normally rainfed, 1-2 irrigation, if irrigated
7. *Duration of crop* — 120-130 days
8. *Parts used* — Grain as human food, stalk for cattle feed
9. *Where grown in India* — U.P., Punjab, Madhya Pradesh, A.P., Karnataka, Maharashtra, Rajasthan, Timil Nadu etc.
10. *Export possibility* — High
11. *Economic yields (q / ha)* — 15-20 (more yield if irrigated)
12. *Economic value (Rs. / ha)* — 15,000 to 20,000
13. *By-product (If any)* — Stalks
14. *Use of by-product* — Cattle feed / fuel / organic matter / mulch / pegs / roofing materials.

231. *CROP* — PEARL MILLET / Bajra

Scientific Name* — *Pennisetum typhoides (Burm.f.) Stapf and CE

Family — Gramineae

Herb / Shrub / Tree — Herb (grass), tall (1.5-2.0 height)

Annual / Biennial / Perennial — Annual

Used as — Food / fodder

1. *Climatic requirements* — Warm season with dry weather

2. *Soil requirements* — Loamy / sandy loam soils
3. *Required land situations* — High / medium lands
4. *Importance of crop* — Grass group with compact inflorescence
5. *Fertilizer management* — 60 kg N, 30 kg P_2O_5 and 30 kg K_2O/ha
6. *Water management* — 2-3 irrigations (normally rainfed)
7. *Duration of crop* — 120-140 days
8. *Parts used* — Seeds for food and plant for cattle feed
9. *Where grown in India* — U.P., Punjab, A.P., Tamil Nadu, Rajasthan, Maharashtra, M.P. etc.
10. *Export possibility* — High
11. *Economic yields (q / ha)* — 15-25
12. *Economic value (Rs. / ha)* — 15,000 to 25,000
13. *By-product (If any)* — Stalk and leaves
14. *Use of by-product* — Cattle feed / fuel / organic matter / mulch

232. *CROP*—FOX TAIL/ITALIAN MILLET Kangni/Kakun/*Kaon/Kangu*

Scientific Name — *Seteria italica* (L.) Beauv

Family — Gramineae

Herb / Shrub / Tree — Herb (grass) (30-80 cm height)

Annual / Biennial / Perennial — Annual

Used as — Food (grains) / fodder

1. *Climatic requirements* — High hot climate with less rainfall
2. *Soil requirements* — Loamy / sandy loam soils
3. *Required land situations* — High land
4. *Importance of crop* — Inflorescence is like tail of a fox / cat
5. *Fertilizer management* — 60 kg N, 30 kg P_2O_5 and 30 kg K_2O/ha
6. *Water management* — 1-2 irrigation if irrigated, normally rainfed
7. *Duration of crop* — 110-130 days
8. *Parts used* — Grains / straw
9. *Where grown in India* — A.P., T.N., Maharashtra, Gujarat, Karnataka, West Bengal (Cooch-behar)
10. *Export possibility* — High
11. *Economic yields (q / ha)* — 15-25
12. *Economic value (Rs. / ha)* — 15,000 to 25,000
13. *By-product (If any)* — Straw
14. *Use of by-product* — Cattle feed / fuel / organic matter / mulch.

233. ***CROP* — FINGER MILLET / RAGI / Mandal / Mandua / *Marua***

Scientific Name — Eleusine coracana (L.) Gaertn

Family — Gramineae

Herb / Shrub / Tree — Herb (grass), small (30-60 cm height)

Annual / Biennial / Perennial — Annual

Used as — Foods / fodder

1. *Climatic requirements* — Moderate temperature with moderate rainfall
2. *Soil requirements* — Sandy loam / loamy soils
3. *Required land situations* — High / medium land
4. *Importance of crop* — Inflorescence like turned finger
5. *Fertilizer management* — 60 kg N, 30 kg P_2O_5 and 30 kg K_2O/ha
6. *Water Management*— 1-3 irrigations, if irrigated, normally rainfed
7. *Duration of crop* — 110-130 days
8. *Parts used* — Grain as food and straw as cattle feed
9. *Where grown in India* — Karnataka, Tamil Nadu, A.P., M.P. and Maharashtra etc.
10. *Export possibility* — High
11. *Economic yields (q / ha)* — 15-20
12. *Economic value (Rs. / ha)* — 20,000 to 30,000
13. *By-product (If any)* — Straw
14. *Use of by-product* — Cattle feed / fuel / organic matter / mulch

234. ***CROP* — *KODO* MILLET / Kodo / Kodra**

Scientific Name — Paspalum scrobiculatum L.

Family — Gramineae

Herb / Shrub / Tree — Herb (50-80 cm height)

Annual / Biennial / Perennial — Annual

Used as — Food / fodder / medicine (decoction of root, for child birth)

1. *Climatic requirements* — Moderate temperature with moderate rainfall
2. *Soil requirements* — Loamy / sandy loam soils
3. *Required land situations* — Medium or high land
4. *Importance of crop* — Wider leaves
5. *Fertilizer management* — 60 kg N, 30 kg P_2O_5 and 30 kg K_2O/ha
6. *Water management* — 1-2 irrigations, if irrigated, normally rainfed

7. *Duration of crop* — 110-130 days
8. *Parts used* — Seed as food / whole plant as fodder
9. *Where grown in India* — All over India, except, too low / too high temperature areas
10. *Export possibility* — High
11. *Economic yields (q / ha)* — 15-20 (grain), 200-250 (green fodder)
12. *Economic value (Rs. / ha)* — 20,000 to 50,000
13. *By-product (If any)* — Straw
14. *Use of by-product* — Cattle feed / fuel / organic matter / mulch

235. *CROP* — LANATUM GRASS / Lanatam ghash

Scientific Name — Pennisetum lanatum Klotzsch

Family — Gramineae

Herb / Shrub / Tree — Herb (0.5-1.0 m height)

Annual / Biennial / Perennial — Annual

Used as — Food / fodder / soil binding plant

1. *Climatic requirements* — Cold climate / weather
2. *Soil requirements* — Loamy / alluvial / sandy soils
3. *Required land situations* — High dry land
4. *Importance of crop* — For small grains (poor man's food)
5. *Fertilizer management* — 100 kg N, 50 kg P_2O_5 and 50 kg K_2O/ha
6. *Water management* — 4-5 irrigations
7. *Duration of crop* — 120-140 days
8. *Parts used* — Grains for man, straw for cattle
9. *Where grown in India* — In hilly areas of India (U.P., H.P., Uttaranchal etc.)
10. *Export possibility* — High
11. *Economic yields (q / ha)* — 15-20 (grain)/ fodder 150-200 (green)
12. *Economic value (Rs. / ha)* — 15,000 to 20,000
13. *By-product (If any)* — Straw
14. *Use of by-product* — Cattle feed / fuel / mulch / organic matter

236. *CROP* — COMMON MILLET / Crow - foot / Makra

Scientific Name — Eleusine aegyptiaca (L.) Desf.

Family — Gramineae

Herb / Shrub / Tree — Herb

Annual / Biennial / Perennial — Annual

Used as — Food / fodder

1. *Climatic requirements* — Warm weather
2. *Soil requirements* — Alluvial / loamy / sandy soils
3. *Required land situations* — High land
4. *Importance of crop* — Both seeds and straw are used
5. *Fertilizer management* — 100 kg N, 50 kg P_2O_5 and 50 kg K_2O/ha
6. *Water management* — 3-4 irrigations
7. *Duration of crop* — 120-150 days
8. *Parts used* — Grain for human food, straw for cattle feed
9. *Where grown in India* — All-over India except, too low / too high temperature areas
10. *Export possibility* — High
11. *Economic yields (q / ha)* — 15-20
12. *Economic value (Rs. / ha)* — 20,000 to 40,000
13. *By-product (If any)* — Straw
14. *Use of by-product* — Cattle feed / fuel / mulch / organic matter

2. PULSES

237. *CROP* — BLACK GRAM / Urd /Kalai / Mashkalai / Beulidal

Scientific Name — Vigna mungo (L.) Hepper

Family — Papilionaceae

Herb / Shrub / Tree — Herb (0.5-1 m long)

Annual / Biennial / Perennial — Annual

Used as — Seeds as pulse/food and plants as cattle feeds

1. *Climatic requirements* — Hot humid climate with high rainfall
2. *Soil requirements* — Loamy / clayey / alluvial soils
3. *Required land situations* — High / medium / low lands
4. *Importance of crop* — Can be used both for pulse, cattle feed or green manure crop. It may be used as *paira* crop
5. *Fertilizer management* — 40 kg N, 60 kg P_2O_5 and 60 kg K_2O/ha
6. *Water management* — 1-2 irrigation, if irrigated, normally rainfed
7. *Duration of crop* — 110-120 days
8. *Parts used* — Grain as pulse, whole plant as cattle feed
9. *Where grown in India* — Punjab, West Bengal, Bihar, A.P., Kerala, Karnataka, Maharashtra, Orissa, T.N. etc.
10. *Export possibility* — High

11. *Economic yields (q / ha)* — Grain : 15-20, Cattle feed : 180-250
12. *Economic value (Rs. / ha)* — 25,000 to 50,000
13. *By-product (If any)* — Haulm or stalk
14. *Use of by-product* — Cattle feed / fuel / organic matter / mulch

238. *CROP* — GREEN GRAM / MUNG BEAN / Mung / Mug / Pessara/ Moong

Scientific Name* — *Vigna mungo (L.) Wilczek

Family — Papilionaceae

Herb / Shrub / Tree — Herb (50-80 cm height)

Annual / Biennial / Perennial — Annual

Used as — Food (pulse) / fodder / cover crop / green manure crop

1. *Climatic requirements* — Hot climate with rainfall
2. *Soil requirements* — Loamy / alluvial / sandy loam soils
3. *Required land situations* — High / medium
4. *Importance of crop* — Seed as food (pulse) and stalk as cattle feeds
5. *Fertilizer management* — 40 kg N, 60 kg P_2O_5 and 60 kg K_2O /ha
6. *Water management* — 1-2 times, if irrigated, normally rainfed
7. *Duration of crop* — 110-130 days
8. *Parts used* — Seeds / stalk
9. *Where grown in India* — Punjab, U.P., W. Bengal, Bihar, Orissa, Andhra Pradesh, Madhya Pradesh, Tamil Nadu, Maharashtra, Kerala, Karnataka, Assam etc.
10. *Export possibility* — High
11. *Economic yields (q / ha)* — 15-30
12. *Economic value (Rs. / ha)* — 30,000 to 60,000
13. *By-product (If any)* — Stalk
14. *Use of by-product* — Cattle feed / fuel / organic matter / mulch.

239. *CROP* — COWPEA / Lobia / Chowli / *Barbati / Lalsa*

Scientific Name* — *Vigna unguiculata (L.) Walp

Family — Papilionaceae

Herb / Shrub / Tree — Herb (0.5-1.5 m long vine)

Annual / Biennial / Perennial — Annual (trailing / spreading)

Used as — Food / vegetable / cattle feed / green manuring crop

1. *Climatic requirements* — High hot temperature with rainfall
2. *Soil requirements* — Loamy / alluvial / sandy soil

3. *Required land situations* — High / medium land
4. *Importance of crop* — Multipurpose uses (including cover crop)
5. *Fertilizer management* — 40 kg N, 60 kg P_2O_5 and 60 kg K_2O/ha
6. *Water management* — 1-2 time irrigations, if irrigated, normally rainfed
7. *Duration of crop* — 120-140 days
8. *Parts used* — Pod for vegetable / vine for fodder
9. *Where grown in India* — Punjab, Haryana, U.P., West Bengal, Orissa, Karnataka, Maharashtra, Tamil Nadu, Andhra Pradesh, Madhya Pradesh, Kerala, etc.
10. *Export possibility* — High
11. *Economic yields (q / ha)* — 40-80 (green pod) / 10-20 (seed) / 150-200 (vine green)
12. *Economic value (Rs. / ha)* — 40,000 to 80,000
13. *By-product (If any)* — Vine / whole plant
14. *Use of by-product* — Cattle feed / fuel / organic matter / mulch

240. *CROP* — RICEBEAN / Sutri / *Gaimung*

Scientific Name* — *Vigna umbellata (Thunb) Ohwi and Ohashi

Family — Papilionanceae

Herb / Shrub / Tree — Herb (0.5 – 1.5 m long vine)

Annual / Biennial / Perennial — Annual (spreading / trailing)

Used as — Fodder / pulse / green manuring crop / cover crop

1. *Climatic requirements* — High hot climate with rainfall (grown in wild)
2. *Soil requirements* — Loamy / alluvial / sandy soils
3. *Required land situations* — High / medium land
4. *Importance of crop* — Trailing habit, controlling weeds, improves soil fertility. Magic crop for barren lands
5. *Fertilizer management* — 40 kg N, 60 kg P_2O_5 and 30 kg K_2O/ha
6. *Water management* — 1-2 times irrigations, if irrigated, normally rainfed
7. *Duration of crop* — 120-140 days
8. *Parts used* — Whole plant (for fodder)
9. *Where grown in India* — All over the country except northern states
10. *Export possibility* — Less
11. *Economic yields (q / ha)* — 200-250

12. *Economic value (Rs. / ha)* — 25,000 to 50,000
13. *By-product (If any)* — Entire crop is taken away for fodder
14. *Use of by-product* — Nil

Note — This crop could be explored for pulse crop also, as it has potentiality.

241. *CROP* — HORSE GRAM / Kulatra

Scientific Name* — *Dolichos falcatus Klein ex Willd

Family — Papilionaceae

Herb / Shrub / Tree — Herb (Trailing) (0.5-1.5 m long vine)

Annual / Biennial / Perennial — Annual

Used as — Fodder / food (seed) / green manure

1. *Climatic requirements* — High temperature with rainfall
2. *Soil requirements* — Loamy / sandy loam soils
3. *Required land situations* — High / medium land
4. *Importance of crop* — Trailing type, hardy in nature
5. *Fertilizer management* — 40 kg N, 60 kg P_2O_5 and 60 kg K_2O/ha
6. *Water management* — 1-2 irrigations, if irrigated, normally rainfed
7. *Duration of crop* — 110-130 days
8. *Parts used* — Seed for food and whole plant for fodder
9. *Where grown in India* — A.P., Karnataka, T.N., Bihar, Orissa etc.
10. *Export possibility* — Less
11. *Economic yields (q / ha)* — 10-15 (grain), 150-200 (green fodder)
12. *Economic value (Rs. / ha)* — 10,000 to 15,000
13. *By-product (If any)* — Stalk (if seed harvested)
14. *Use of by-product* — Cattle feed / fuel / organic matter / mulch.

3. OIL SEEDS

242. *CROP* — SESAME / GINGELLY / *Til*

Scientific Name* — *Sesamum indicum L.

Family — Pedaliaceae

Herb / Shrub / Tree — Herb (50-80 cm height)

Annual / Biennial / Perennial — Annual

Used as — Oil seeds / Industrial uses – soap, oil, cosmetics

1. *Climatic requirements* — High temperature with rainfall

2. *Soil requirements* — Loamy / alluvial / sandy loam soils
3. *Required land situations* — High / medium land
4. *Importance of crop* — C4 type crop, can be grown at all times
5. *Fertilizer management* — 60 kg N, 30 kg P_2O_5 and 30 kg K_2O/ha
6. *Water management* — 1-2 irrigation, if irrigated, normally rainfed
7. *Duration of crop* — 110-130 days
8. *Parts used* — Seeds
9. *Where grown in India* — M.P., U.P., Orissa, Gujarat, A.P., T.N., Bihar, Maharashtra, West Bengal etc.
10. *Export possibility* — High
11. *Economic yields (q / ha)* — 10-15
12. *Economic value (Rs. / ha)* — 20,000 to 30,000
13. *By-product (If any)* — Stalk
14. *Use of by-product* — Fuel / organic matter / pegs (for vegetable staking)

243. *CROP* — CASTOR-SEED / CASTOR / Arandi / Veranda / Bherenda / Reri / Castor oil

Scientific Name* — *Ricinus communis L.

Family — Euphorbiaceae

Herb / Shrub / Tree — Shrub / tree (small)

Annual / Biennial / Perennial — Perennial

Used as — Oil seeds (as lubricant in machine) / purgative / bio-fuel)

1. *Climatic requirements* — Moderate hot climate with rainfall
2. *Soil requirements* — Loamy / sandy loam / alluvial soils
3. *Required land situations* — High / medium land
4. *Importance of crop* — Oils / lubricant (for cars) for industries / purgative
5. *Fertilizer management* — 60 kg N, 30 kg P_2O_5 and 30 kg K_2O/ha
6. *Water management* — 1-2 irrigations, if irrigated; normally rainfed
7. *Duration of crop* — 2-3 years
8. *Parts used* — Seeds (bio fuel)
9. *Where grown in India* — A.P., Maharashtra, Karnataka, Orissa, West Bengal, Madhya Pradesh, etc.
10. *Export possibility* — High / very high
11. *Economic yields (q / ha)* — 10-15 (seed)
12. *Economic value (Rs. / ha)* — 40,000 to 60,000

13. *By-product (If any)* — Stalk and leaves
14. *Use of by-product* — Fuel / organic matter / cattle feed

244. *CROP* — SUNFLOWER / Surajmukhi / *Suryamukhi*

Scientific Name* — *Helianthus annuus L.

Family — Compositae

Herb / Shrub / Tree — Herb (60-120 cm height)

Annual / Biennial / Perennial — Annual

Used as — Oil seeds / ornamental / industrial use

1. *Climatic requirements* — Hot, high or moderate temperature with rainfall
2. *Soil requirements* — Loamy / sandy loam / alluvial soils
3. *Required land situations* — High / medium lands
4. *Importance of crop* — Head (big) moves according to the sun's movement
5. *Fertilizer management* — 60 kg N, 30 kg P_2O_5 and 30 kg K_2O/ha
6. *Water management* — 1-2 irrigations, if irrigated; normally rainfed
7. *Duration of crop* — 110-120 days
8. *Parts used* — Seeds
9. *Where grown in India* — All over India (in different seasons) except too low temperature areas.
10. *Export possibility* — Very high
11. *Economic yields (q / ha)* — 15-20
12. *Economic value (Rs. / ha)* — 45,000 to 60,000
13. *By-product (If any)* — Stalk
14. *Use of by-product* — Fuel / organic matter / pegs / mulch / cattle feed

245. *CROP* — GROUNDNUT / PEANUT / Mungphali / Badam

Scientific Name* — *Arachis hypogaea L.

Family — Papilionaceae

Herb / Shrub / Tree — Herb (erect trailing), (50-80 cm height)

Annual / Biennial / Perennial — Annual

Used as — Oil seeds / fodder / green manure / food

1. *Climatic requirements* — Hot, high temperature with rainfall
2. *Soil requirements* — Loamy / sandy loam / alluvial soils
3. *Required land situations* — High / medium lands

4. *Importance of crop* — Pods are developed below ground, but flowers develop above ground
5. *Fertilizer management* — 40 kg N, 60 kg P_2O_5 and 60 kg K_2O/ha
6. *Water management* — 3-4 irrigations
7. *Duration of crop* — 120-160 days
8. *Parts used* — Nut (for oils) / haulm (for fodder) or green manure
9. *Where grown in India* — Gujarat, Maharashtra, T.N., Punjab, Rajasthan, M.P., Orissa, West Bengal etc.
10. *Export possibility* — Very high
11. *Economic yields (q / ha)* — 15-25
12. *Economic value (Rs. / ha)* — 45,000 to 75,000
13. *By-product (If any)* — Haulm
14. *Use of by-product* — Fuel / green manure in WB / organic manure / cattle feed.

246. *CROP* — PHYSIC NUT / Safed arand / Bagbherenda / Sadabherenda

Scientific Name* — *Jatropha curcas L.

Family — Euphorbiaceae

Herb / Shrub / Tree — Shrub / tree (small)

Annual / Biennial / Perennial — Perennial

Used as — Oil / bio-energy / fuel

1. *Climatic requirements* — Warm weather
2. *Soil requirements* — Loamy / sandy loam / alluvial soils
3. *Required land situations* — High / medium lands
4. *Importance of crop* — With the oil from seed, petrol may be substituted partly
5. *Fertilizer management* — 80 kg N, 60 kg P_2O_5 and 30 kg K_2O/ha
6. *Water management* — 2-3 irrigations, annually
7. *Duration of crop* — 2-3 years
8. *Parts used* — Seeds
9. *Where grown in India* — Kerala, T.N., A.P., W.B., etc. (mainly in Coromandel Coast and Travancore in T.N.)
10. *Export possibility* — Very high
11. *Economic yields (q / ha)* — 25-30
12. *Economic value (Rs. / ha)* — 60,000 to 80,000
13. *By-product (If any)* — Plant and leaves
14. *Use of by-product* — Fuel / cattle feed / organic matter / pegs.

247. *CROP* — Jatropha / Jangliarandi / *Banbherenda* / Safed arand

Scientific Name* — *Jatropha glandulifera Roxb./ ***J. curcas*** L.

Family — Euphorbiaceae

Herb / Shrub / Tree — Shrub / tree (small)

Annual / Biennial / Perennial — Perennial

Used as — Seed-oil / bio-energy / medicine / bio-fuel

1. *Climatic requirements* — Warm weather
2. *Soil requirements* — Loamy / sandy loam / alluvial soils
3. *Required land situations* — High / medium lands
4. *Importance of crop* — Some engine may be run with this oil, substituted partly with petrol
5. *Fertilizer management* — 80 kg N, 60 kg P_2O_5 and 40 kg K_2O/ha
6. *Water management* — 2-3 irrigations annually
7. *Duration of crop* — 2-3 years
8. *Parts used* — Seeds (for oil)
9. *Where grown in India* — Kerala, T.N., A.P., Karnataka, West Bengal, Assam, Bihar, etc.
10. *Export possibility* — Very high
11. *Economic yields (q / ha)* — 20-25
12. *Economic value (Rs. / ha)* — 40,000 to 60,000
13. *By-product (If any)* — Stem and leaves
14. *Use of by-product* — Pegs / fuel / organic matter / cattle feed

4. FIBRES

248. *CROP* — JUTE / WHITE JUTE /Pat / Titapat / Mithapat / Nalita / Narcha/ Pattam

Scientific Name* — *Corchorus capsularis* / *C. olitorius L.

Family — Tiliaceae

Herb / Shrub / Tree — Shrub (tall), (1.0 to 3.0 m height)

Annual / Biennial / Perennial — Annual

Used as — Fibre / Green manuring for availability of P / vegetable

1. *Climatic requirements* — Hot, high temperature with high humidity and high rainfall
2. *Soil requirements* — Alluvial / loamy / sandy loam soils
3. *Required land situations* — High / medium / low lands
4. *Importance of crop* — Bast fibre from bark / stem. *Capsularis* for low and *Olitorius* for high lands

5. *Fertilizer management* — 80 kg N, 30 kg P_2O_5 and 60 kg K_2O/ha
6. *Water management* — 2-3 irrigations, if irrigated (In W.B. rainfed)
7. *Duration of crop* — 120-140 days
8. *Parts used* — Stem as fibre / leaves for vegetable and soil fertility
9. *Where grown in India* — West Bengal, Assam, Orissa, Bihar, A.P., etc. (mostly in East India)
10. *Export possibility* — Very high
11. *Economic yields (q / ha)* — 15-25
12. *Economic value (Rs. / ha)* — 45,000 to 75,000
13. *By-product (If any)* — Stick / leaves
14. *Use of by-product* — Stick for boards / papers and leaves for manure (leaves contain nearly 0.5% N).

249. *CROP — MESTA / KENAF / ROSELLE* / Patsan / Ambari / *Patwa*

Scientific Name — Hibiscus cannabinus L. / ***H. sabdariffa*** L.

Family — Malvaceae

Herb / Shrub / Tree — Shrub (tall) (1.0 to 3.0m height)

Annual / Biennial / Perennial — Annual

Used as — Fibre / paper-pulp

1. *Climatic requirements* — High temperature with high / moderate rainfall with high humidity
2. *Soil requirements* — Loamy / alluvial / sandy loam soils
3. *Required land situations* — High / medium low lands
4. *Importance of crop* — Bast fibre (fibre from stem)
5. *Fertilizer management* — 60 kg N, 30 kg P_2O_5 and 30 kg K_2O/ha
6. *Water management* — Normally rainfed; 1-2 times irrigations, if irrigated
7. *Duration of crop* — 110-140 days
8. *Parts used* — Fibre from stem
9. *Where grown in India* — West Bengal, Orissa, Assam, Bihar, Tamil Nadu, Uttar Pradesh, Punjab, M.P., A.P. etc.
10. *Export possibility* — High
11. *Economic yields (q / ha)* — 15-25
12. *Economic value (Rs. / ha)* — 30,000 to 50,000
13. *By-product (If any)* — Stick and leaves
14. *Use of by-product* — Stick is used as fuel / pegs and leaves as organic manure / mulch / cattle feed.

250. CROP — SUNNHEMP / Sunn / Shan / Shanpat / San

Scientific Name — Crotalaria juncea L.

Family — Papilionaceae

Herb / Shrub / Tree — Shrub (tall), 1.0 – 3.0 m length

Annual / Biennial / Perennial — Annual

Used as — Fibre (from stem) / green manure

1. *Climatic requirements* — High temperature with high / moderate rainfall with humidity
2. *Soil requirements* — Alluvial / loamy / sandy loam soils
3. *Required land situations* — High / medium / low lands
4. *Importance of crop* — Bast fibre from stem
5. *Fertilizer management* — 80 kg N, 60 kg P_2O_5 and 40 kg K_2O/ha
6. *Water management* — Normally rainfed; 1-2 irrigations, if irrigated
7. *Duration of crop* — 110-130 days for fibre in summer, 130-150 days for seed in winter season
8. *Parts used* — Fibre from the stem (green manure crop also)
9. *Where grown in India* — West Bengal, Orissa, Bihar, Assam, Uttar Pradesh, Punjab, Maharashtra, Madhya Pradesh, Andhra Pradesh, Tamil Nadu.
10. *Export possibility* — High
11. *Economic yields (q / ha)* — Fibre 15-20
12. *Economic value (Rs. / ha)* — 30,000 to 40,000
13. *By-product (If any)* — Stick and leaves
14. *Use of by-product* — Stick as fuel / pegs / leaves for organic manure

251. *CROP* — RAMIE / Kankhura / *Maslin / Remi*

Scientific Name — Boehmeria nivea (L.) Gaud

Family — Urticaceae

Herb / Shrub / Tree — Shrub (tall), (1.0 to 2.0 m height)

Annual / Biennial / Perennial — Perennial

Used as — Fibre (from stem)

1. *Climatic requirements* — High temperature with high rainfall and high humidity
2. *Soil requirements* — Loamy / alluvial / sandy loam soils
3. *Required land situations* — High / medium lands
4. *Importance of crop* — Strong fibre from stem obtained

5. *Fertilizer management* — 80 kg N, 40 kg P_2O_5 and 40 kg K_2O/ha
6. *Water management* — 1-3 irrigations annually
7. *Duration of crop* — 2-3 years
8. *Parts used* — Fibre
9. *Where grown in India* — Assam, W.B., Bihar, Orissa, M.P. etc.
10. *Export possibility* — High (due to strong fibre used in tyres etc.)
11. *Economic yields (q / ha)* — 10-15
12. *Economic value (Rs. / ha)* — 30,000 to 45,000
13. *By-product (If any)* — Stick and leaves
14. *Use of by-product* — Stick for fuel / pegging materials and leaves as organic matter / cattle feed

252. *CROP* — AGAVE / sisal / *Sisalgachh*

Scientific Name* — *Agave sisalana Perr. Ex Engelm.

Family — Agavaceae

Herb / Shrub / Tree — Shrub (50-150 cm height)

Annual / Biennial / Perennial — Perennial

Used as — Fibre (from leaves)

1. *Climatic requirements* — Dry conditions (high temperature, with less rainfall and less humidity)
2. *Soil requirements* — Sandy / oxisol with rocks / sandy loam soils
3. *Required land situations* — High land
4. *Importance of crop* — Fibre from the leaves / xerophytic plant
5. *Fertilizer management* — 60 kg N, 0 kg P_2O_5 and 30 kg K_2O/ha
6. *Water management* — 1-2 irrigations, if irrigated, normally rainfed
7. *Duration of crop* — 2-3 years
8. *Parts used* — Leaves
9. *Where grown in India* — Dry tracks of Bengal, Orissa, Bihar, M.P., Maharashtra, T.N., Karnataka
10. *Export possibility* — High
11. *Economic yields (q / ha)* — 10-15
12. *Economic value (Rs. / ha)* — 30,000 to 45,000
13. *By-product (If any)* — Other leaves materials and stem
14. *Use of by-product* — Fibre manufacturing (from leaves) products sisal wax / fuel

253. ***CROP*** **— UPLAND COTTON / ASIATIC COTTON / EGYPTIAN COTTON / Cotton / Rui / *Kapash / Tula***

Scientific Name* — *Gossypium hirsutum L./ ***arboreum*** L. / ***herbaceum*** L.

Family — Malvaceae

Herb / Shrub / Tre —Much branched shrub (bushy) (0.5 to 1.5 m. height)

Annual / Biennial / Perennial — Annual

Used as — Cotton fibre (from bolls)

1. *Climatic requirements* — Warm season / weather
2. *Soil Requirements*—Black cotton / loamy / sandy / sandy loam soils
3. *Required land situations* — High land
4. *Importance of crop* — Boll from flower is formed, where from cotton is obtained
5. *Fertilizer management* — 100 kg N, 50 kg P_2O_5 and 50 kg K_2O/ha
6. *Water management* — 5-6 irrigations
7. *Duration of crop* — 120-140 days
8. *Parts used* — Bolls (from flowers)
9. *Where grown in India* — Pubjab, Gujarat, Rajasthan, T.N., Karnataka, Maharashtra, M.P. and A.P.
10. *Export possibility* — High
11. *Economic yields (q / ha)* — 10-20
12. *Economic value (Rs. / ha)* — 40,000 to 80,000
13. *By-product (If any)* — Branches and leaves
14. *Use of by-product* — Fuel / cattle feed / organic matter (vine crops may be trailed on the bushy plant after harvest of boll)

5. TUBERS / RHIZOMES

254. ***CROP*** **— ASPARAGUS / *Satmuli* / Satawar / Phusar / Satamuli**

Scientific Name* — *Asparagus racemosus Willd

Family — Liliaceae

Herb / Shrub / Tree — Shrub (tall)

Annual / Biennial / Perennial — Perennial

Used as — Vegetable / food / medicine

1. *Climatic requirements* — Dry conditions with less rainfall
2. *Soil requirements* — Sandy / rocky / sandy loam soils

3. *Required land situations* — High / medium land
4. *Importance of crop* — Spinous shrub and swollen roots
5. *Fertilizer management* — 60 kg N, 30 kg P_2O_5 and 30 kg K_2O/ha
6. *Water management* — Normally rainfed, 1-2 times irrigations annually
7. *Duration of crop* — 2-3 years
8. *Parts used* — Roots / twigs
9. *Where grown in India* — Dry tropical / subtropical areas of A.P., Kerala, Pondicherry, Maharashtra, M.P., T.N., etc.
10. *Export possibility* — High
11. *Economic yields (q / ha)* — 10-15
12. *Economic value (Rs. / ha)* — 10,000 to 15,000
13. *By-product (If any)* — Stem
14. *Use of by-product* — Fuel / pegs / organic matter / veterinary medicine

255. *CROP* — GLOBE ARTICHOKE / Hathi chuk

Scientific Name* — *Cynara scolymus L.

Family — Composiae

Herb / Shrub / Tree — Herb

Annual / Biennial / Perennial — Perennial

Used as — Vegetable / medicine (diabetics)

1. *Climatic requirements* — Moderate temperature with moderate rainfall (Mediterranean climate)
2. *Soil requirements* — Loamy / sandy loam soils
3. *Required land situations* — High / medium
4. *Importance of crop* — Flower is eaten
5. *Fertilizer management* — 100 kg N, 60 kg P_2O_5 and 50 kg K_2O/ha
6. *Water management* — 4-5 irrigations
7. *Duration of crop* — 120-130 days
8. *Parts used* — Soft and fleshy receptacle of flower heads
9. *Where grown in India* — Throughout India except, too low / too high temperature areas
10. *Export possibility* — High
11. *Economic yields (q / ha)* — 10-15
12. *Economic value (Rs. / ha)* — 30,000 to 45,000
13. *By-product (If any)* — Stalks and leaves

14. *Use of by-product*— Fuel / pegs / organic matter / medicine (with leaves).

256. *CROP* — JERUSALEM ARTICHOKE / *Hathi chuk*

Scientific Name* — *Helianathus tuberosus L.

Family — Compositae

Herb / Shrub / Tree — Shrub

Annual / Biennial / Perennial — Annual

Used as — Food / medicine / vegetable (green tops)

1. *Climatic requirements* — Moderate temperature with moderate rainfall
2. *Soil requirements* — Loamy / sandy loam soil
3. *Required land situations* — High land
4. *Importance of crop* — A tuberous under shrub (i.e. edible, subterranean tuberous stems)
5. *Fertilizer management* — 60 kg N, 30 kg P_2O_5 and 30 kg K_2O/ha
6. *Water management* — 1-2 irrigations, if irrigated; normally rainfed
7. *Duration of crop* — 110-140 days
8. *Parts used* — Tuberous stem
9. *Where grown in India* — West Bengal, Assam, Maharashtra, U.P., Andhra Pradesh etc.
10. *Export possibility* — High
11. *Economic yields (q / ha)* — 15-20
12. *Economic value (Rs. / ha)* — 30,000 to 20,000
13. *By-product (If any)* — Green tops and stems (stalks)
14. *Use of by-product* — Used as feed of cattle / fuel / organic matter

257. *CROP* — ARUM / TARO / Colocasia / *Ganti kachu* / Arvi / Kachalu/ Ghuiyan / *Jalkachu*

Scientific Name* — *Colocasia esculenta L. Schott.

Family — Araceae

Herb / Shrub / Tree — Herb (rhizomatous), 0.5 m height)

Annual / Biennial / Perennial — Perennial

Used as — Food (Rhizomes / petiole / runners / leaves)

1. *Climatic requirements* — Moderate temperature with moderate rain falls
2. *Soil requirements* — Loamy / alluvial / sandy loam / clayey soils (according to varieties)

3. *Required land situations* — Low / medium
4. *Importance of Crop*—Rhizomes, runners, tubers and leaves are eaten
5. *Fertilizer management* — 60 kg N, 30 kg P_2O_5 and 30 kg K_2O/ha
6. *Water management* — Normally rainfed, if necessary 1-2 times irrigations (some in waterlogged soils)
7. *Duration of crop* — 110-150 days
8. *Parts used* — Petiole, rhizomatious stems and runners
9. *Where grown in India* — West Bengal, Kerala, Assam, Bihar, Orissa, Sikkim, Maharashtra, Karnataka, Tamil Nadu, Madhya Pradesh, A.P., etc.
10. *Export possibility* — High
11. *Economic yields (q / ha)* — 20-25
12. *Economic value (Rs. / ha)* — 20,000 to 25,000
13. *By-product (If any)* — Leaves
14. *Use of by-product* — Vegetables / organic matter (sap of stem is used for staining cloths)

258. *CROP* — ARAI KACHU / *Dudhkachu* / Arum

Scientific Name* — *Colocasia anatiquorum Schott. var. ***esculenta*** Schott. / ***Arum esculentum*** L.

Family — Araceae

Herb / Shrub / Tree — Rhizomatous herb, (50-150 cm height)

Annual / Biennial / Perennial — Perennial

Used as — Food (starch) / vegetable

1. *Climatic requirements* — Warm weather with high rainfall
2. *Soil requirements* — Alluvial / laterite / sandy / sandy loam soils
3. *Required land situations* — High land
4. *Importance of crop* — White milk comes from leaf when torn
5. *Fertilizer management* — 100 kg N, 50 kg P_2O_5 and 50 kg K_2O/ha
6. *Water management* — 4-5 irrigations, if irrigated
7. *Duration of crop* — 2-3 years
8. *Parts used* — Rhizomes / leaves / petiole / runners
9. *Where grown in India* — All over India except too dry or too low temperature areas
10. *Export possibility* — High
11. *Economic yields (q / ha)* — 150-200
12. *Economic value (Rs. / ha)* — 20,000 to 40,000

13. *By-product (If any)* — Nil
14. *Use of by-product* — Nil

6. SUGARS / STARCHES

259. *CROP* — SUGARCANE / Nobelcane / *Ganna* / Paunda / *Akh* / *Ikshu*

Scientific Name* — *Saccharum officinarum L.

Family — Gramineae

Herb / Shrub / Tree — Tall herb (grass), (1.0 to 3.0 m height)

Annual / Biennial / Perennial — Perennial

Used as — Sugars (juice from cane and sugar from juice)

1. *Climatic requirements* — Warm climate
2. *Soil requirements* — Loamy / sandy loam / alluvial soils
3. *Required land situations* — Low / medium / high (according to varieties)
4. *Importance of crop* — Sweet cane (one variety of sugarcane) can be chewed
5. *Fertilizer management* — 100 kg N, 50 kg P_2O_5 and 50 kg K_2O/ha
6. *Water management* — 5-6 irrigations annually
7. *Duration of crop* — 2-3 years
8. *Parts used* — Entire stem (nodes and inter nodes)
9. *Where grown in India* — U.P., Bihar, Punjab, T.N., Karnataka and Maharashtra etc.
10. *Export possibility* — High
11. *Economic yields (q / ha)* — 10-15 (sugar)
12. *Economic value (Rs. / ha)* — 40,000 to 60,000
13. *By-product (If any)* — Bagasse and leaf
14. *Use of by-product* — Fuel / mulching material / organic matter / board / cattle feed

260. *CROP* — SWEET POTATO / Shakarkand / *Mitha alu* / *Misti Alu*

Scientific Name* — *Ipomoea batatas (L) Lamk

Family — Convolvulaceae

Herb / Shrub / Tree — Herb (spreading), (0.5 to 1.5 length vine)

Annual / Biennial / Perennial — Annual

Used as — Starch (from swollen / tubers) / vegetable / fodder

1. *Climatic requirements* — Warm weather conditions

2. *Soil requirements* — Sandy / loamy / alluvial soils
3. *Required land situations* — High / medium lands
4. *Importance of crop* — Rooting at the nodes, tubers inside soils
5. *Fertilizer management* — 50 kg N, 30 kg P_2O_5 and 50 kg K_2O/ha
6. *Water management* — 2-3 irrigations, normally rainfed
7. *Duration of crop* — 120-140 days
8. *Parts used* — Swollen tubers as food; vines as fodder
9. *Where grown in India* — All drier places in India except, too low temperature and too hot areas
10. *Export possibility* — High
11. *Economic yields (q / ha)* — 150-200 (green)
12. *Economic value (Rs. / ha)* — 15,000 to 40,000
13. *By-product (If any)* — Vine / stem / leaves
14. *Use of by-product* — Cattle feed / organic matter / mulch / fuel.

261. *CROP* — TAPIOCA / Manioc / Sakarkanda / *Maravuli / Cassava*

Scientific Name* — *Manihot esculenta Crantz

Family — Euphorbiaceae

Herb / Shrub / Tree — Shrub (small), (1.0-2.0 m height)

Annual / Biennial / Perennial — Perennial

Used as — Starch (sago / semolina / flour etc. produced)

1. *Climatic requirements* — Warm weather
2. *Soil requirements* — Sandy / loamy / alluvial soils
3. *Required land situations* — High lands
4. *Importance of crop* — Starch is from tubers
5. *Fertilizer management* — 80 kg N, 30 kg P_2O_5 and 30 kg K_2O/ha
6. *Water management* — 2-3 irrigations annually
7. *Duration of crop* — 120-140 days
8. *Parts used* — Tubers
9. *Where grown in India* — Tamil Nadu, Karnataka Andhra Pradesh, Madhya Pradesh, Kerala etc.
10. *Export possibility* — High
11. *Economic yields (q / ha)* — 100-200
12. *Economic value (Rs. / ha)* — 15,000 to 25,000
13. *By-product (If any)* — Stems and leaves
14. *Use of by-product* — Fuel and organic matter / cattle feed / pegs

7. VEGETABLES

262. (as in pre-*kharif* season)

8. FODDER / FORAGE
GRASSY

263. (as in pre-*kharif* season)

NON-GRASSY

264. (as in pre-*kharif* season)

9. GREEN MANURING CROPS
LEGUMES

265. (as in pre-*kharif* season)

NON-LEGUMES

266. (as in pre-*kharif* season)

10. MEDICINAL PLANTS

267. *CROP* — ROSE GERANIUM

Scientific Name* — *Pelargonium graveolens L'Herit

Family — Geraniaceae

Herb / Shrub / Tree — Shrub

Annual / Biennial / Perennial — Perennial

Used as — Geranium oil / Toilet soap preparation

1. *Climatic requirements* — Warm weather
2. *Soil requirements* — High land
3. *Required land situations* — Loamy / sandy loam soils
4. *Importance of crop* — It is antibacterial (in tooth paste / powder etc.). It is used for flavouring tobacco products
5. *Fertilizer Management*—80 kg N, 60 kg P_2O_5 and 30 kg K_2O/ha annually.
6. *Water management* — 2-3 irrigations annually

7. *Duration of crop* — 2-3 years
8. *Parts used* — Leaves and twigs (for Geranium oil for toilet soap preparation)
9. *Where grown in India* — T.N. (Shevaroy areas)
10. *Export possibility* — High
11. *Economic yields (q / ha)* — 10-15
12. *Economic value (Rs. / ha)* — 40,000 to 80,000
13. *By-product (If any)* — Stems / trunk
14. *Use of by-product* — Timber / fuel.

268. *CROP* — PATCHOULI

Scientific Name — Pogostemon cablin (Blanco) Benth

Family — Labiatae

Herb / Shrub / Tree — Herb

Annual / Biennial / Perennial — Annual

Used as — Patchoulic oil (antibacterial)

1. *Climatic requirements* — Warm weather
2. *Soil requirements* — Loamy / sandy loam soils
3. *Required land situations* — High lands
4. *Importance of Crop*— Oil is used in perfume industries (for cosmetics etc.)
5. *Fertilizer management* — 80 kg N, 60 kg P_2O_5 and 30 kg K_2O/ha
6. *Water management* — 2-3 irrigations
7. *Duration of crop* — 140-160 days
8. *Parts used* — All parts
9. *Where grown in India* — T.N., Maharashtra, etc.
10. *Export possibility* — High
11. *Economic yields (q / ha)* — 15-25
12. *Economic value (Rs. / ha)* — 50,000 to 60,000
13. *By-product (If any)* — Nil
14. *Use of by-product* — Nil

11. SPICES

269. (as in pre-*kharif* season)

12. FRUITS

270. (as in pre-*kharif* season)

13. FLOWERS / SUCCULENTS / ORNAMENTALS

271. (as in pre-*kharif* season)

14. BEVERAGE AND NARCOTICS

272. (as in pre-*kharif* season)

Chapter - 4

WEEDS IN CROPS OF DIFFERENT SEASONS

INTRODUCTION

Chapter 4 deals with different weeds both under broad and narrow leaved weeds in different seasons such as winter (*rabi*), summer (pre-*kharif*) and rainy / wet (*kharif*) season.

Everybody is aware that weed is a 'menace' in crop production in India and it grows along with the crops and as a result production of crop is hampered. But in some cases weed may be a boon to the crop production. In case of *paira* crop (no-tillage relay crops) weed is very helpful for crop production. Some vine crops (pea / lathyrus / lentil etc.) grow well, in winter season (in West Bengal), over the weeds like *Bengalvetch / Cyperus / Blumea* etc. and as a result the productions of main crops are high.

Though weed, in general, is a menace, but these weeds may be utilized for other important purposes, such as medicines, fodders, fruits, vegetables, narcotics, beverage or even mulching purposes for conserving moisture. In weeds, of course, the production and value aspects, have been omitted sometimes in the format. 'Weed', being a plant, may be utilized for other important purposes, it also may be called as crop. Hence, it has been incorporated here.

Note : Weed is a 'menace' in crop production in India and it grows along with the crop, as a result crop production is hampered. Therefore, production and value aspects have been omitted. But these weeds may be utilized for other purposes (medicine / fruits / vegetable / narcotics etc.)

1. WINTER (*Rabi*) Season

BROAD LEAVED WEEDS

273. *CROP*—BENGAL VETCH /Purple vetch / *Lalhetka / Hetka*

Scientific Name — Vicia benghalensis L.

Family — Papilionaceae

Herb / Shrub / Tree — Herb (30-50 cm height)

Annual / Biennial / Perennial — Annual

Used as — Fodder, though weed in *Lathyrus*

1. *Climatic requirements* — Cold weather
2. *Soil requirements* — Alluvial / loamy / sandy clayey soils
3. *Required land situations* — Medium / low land
4. *Importance of crop* — It is the weed in sunnhemp or *Lathyrus*
5. *Fertilizer management* — 40 kg N, 60 kg P_2O_5 and 30 kg K_2O/ha
6. *Water management* — 2-3 irrigation, if needed
7. *Duration of crop* — 120-140 days
8. *Parts used* — Entire plant for fodder
9. *Where grown in India* — In eastern part of India (i.e. West Bengal, Assam, Orissa, U.P., Bihar, M.P. etc.)
10. *Export possibility* — High
11. *Economic yields (q / ha)* — 150-200 (if grown as fodder)
12. *Economic value (Rs. / ha)* — 15,000-20,000
13. *By-product (If any)* — Nil
14. *Use of by-product* — Weed can be used as fodder, fuel, mulching materials, organic matter

274. *CROP* — WHITE SWEET CLOVER / Huban clover / *Senji methi*

Scientific Name — *Melilotus alba* Medik

Family — Papilionaceae

Herb / Shrub / Tree — Herb (30-40 cm height)

Annual / Biennial / Perennial — Annual

Used as — Fodder / Weed in *Lathyrus*

1. *Climatic requirements* — Cold weather
2. *Soil requirements* — Alluvial / loamy / sandy / clayey soils
3. *Required land situations* — Medium or low lands
4. *Importance of crop* — It is a good fodder
5. *Fertilizer management* — 40 kg N, 60 kg P_2O_5 and 30 kg K_2O/ha
6. *Water management* — 2-3 irrigation, if needed (for fodder)
7. *Duration of crop* — 120-140 days
8. *Parts used* — Though weed, entire plant is used as fodder
9. *Where grown in India* — Assam, West Bengal, Orissa, Bihar, Madhya Pradesh, U.P., etc.

10. *Export possibility* — High
11. *Economic yields (q / ha)* — Nil
12. *Economic value (Rs. / ha)* — As weed it is very detrimental but as fodder it is good. As weed it can be utilized as fuel, fodder, mulching materials and organic matter.
13. *By-product (If any)* — Nil
14. *Use of by-product* — Nil

275. *CROP* — YELLOW SWEET CLOVER / *Banmethi / Senji*

Scientific Name* — *Melilotus indica All.

Family — Papilionaceae

Herb / Shrub / Tree — Herb (30-50 cm height)

Annual / Biennial / Perennial — Annual

Used as — Foods / weed in grass pea/ fodder

1. *Climatic requirements* — Cold weather, dry conditions
2. *Soil requirements* — Alluvial / loamy / sandy / clayey soils
3. *Required land situations* — Medium or low lands
4. *Importance of crop* — It is a good fodder also, though it is the winter season weed. Flower yellow.
5. *Fertilizer management* — As fodder 40-60-30 (N, P_2O_5 and K_2O/ha respectively) but as weed it is detrimental for crops
6. *Water management* — 3-4 irrigations, if grown as fodder
7. *Duration of crop* — 120-140 days
8. *Parts used* — Whole plant as fodder
9. *Where grown in India* — All over India, except southern states
10. *Export possibility* — High
11. *Economic yields (q / ha)* — Nil (as weed)
12. *Economic value (Rs. / ha)* — As weed, it is detrimental to the winter crop production, but as fodder it has much value. It is used as medicine also (clover)
13. *By-product (If any)* — Nil
14. *Use of by-product* — Nil

276. *CROP* — SWEET CLOVER / Bunomethi / *Jatamethi*

Scientific Name* — *Melilotus parvifloa Desf.

Family — Papilionaceae

Herb / Shrub / Tree — Herb (30-50 height)

Annual / Biennial / Perennial — Annual

Used as — Fodder (Weed in grass pea)

1. *Climatic requirements* — Cool season, dry conditions
2. *Soil requirements* — Alluvial / loamy / sandy soils
3. *Required land situations* — Medium or low lands
4. *Importance of crop* — It is a good fodder also, though it is the winter season weed. Flower white.
5. *Fertilizer management* — As fodder 40-60-30 (N, P_2O_5 and K_2O/ha respectively) but as weed it is detrimental for crops
6. *Water management* — 3-4 irrigation, if grown as fodder
7. *Duration of crop* — 120-140 days
8. *Parts used* — Whole plant as fodder
9. *Where grown in India* — All over India, except southern states
10. *Export possibility* — High
11. *Economic yields (q / ha)* — Nil
12. *Economic value (Rs. / ha)* — As weed, it is detrimental to the winter crop production, but as fodder it has much value. It is used as medicine also
13. *By-product (If any)* — Nil
14. *Use of by-product* — Nil

277. CROP — FUMITORY /Pit-Papara / *Sulpha*

Scientific Name — Fumaria officinalis L.

Family — Fumariaceae

Herb / Shrub / Tree — Herb (30-40 cm height)

Annual / Biennial / Perennial — Annual

Used as — It is a weed in winter crop but is used as medicine (hepatic ailments)

1. *Climatic requirements* — Cool weather condition
2. *Soil requirements* — Alluvial / loamy / sandy loam soils
3. *Required land situations* — High / medium land
4. *Importance of crop* — Smell in leaves
5. *Fertilizer management* — 60 kg N, 60 kg P_2O_5 and 30 kg K_2O/ha
6. *Water management* — 2-3 irrigations if grown as medicinal crop but it is detrimental to winter crops
7. *Duration of crop* — 120-140 days
8. *Parts used* — Leaves for medicine / stem for cattle fed
9. *Where grown in India* — Cool areas in India except southern states

10. *Export possibility* — High
11. *Economic yields (q / ha)* — 10-15 (if medicine)
12. *Economic value (Rs. / ha)* — As weed it is detrimental but as medicine it is highly valuable
13. *By-product (If any)* — Nil
14. *Use of by-product* — Nil

278. *CROP* — FENNEL / Saunf / Sonp / *Mauri*

Scientific Name* — *Foeniculum vulgare Mill

Family — Umbeliferae

Herb / Shrub / Tree — Herb (Aromatic)

Annual / Biennial / Perennial — Annual (50-90 cm height)

Used as — As medicine (rheumatism) and spice (seed) but observed as weed in winter crops

1. *Climatic requirements* — Cold weather conditions / partially warm
2. *Soil requirements* — Alluvial / loamy and sandy loam soils
3. *Required land situations* — High / medium land
4. *Importance of crop* — Aromatic herb used as spice
5. *Fertilizer management* — 60 kg N, 50 kg P_2O_5 and 30 kg K_2O/ha
6. *Water management* — 2-3 irrigations, if grown as spice
7. *Duration of crop* — 120-140 days
8. *Parts used* — Leaves and seeds
9. *Where grown in India* — Punjab, Assam, Maharashtra, Orissa, Bihar, West Bengal and Karnataka
10. *Export possibility* — High
11. *Economic yields (q / ha)* — 5-10 (seed) when grown for seeds
12. *Economic value (Rs. / ha)* — 25,000 to 50,000
13. *By-product (If any)* — Stalk and as weed
14. *Use of by-product* — Cattle feed / organic matter / mulch.

279. *CROP* — ANAGALIS / *Krishnanil*

Scientific Name* — *Anagallis arvensis L.

Family — Primulaceae

Herb / Shrub / Tree — Herb (10-30 cm height)

Annual / Biennial / Perennial — Annual

Used As — As fodder but found as weeds in winter crops particularly in wheat

1. *Climatic requirements* — Cold climatic condition

2. *Soil requirements* — Alluvial, sandy, loamy soils
3. *Required land situations* — High / land
4. *Importance of crop* — Flowers are blue (deep)
5. *Fertilizer management* — 60 kg N, 30 kg P_2O_5 and 30 kg K_2O/ha
6. *Water management* — 2-3 irrigations
7. *Duration of crop* — 120-140 days
8. *Parts used* — Whole plant for fodder, flower for medicine (stomach)
9. *Where grown in India* — All over India (under cool condition)
10. *Export possibility* — High
11. *Economic yields (q / ha)* — as weed nil
12. *Economic value (Rs. / ha)* — As fodder it is good, but as weed, it is very bad for crop (wheat) production
13. *By-product (If any)* — Nil as weed
14. *Use of by-product* — Nil as weed

280. *CROP* — TINY VETCH / Munmana / *Hetkachhoto*

Scientific Name* — *Vicia hirsuta (L.) S.F. Gray

Family — Papilionaceae

Herb / Shrub / Tree — Herb (30-40 cm height)

Annual / Biennial / Perennial — Annual

Used as — As fodder but it is found in *Lathyrus* as weed.

1. *Climatic requirements* — Cool climate conditions
2. *Soil requirements* — Alluvial / sandy / sandy loamy soils
3. *Required land situations* — High / medium / low land
4. *Importance of crop* — Smaller leaves, plant trailing type
5. *Fertilizer management* — 40 kg N, 60 kg P_2O_5 and 30 kg K_2O/ha
6. *Water management* — 2-3 irrigations, if grown as fodder
7. *Duration of crop* — 120-140 days
8. *Parts used* — Whole plant for fodder
9. *Where grown in India* — Punjab, Assam, U.P., M.P., West Bengal, Nilgiris (TN)
10. *Export possibility* — High
11. *Economic yields (q / ha)* — As fodder, 150-200
12. *Economic value (Rs. / ha)* — 15,000 to 20,000
13. *By-product (If any)* — As weed it is dangerous for the winter crop specially sunnhemp, Lathyrus etc.

14. *Use of by-product* — Nil

281. ***CROP*** **— COMMON VETCH /** ***Chhoto*** **/ Khesari – Hetka / Small lathyrus**

Scientific Name **—** ***Vicia indica*** L.

Family **—** Papilionaceae

Herb / Shrub / Tree **—** Herb (30-40 cm height)

Annual / Biennial / Perennial **—** Annual

Used as **—** Fodder but found in lathyrus (grass pea / *khesari*) as weed

1. *Climatic requirements* — Cool climatic conditions
2. *Soil requirements* — Loamy / sandy and alluvial soils
3. *Required land situations* — High / medium / low lands
4. *Importance of crop* — Trailing type on the crops
5. *Fertilizer management* — 40 kg N, 60 kg P_2O_5 and 30 kg K_2O/ha
6. *Water management* — 2-3 irrigations, if grown as fodder
7. *Duration of crop* — 120-140 days
8. *Parts used* — Whole plant as fodder
9. *Where grown in India* — Orissa, West Bengal, Bihar, Assam, U.P. and Madhya Pradesh
10. *Export possibility* — High
11. *Economic yields (q / ha)* — 150-200 (green fodder)
12. *Economic value (Rs. / ha)* — 15,000 to 20,000
13. *By-product (If any)* — Entire crop is used as fodder, but as weed in winter it damages the crops
14. *Use of by-product* — Nil

282. ***CROP*** **— SUNBERRY / Papotan /** ***Bantepari*** **/ Tulatipati**

Scientific Name **—** ***Physalis minima*** L.

Family **—** Solanaceae

Herb / Shrub / Tree **—** Herb (50-60 cm height)

Annual / Biennial / Perennial **—** Annual

Used as **—** Fruits but weeds in some winter crops

1. *Climatic requirements* — Cool climatic condition, can also grow in hot climate
2. *Soil requirements* — Loamy / sandy and alluvial soils
3. *Required land situations* — High / medium / low lands
4. *Importance of crop* — Fruits are in the sacks, yellow in colour
5. *Fertilizer management* — 100 kg N, 50 kg P_2O_5 and 30 kg K_2O/ha

6. *Water management* — 4-5 irrigations, if grown as fruit, but no irrigation when grown as weed in winter crops
7. *Duration of crop* — 120-140 days
8. *Parts used* — Fruits / entire plant (for fodder)
9. *Where grown in India* — U.P., W.B., Assam, Bihar and Orissa etc.
10. *Export possibility* — High
11. *Economic yields (q / ha)* — 100-150 (fruits, green)
12. *Economic value (Rs. / ha)* — 10,000 to 15,000
13. *By-product (If any)* — If it is grown as fruits, entire plant is used for cattle feed, but most of the time, it is the weed of the winter crops, particularly in wheat, grasspea, linseed, potato etc.
14. *Use of by-product* — Nil

283. *CROP* — YELLOW VETCHLING / Jangli matar / *Banmator*

Scientific Name* — *Lathyrus aphaca L.

Family — Papilionaceae

Herb / Shrub / Tree — Herb (40-60 cm height)

Annual / Biennial / Perennial — Annual

Used as — Fodder but found as weed in *rabi* crops particularly in grasspea, sunnhemp (seed) etc.

1. *Climatic requirements* — Cool climatic conditions
2. *Soil requirements* — Loamy / alluvial or sandy soils
3. *Required land situations* — Low / medium lands
4. *Importance of crop* — Much-branched trailing type plant
5. *Fertilizer management* — 40 kg N, 60 kg P_2O_5 and 30 kg K_2O/ha
6. *Water management* — 2-3 irrigation, if grown for fodder
7. *Duration of crop* — 120-140 days
8. *Parts used* — Whole plant, if fodder
9. *Where grown in India* — Punjab, U.P., Bengal, M.P., etc.
10. *Export possibility* — High
11. *Economic yields (q / ha)* — 150-200 (green fodder)
12. *Economic value (Rs. / ha)* — 15,000 to 20,000
13. *By-product (If any)* — If grown as fodder, entire plant is harvested, but as weed it is found in grasspea etc.
14. *Use of by-product* — Nil

284. *CROP* — LAUNAEA / *Lania*

Scientific Name* — *Launaea pinnatifida Cass / ***L. sarmentosa*** (Willd) Alston

Family — Compositae

Herb / Shrub / Tree — Herb (30-40 cm height)

Annual / Biennial / Perennial — Annual

Used as — Soil-binder / medicine / fodder, besides weed in winter crop (potato)

1. *Climatic requirements* — Cool climatic conditions
2. *Soil requirements* — Alluvial / sandy or loamy soils
3. *Required land situations* — High / medium / low land
4. *Importance of crop* — Yellow star-like beautiful flower with long roots and stalk, but plant is small.
5. *Fertilizer management* — 60 kg N, 30 kg P_2O_5 and 30 kg K_2O/ha
6. *Water management* — 2-3 irrigations, if planted for medicine / fodder
7. *Duration of crop* — 120-150 days
8. *Parts used* — Leaves for medicine, whole plant for fodder
9. *Where grown in India* — All over India except snow areas
10. *Export possibility* — High
11. *Economic yields (q / ha)* — Fodder : 150-200 (green), Medicine 5-10
12. *Economic value (Rs. / ha)* — Fodder : 150,000 to 20,000
    Medicine : 50,000 to 80,000
13. *By-product (If any)* — When grown as medicine some stalk will be there but as weed, crops are damaged
14. *Use of by-product* — Nil

285. *CROP* — ECLIPTA / Kesut / Bhangra / Mochkand

Scientific Name* — *Eclipta alba L.

Family — Compositae

Herb / Shrub / Tree — Herb (30-40 cm)

Annual / Biennial / Perennial — Annual

Used as — Medicine (hair fall and jaundice) / fodder as well as weed in crops

1. *Climatic requirements* — Cool or warm climatic conditions
2. *Soil requirements* — Alluvial / sandy or loamy soils
3. *Required land situations* — High / medium / low lands

4. *Importance of crop* — Bending type inflorescence / Hair oil is produced
5. *Fertilizer management* — 60 kg N, 30 kg P_2O_5 and 30 kg K_2O/ha
6. *Water management* — 2-3 irrigations, if grown for fodder or medicine
7. *Duration of crop* — 120-140 days
8. *Parts used* — For medicine leaves and for fodder whole plant
9. *Where grown in India* — All over India except too low / high temperature
10. *Export possibility* — High
11. *Economic yields (q / ha)* — 10-15 (for medicine)
12. *Economic value (Rs. / ha)* — Nearly 30,000 to 60,000
13. *By-product (If any)* — Stalk
14. *Use of by-product* — Fuel / cattle feed. Otherwise, as weed it damages the crops.

286. CROP — LITTLE CROTON / Lankeswar / Chuka

Scientific Name* — *Croton roxburghii Bal

Family — Euphorbiaceae

Herb / Shrub / Tree — Herb (30-40 cm height)

Annual / Biennial / Perennial — Annual

Used as — Medicine (Purgative / toothache) / insecticide; weeds in crops

1. *Climatic requirements* — Cool or warm climatic conditions
2. *Soil requirements* — Alluvial, sandy or loamy soil
3. *Required land situations* — High or medium land
4. *Importance of crop* — Capscicum leaf-like leaves, though small
5. *Fertilizer management* — 60 kg N, 50 kg P_2O_5 and 30 kg K_2O/ha, if grown for medicine
6. *Water management* — 2-3 irrigations, if grown for medicine
7. *Duration of crop* — 120-150 days
8. *Parts used* — Leaves and seeds for medicine
9. *Where grown in India* — All over India, except too low / too high temperature areas
10. *Export possibility* — High
11. *Economic yields (q / ha)* — For medicine, highly valuable
12. *Economic value (Rs. / ha)* — Nearly 50,000 to 60,000

13. *By-product (If any)* — If grown for medicine, stalk is available for cattle feed etc., otherwise it is the weed, which damages the crops.

14. *Use of by-product* — Nil

287. *CROP* — PRICKLY AMARANTH / Kantewali chaulai / *Kantanate* / Goja

Scientific Name* — *Amaranthus spinosus L.

Family — Amaranthaceae

Herb / Shrub / Tree — Herb (30-40 cm height)

Annual / Biennial / Perennial — Annual

Used as — Vegetables / fodder; but as a weed in crops/or in waste lands

1. *Climatic requirements* — Cool and warm climates
2. *Soil requirements* — Sandy / loamy / alluvial / laterite / clayey soils
3. *Required land situations* — High / medium lands
4. *Importance of crop* — Spines on the stems
5. *Fertilizer management* — 60 kg N, 30 kg P_2O_5 and 30 kg K_2O/ha if sown for vegetable / fodder
6. *Water management* — 2-3 irrigations (normally rainfed), if vegetable
7. *Duration of crop* — 120-140 days
8. *Parts used* — Entire plants (for fodder / vegetable)
9. *Where grown in India* — All over India, except extreme cold or too hot areas
10. *Export possibility* — High
11. *Economic yields (q / ha)* — 150-200
12. *Economic value (Rs. / ha)* — It is valuable for fodder or vegetable but damaging as weed in crops
13. *By-product (If any)* — No product, if grown as vegetable
14. *Use of by-product* — Nil

288. *CROP* — GOLDEN DOCK / Jangli palak / *Banpalang* / Jub-palum

Scientific Name* — *Rumex maritimus L.

Family — Polygonaceae

Herb / Shrub / Tree — Herb (50-60 cm height)

Annual / Biennial / Perennial — Annual

Used as — Medicine (cathartic/purgative) but weeds in low land winter crops

1. *Climatic requirements* — Cool climatic conditions

2. *Soil requirements* — Loamy / sand / alluvial soils
3. *Required land situations* — Low / medium lands
4. *Importance of crop* — This plant is like spinach / roots are purgative
5. *Fertilizer management* — 100 kg N, 50 kg P_2O_5 and 50 kg K_2O/ha, if grown as medicine
6. *Water management* — 4-5 irrigations
7. *Duration of crop* — 120-140 days
8. *Parts used* — Leaves and roots
9. *Where grown in India* — Hilly regions, West Bengal, Assam, U.P. and Karnataka etc.
10. *Export possibility* — High
11. *Economic yields (q / ha)* — 10-15
12. *Economic value (Rs. / ha)* — This is highly valuable when sown for medicine but as weed, it is damaging to crops.
13. *By-product (If any)* — Stock
14. *Use of by-product* — Fuel / organic matter

289. *CROP* — Nicotiana weed / Wild tobacco / *Bantamak / Janglitamak*

Scientific Name — Nicotiana plumbaginifolia L.

Family — Solanaceae

Herb / Shrub / Tree — Herb (50-80 cm height)

Annual / Biennial / Perennial — Annual

Used as — Medicine (narcotics) but weed in winter crops particularly in wheat or potato in West Bengal

1. *Climatic requirements* — Cool climatic conditions
2. *Soil requirements* — Alluvial / loamy / sandy soils
3. *Required land situations* — Low / medium lands
4. *Importance of crop* — Leaves like tobacco, bigger at the base
5. *Fertilizer management* — 50 kg N, 30 kg P_2O_5 and 30 kg K_2O/ha (if grown as medicine)
6. *Water management* — 2-3 irrigations if cultivated
7. *Duration of crop* — 120-140 days
8. *Parts used* — Leaves
9. *Where grown in India* — U.P., Bihar, West Bengal, Assam etc.
10. *Export possibility* — High
11. *Economic yields (q / ha)* — 10-15

12. *Economic value (Rs./ha)* — It is valuable when grown as medicinal plant but normally grows as weed in winter crops
13. *By-product (If any)* — Stems
14. *Use of by-product* — Fuel / mulch / organic matter

290. ***CROP*** **— NIGHTSHADE / Makoi /** ***Banbegun***

Scientific Name — Solanum nigrum L.

Family — Solanaceae

Herb / Shrub / Tree — Herb (80-90 cm height)

Annual / Biennial / Perennial — Annual

Used as — Medicine (for anthrax pustules and enema in infant); but grown wildly as weed in winter / warm season crops

1. *Climatic requirements* — Cool / warm weather conditions
2. *Soil requirements* — Alluvial / loamy / sandy soils
3. *Required land situations* — High / medium land
4. *Importance of crop* — Like smaller brinjals, but with open small fruits
5. *Fertilizer management* — 60 kg N, 30 kg P_2O_5 and 30 kg K_2O/ha
6. *Water management* — 2-3 irrigations (normally rainfed) for medicine
7. *Duration of crop* — 120-140 days
8. *Parts used* — Fruits
9. *Where grown in India* — All over India except extreme cold and hot areas
10. *Export possibility* — High
11. *Economic yields (q/ha)* — 10-15
12. *Economic value (Rs./ha)* — It is a valuable crop when grown as medicine but as weed, it is damaging to crops
13. *By-product (If any)* — Stems / leaves
14. *Use of by-product* — Cattle feed / organic manure / fuel

291. ***CROP*** **— PIGWEED /** ***Bathua / Betho*** **/ Lambs quarters**

Scientific Name — Chenopodium album L.

Family — Chenopodiaceae

Herb / Shrub / Tree — Herb (50-80 cm height)

Annual / Biennial / Perennial — Annual

Used as — Fodder and vegetable, though it is a weed in wheat

1. *Climatic requirements* — Cool climatic conditions
2. *Soil requirements* — Alluvial / loamy / sandy soil
3. *Required land situations* — High / medium / low lands
4. *Importance of crop* — Leaves used as vegetable / fodder
5. *Fertilizer management* — 60 kg N, 30 kg P_2O_5 and 30 kg K_2O/ha (if grown as fodder)
6. *Water management* — 2-3 irrigations
7. *Duration of crop* — 120-140 days
8. *Parts used* — Leaves and whole plant
9. *Where grown in India* — All over India except too low / too high temperature areas
10. *Export possibility* — High
11. *Economic yields (q / ha)* — 50-100 (green leaves and twigs)
12. *Economic value (Rs. / ha)* — It has got high value when it grows as fodder, but it is damaging wheat crop
13. *By-product (If any)* — Nil
14. *Use of by-product* — Nil

NARROW LEAVED WEEDS

292. *CROP* — NUTGRASS / Motha / *Mutha*

Scientific Name* — *Cyperus rotundus L.

Family — Cyperaceae

Herb / Shrub / Tree — Sedge (Herb)

Annual / Biennial / Perennial — Perennial

Used as—Obnoxious weeds; Industrial uses (Rhizomes or nut is used as perfumery / soap making / insect repellent, due to essential oils)

1. *Climatic requirements* — Cool / warm climatic conditions
2. *Soil requirements* — Alluvial / loamy / sandy or clayey soil
3. *Required land situations* — High / medium / low lands
4. *Importance of crop* — Rhizomatous plant, nut inside soils
5. *Fertilizer management* — N, P_2O_5 and K_2O @ 80-50-30 kg/ha respectively, when grown as medicine
6. *Water management* — 2-3 irrigations (if needed for industrial use)
7. *Duration of crop* — 2-3 years
8. *Parts used* — Nuts (insides soils)

9. *Where grown in India* — All over India both in cool and warm conditions
10. *Export possibility* — High
11. *Economic yields (q / ha)* — 50-100 (green) 10-15
12. *Economic value (Rs. / ha)* — The rhizomes are used for extracting essential oils other than industrial uses, though it is the obnoxious weed in crops throughout the year and damages the crop production
13. *By-product (If any)* — Leaves and stems
14. *Use of by-product* — Cattle feed, fuel, organic matter

293. *CROP* — BERMUDA GRASS / Doop / *Durba* / *Bahamagrass*

Scientific Name* — *Cynodon dactylon (L.) Pers.

Family — Gramineae

Herb / Shrub / Tree — Herb (spreading just above soil)

Annual / Biennial / Perennial — Perennial

Used as — Fodder, but weed in crops / waste land and a good pasture crop

1. *Climatic requirements* — Cool or warm climate
2. *Soil requirements* — Alluvial / loamy / sandy / clayey / laterite soils
3. *Required land situations* — High / medium / low lands
4. *Importance of crop* — Rapid multiplication in the soil surface and soil binder
5. *Fertilizer management* — For fodder / lawn, 100 kg N, 50 kg P_2O_5 and 50 kg K_2O/ha
6. *Water management* — 3-4 irrigations annually
7. *Duration of crop* — 2-4 years
8. *Parts used* — Entire plant for fodder
9. *Where grown in India* — All over India, in different seasons except too low/too high temperature areas
10. *Export possibility* — High
11. *Economic yields (q / ha)* — 50 – 100 (green)
12. *Economic value (Rs. / ha)* — It is highly valuable for fodder / lawn, but it is a damaging weed in crops
13. *By-product (If any)* — Nil
14. *Use of by-product* — Nil

294. ***CROP*** **— CHIRIA-BAJRA / FALAGHASH / Cannary grass / *Chiriakadana***

Scientific Name — Phalaris minor Retz

Family — Gramineae

Herb / Shrub / Tree — Herb (80-90 cm height)

Annual / Biennial / Perennial — Annual

Used as — Fodder but an obnoxious weed in wheat

1. *Climatic requirements* — Cool climate / weather
2. *Soil requirements* — Alluvial / sandy / loamy soils
3. *Required land situations* — High / medium land
4. *Importance of crop* — Fox tail inflorescence (compact) / flowers earlier than wheat
5. *Fertilizer management* — 100 kg N, 50 kg P_2O_5 and 50 kg K_2O /ha when grown as fodder
6. *Water management* — 4-5 irrigations
7. *Duration of crop* — 110-120 days (cut at flower)
8. *Parts used* — For fodder, entire plant
9. *Where grown in India* — Uttar Pradesh Rajasthan, Haryana, Madhya Pradesh, Bihar, Punjab, West Bengal, etc.
10. *Export possibility* — High
11. *Economic yields (q / ha)* — 150-200 (green)
12. *Economic value (Rs. / ha)* — It is highly valuable when grown as fodder, but as weed in wheat it is highly damaging to the crop
13. *By-product (If any)* — Nil
14. *Use of by-product* — Nil

295. ***CROP*** **— WILD OAT / *Banjjai* or Bunojjai**

Scientific Name — Avena ludoviciana Durien / ***A. fatua*** L.

Family — Gramineae

Herb / Shrub / Tree — Herb (80-90 cm height)

Annual / Biennial / Perennial — Annual

Used as — Fodder, but it is the obnoxious weed in wheat

1. *Climatic requirements* — Cool climatic conditions
2. *Soil requirements* — Alluvial, loamy, sandy soils
3. *Required land situations* — High / medium land
4. *Importance of crop* — Spreading inflorescence (like fire works)
5. *Fertilizer management* — 100 kg N, 50 kg P_2O_5 and 50 kg K_2O/ha. when grown as fodder

6. *Water management* — 4-5 irrigations
7. *Duration of crop* — 120-140 days
8. *Parts used* — For fodder entire plant
9. *Where grown in India* — Punjab, Haryana, Rajasthan, M.P., Bihar and West Bengal etc.
10. *Export possibility* — High
11. *Economic yields (q / ha)* — 150-200
12. *Economic value (Rs. / ha)* — It is highly valuable when grown as fodder but it is damaging in wheat crop when grown as weed
13. *By-product (If any)* — Nil
14. *Use of by-product* — Nil

296. CROP — DARNEL GRASS / Karatghash / Machni

Scientific Name — Lolium temulentum L.

Family — Gramineae

Herb / Shrub / Tree — Herb (30-60 cm height)

Annual / Biennial / Perennial — Annual

Used as — Fodder, but it is the obnoxious weed in wheat

1. *Climatic requirements* — Cool climate condition
2. *Soil requirements* — Alluvial / sand / loamy and clayey soils
3. *Required land situations* — High / medium lands
4. *Importance of crop* — The long inflorescence is like sickle, smaller than wheat
5. *Fertilizer management* — 100 kg N, 50 kg P_2O_5 and 50 kg K_2O/ha, when grown as fodder
6. *Water management* — 4-5 irrigation
7. *Duration of crop* — 120-140 days
8. *Parts used* — Whole plant for fodder
9. *Where grown in India* — All over north, west and east India during winter
10. *Export possibility* — High
11. *Economic yields (q / ha)* — 150-200
12. *Economic value (Rs. / ha)* — When grown as fodder, it is highly valuable, but it is the obnoxious weed in wheat and then it is damaging to the crop
13. *By-product (If any)* — Nil
14. *Use of by-product* — Nil

2. SUMMER (PRE-*KHARIF*) SEASON

BROAD LEAVED WEEDS

297. ***CROP* — JANGLI AMLI / Bhuinanvalah / Hazardana / Jaramla / *Bhuiamla***

Scientific Name* — *Phyllanthus niruri Hook / ***P. fraternus*** Webster

Family — Euphorbiaceae

Herb / Shrub / Tree — Herb (30-40 cm height)

Annual / Biennial / Perennial — Annual

Used as — Medicinal plants (stomach ache / diarrhoea); fresh roots for jaundice, but a weed in pre-*kharif* season

1. *Climatic requirements* — Warm / cool climatic conditions
2. *Soil requirements* — Alluvial / sandy or loamy soils
3. *Required land situations* — High / medium lands
4. *Importance of crop* — Small bead-like structure just below the surface of the leaves of small plant
5. *Fertilizer management* — 50 kg N, 30 kg P_2O_5 and 30 kg K_2O/ha, if grown as medicinal crop
6. *Water management* — 2-3 irrigations
7. *Duration of crop* — 120-130 days
8. *Parts used* — For medicine, the leaves and beads
9. *Where grown in India* — All over of India in pre-*kharif* season
10. *Export possibility* — High
11. *Economic yields (q / ha)* — 10-15
12. *Economic value (Rs. / ha)* — As the medicinal crop, it is highly valuable but as weed, it is detrimental to crops
13. *By-product (If any)* — Stems
14. *Use of by-product* — Fuel / Cattle feed / organic matter

298. ***CROP* — LALDUDHI / *Dudhia***

Scientific Name* — *Euphorbia hirta L.

Family — Euphorbiaceae

Herb / Shrub / Tree —Herb (30-40 cm height)

Annual / Biennial / Perennial — Annual

Used as — Medicine (bronchial infection), asthma; but a weed in warm season

1. *Climatic requirements* — Warm climate
2. *Soil requirements* — Alluvial / loamy and sandy soils
3. *Required land situations* — High / medium / low land
4. *Importance of crop* — White milk-like latex comes out of the stem / leaves if pierced
5. *Fertilizer management* — 60 kg N, 30 kg P_2O_5 and 30 kg K_2O/ha, if grown for medicine
6. *Water management* — 3-4 irrigations
7. *Duration of crop* — 120-150 days
8. *Parts used* — Entire plant at flowering / fruiting stage
9. *Where grown in India* — All over India except too low / too high temperature areas
10. *Export possibility* — High
11. *Economic yields (q / ha)* — 150-160
12. *Economic value (Rs. / ha)* — It is valuable when grown as medicine; but it is very damaging to crops, when grown as weed in warm climate
13. *By-product (If any)* — Nil
14. *Use of by-product* — Nil

299. *CROP* — BLACK NIGHT SHADE / *Makoi / Banbegun*

Scientific Name — Solanum nigram L.

Family — Solanaceae

Herb / Shrub / Tree — Herb (80-90 cm height)

Annual / Biennial / Perennial — Annual

Used as — Medicine (abdominal upsets), but acts as weed in warm season crops

1. *Climatic requirements* — Warm / cool climatic conditions
2. *Soil requirements* — Alluvial / sandy / loamy and clayey soils
3. *Required land situations* — High / medium / low lands
4. *Importance of crop* — Some berry but small brinjal like fruits are attached to the plant which is spiny
5. *Fertilizer management* — 60 kg N, 30 kg P_2O_5 and 30 kg K_2O/ha, if grown for medicinal plant
6. *Water management* — 2-3 irrigation
7. *Duration of crop* — 120-140 days
8. *Parts used* — Leaves and seeds
9. *Where grown in India* — All over India except too low / too high temperature

10. *Export possibility* — High
11. *Economic yields (q / ha)* — 15-20
12. *Economic value (Rs. / ha)* — It is highly valuable when grown as medicine, but very damaging when grown in crops
13. *By-product (If any)* — Stems
14. *Use of by-product* — Fuel / organic matter

300. *CROP* — CROTON WEED / Lankeswar

Scientific Name* — *Croton sperciflora L. / ***C. tiglium*** L.

Family — Euphorbiaceae

Herb / Shrub / Tree — Herb (40-60 cm height)

Annual / Biennial / Perennial — Annual

Used as — Medicine (purgative from croton oil)

1. *Climatic requirements* — Warm / cool climate
2. *Soil requirements* — Alluvial / loamy / sandy and clayey soils
3. *Required land situations* — High / medium / low lands
4. *Importance of crop* — Chilly-like leaves is small plant with small fruits
5. *Fertilizer management* — 60 kg N, 30 kg P_2O_5 and 30 kg K_2O/ha if grown as medicinal plant
6. *Water management* — 3-4 irrigation
7. *Duration of crop* — 120-140 days
8. *Parts used* — Seeds (for croton oil)
9. *Where grown in India* — Assam, West Bengal, South India etc.
10. *Export possibility* — High
11. *Economic yields (q / ha)* — 10-15
12. *Economic value (Rs. / ha)* — It is a valuable plant when grown as a medicinal plant but very damaging, if grown as weed in crops
13. *By-product (If any)* — Stems and leaves
14. *Use of by-product*— Fuel / mulch / cattle feed / organic matter

301. *CROP* — SUNBERRY / Tulatipati / Bantepari / Papotan

Scientific Name* — *Physalis minima L.

Family — Solanaceae

Herb / Shrub / Tree — Herb (30-90 cm height)

Annual / Biennial / Perennial — Annual

Used as — Fruits, but grows as weed in summer crops

1. *Climatic requirements* — Warm and cool climate

2. *Soil requirements* — Alluvial / loamy / sandy and clayey soils
3. *Required land situations* — High / medium / low lands
4. *Importance of crop* — Fruits (small) are in the sack-like structure
5. *Fertilizer management* — 100 kg N, 50 kg P_2O_5 and 50 kg K_2O/ha, if grown as fruits
6. *Water management* — 3-4 irrigations
7. *Duration of crop* — 120-140 days
8. *Parts used* — Fruits
9. *Where grown in India* — Assam, West Bengal, Bihar, Orissa etc.
10. *Export possibility* — High
11. *Economic yields (q / ha)* — 50-80
12. *Economic value (Rs. / ha)* — It is a valuable crop when grown for fruits but when it grows as weed, it is damaging to crop production
13. *By-product (If any)* — Stems and leaves
14. *Use of by-product* — Cattle feed / fuel / organic matter / mulch

302. *CROP* — CORCHORUS WEED / *Ban pat*

Scientific Name* — *Corchorus acquitangular L.

Family — Tiliaceae

Herb / Shrub / Tree — Herb (50-90 cm height)

Annual / Biennial / Perennial — Annual

Used as — Medicine (worms), but grows as weed in warm season

1. *Climatic requirements* — Warm climate
2. *Soil requirements* — Alluvial, loamy and sandy soils
3. *Required land situations* — High / medium land
4. *Importance of crop* — Small capsules like *capsularis* jute
5. *Fertilizer management* — 60 kg N, 40 kg P_2O_5 and 60 kg K_2O/ha, when grown as medicine
6. *Water management* — 2-3 irrigations
7. *Duration of crop* — 120-150 days
8. *Parts used* — Leaves and seeds
9. *Where grown in India* — West Bengal, Assam, Orissa, Bihar etc.
10. *Export possibility* — High
11. *Economic yields (q / ha)* — 10-15
12. *Economic value (Rs. / ha)* — It is a valuable crop when grown as medicinal plant but if it grows as weed in pre-*kharif* crops, it is damaging

13. *By-product (If any)* — Stems
14. *Use of by-product* — Fuel / Cattle feed / organic matter

303. *CROP* — CASSIA / Sophera senna / Sickle senna / Pammar / Chakunda / *Kalkasunde / Kasaundi*

***Scientific Name* — *Cassia sophera* / *C. tora* L.**

Family — Caesalpiniaceae

Herb / Shrub / Tree — Shrub (60-90 cm or more height)

Annual / Biennial / Perennial — Perennial

Used as — Medicine (skin infection / ophthalmia), but grows as weed in warm season on road side

1. *Climatic requirements* — Warm seasons (pre-*kharif* and *kharif*)
2. *Soil requirements* — Alluvial / loamy sandy etc. soils
3. *Required land situations* — High / medium land
4. *Importance of crop* — Inflorescence is ball-like in *sophera* but like sickle in *tora*. Seeds are used in preparation of sweets.
5. *Fertilizer management* — 60 kg N, 60 kg P_2O_5 and 30 kg K_2O/ha, if grown as medicine and other purposes.
6. *Water management* — 2-3 irrigations
7. *Duration of crop* — 1-3 years
8. *Parts used* — Seeds
9. *Where grown in India* — All over India in warm season except too low / too high temperature areas
10. *Export possibility* — High
11. *Economic yields (q / ha)* — 10-20
12. *Economic value (Rs. / ha)* — It is a valuable plant as medicine / sweets but as weed, it is damaging to the crops
13. *By-product (If any)* — Stem and leaves
14. *Use of by-product* — Fuel / organic matter / pegs

304. *CROP* — ECLIPTA / Bhangra / *Kesut*

Scientific Name* — *Eclipta alba (L.) / ***E. prostrata*** L.

Family — Compositae

Herb / Shrub / Tree — Herb (40-50 m cm height)

Annual / Biennial / Perennial — Annual

Used as — Medicine (jaundice) and used as hair oil (cosmetics), but grows as weeds in warm season

1. *Climatic requirements* — Warm climate

2. *Soil requirements* — Alluvial / loamy and sandy soils
3. *Required land situations* — High / medium / low lands
4. *Importance of crop* — Sickle like bent stalk of the inflorescence with star like flower
5. *Fertilizer management* — 60 kg N, 30 kg P_2O_5 and 30 kg K_2O/ha, if grown as medicine
6. *Water management* — 2-3 irrigation, if grown for medicinal plant
7. *Duration of crop* — 120-140 days
8. *Parts used* — Seeds
9. *Where grown in India* — All over in India in warm climate except too low / high temperature
10. *Export possibility* — High
11. *Economic yields (q / ha)* — 10-12
12. *Economic value (Rs. / ha)* — It is a valuable plant when grown as medicine, but it is damaging to crop when it grows as weed
13. *By-product (If any)* — Stems and leaves
14. *Use of by-product* — Cattle feed / fuel / organic matter

305. *CROP* — COMMELINA / Kanchara / Kanteri / *Nilsechi*

Scientific Name* — *Commelina benghalensis L.

Family — Commelinaceae

Herb / Shrub / Tree — Herb (30-40 cm height)

Annual / Biennial / Perennial — Annual

Used as —Vegetable / fodder / medicine (carbuncle), but grows in warm season as weed in crops. It produces blue ear – like flowers

1. *Climatic requirements* — Warm climate
2. *Soil requirements* — Loamy / sandy / clayey / alluvial soils
3. *Required land situations* — High / medium / low lands
4. *Importance of crop* — Blue coloured flowers on leaf axils. It grows in water also.
5. *Fertilizer management* — 60 kg N, 30 kg P_2O_5 and 30 kg K_2O/ha when grown as vegetable or medicine
6. *Water management* — 2-3 irrigations
7. *Duration of crop* — 120-150 days
8. *Parts used* — Whole plant
9. *Where grown in India* — All over in India, in warm seasons except too low / too high temperature areas
10. *Export possibility* — High

11. *Economic yields (q / ha)* — 150-200 (green)
12. *Economic value (Rs. / ha)* — It is a valuable plant, if it is grown as medicine / fodder / vegetable, but when it grows as weed, it is the enemy to the crops
13. *By-product (If any)* — Nil
14. *Use of by-product* — Nil.

306. *CROP* — HORSE PURSLANE / Santhi / Lalsabuni

Scientific Name — Triantheme monogyna L. / ***T. portulacastrum*** L.

Family — Aizoaceae

Herb / Shrub / Tree — Succulent herb (30-60 cm height)

Annual / Biennial / Perennial — Annual

Used as — Medicine (disorder of liver or kidney) but grows in warm season as weeds

1. *Climatic requirements* — Warm climate
2. *Soil requirements* — Alluvial loamy / sandy and clayey soil
3. *Required land situations* — High / Medium / Low lands
4. *Importance of crop* — Succulent plant
5. *Fertilizer management* — 60 kg N, 30 kg P_2O_5 and 30 kg K_2O/ha if grown as medicine
6. *Water management* — 2-3 irrigations
7. *Duration of crop* — 120-140 days
8. *Parts used* — Whole plant
9. *Where grown in India* — All over India (in warm season) except too hot or snowy areas
10. *Export possibility* — High
11. *Economic yields (q / ha)* — 50 – 100 (green)
12. *Economic value (Rs. / ha)* — It is a valuable plant if grown for medicinal use, but as weed, it is the harmful plant to the other crops.
13. *By-product (If any)* — Nil
14. *Use of by-product* — Nil

307. *CROP* — ALTERNANTHERA / KANCHARI / *Senchisag* / Nadarang / *Water spinach*

Scientific Name — Alternanthera sessilis / A. amoena Voss

Family — Amaranthaceae

Herb / Shrub / Tree — Trailing herb (2-3 m in length, height 50 cm)

Annual / Biennial / Perennial — Annual

Used as — Vegetable but weed in warm season crops

1. *Climatic requirements* — Warm climate
2. *Soil requirements* — Alluvial / loamy / sandy and clayey soils
3. *Required land situations* — High / medium / low land also
4. *Importance of crop* — Grows in marshy land / star – like small flowers.
5. *Fertilizer Management*—60 kg N, 30 kg P_2O_5 and 30 kg K_2O/ha, if grown as vegetable
6. *Water management* — 2-3 irrigations (if in high / medium land), it can also be grown in water logged conditions
7. *Duration of crop* — 4-8 months
8. *Parts used* — Entire twig (for vegetable)
9. *Where grown in India* — Bengal, Orissa, Tamil Nadu, Bihar, Assam, etc.
10. *Export possibility* — High
11. *Economic yields (q / ha)* — 100-150 (green)
12. *Economic value (Rs. / ha)* — As vegetable it is very valuable; but in warm season crops (like rice etc.) it is very damaging to the crops
13. *By-product (If any)* — Lower soft stems
14. *Use of by-product* — Cattle feed / fuel /organic matters

308. *CROP* — BAN BEGUN / Wild brinjal / Khasia begun

Scientific Name — Solanum khasianum L.

Family — Solanaceae

Herb / Shrub / Tree — Shrub (spiny)

Annual / Biennial / Perennial — Perennial (0.8-1.2 m height)

Used as — Medicine (for curing sterility), but grows as weed in warm season

1. *Climatic requirements* — Warm climate
2. *Soil requirements* — Alluvial / loamy / sandy and clayey soils
3. *Required land situations* — High / Medium land
4. *Importance of Crop*— Spines on the plant / Half-ripe fruits used in *Chatni*
5. *Fertilizer management* — 100 kg N, 50 kg P_2O_5 and 50 kg K_2O/ha, if grown as medicinal plant
6. *Water management* — 2-3 irrigations
7. *Duration of crop* — 2-3 years

8. *Parts used* — Fruits / seeds
9. *Where grown in India* — All over India except too cold or too hot areas
10. *Export possibility* — High
11. *Economic yields (q / ha)* — 10-15
12. *Economic value (Rs. / ha)* — It is a good medicinal plant and is highly valuable but as weed, it is the obnoxious and deadly damaging to the crops
13. *By-product (If any)* — Stem and leaves
14. *Use of by-product* — Fuel / organic matter / fence

309. *CROP* — SWAMP CABBAGE / Water spinach / Kalmi sag / Sarnali / Ganthian

Scientific Name* — *Ipomoea reptans (L.) Poir

***Family* —** Convolvulaceae

***Herb / Shrub / Tree* —** Trailing aquatic herb

***Annual / Biennial / Perennial* —** Perennial (2-3 m length)

***Used as* —** Vegetable and fodder, but grows as weed in marshy land

1. *Climatic requirements* — Warm climate conditions
2. *Soil requirements* — Alluvial, loamy, sandy and clayey soils
3. *Required land situations* — Medium / low lands (marshy land)
4. *Importance of Crop*— Aquatic / semi aquatic trailing plant with white flowers
5. *Fertilizer management* — 50 kg N, 30 kg P_2O_5 and 30 kg K_2O/ha, (through pellet in water) when grown for vegetable
6. *Water management* — In dry condition, 2-3 irrigations should be applied
7. *Duration of crop* — 2-3 years
8. *Parts used* — Young twigs (for vegetable) and vines (for fodder)
9. *Where grown in India* — Maharashtra, Bihar, West Bengal, Orissa, Kerala, Tamil Nadu, Assam, A.P. etc.
10. *Export possibility* — High
11. *Economic yields (q / ha)* — 50-100 (green)
12. *Economic value (Rs. / ha)* — It is highly valuable, if it is grown for vegetable, but in marshy land crops, it is the obnoxious weed
13. *By-product (If any)* — Stem
14. *Use of by-product* — Fuel / organic matter / cattle feed

310. *CROP* — DHOLKALMI / Dudhia kalmi / Moon flower

Scientific Name* — *Ipomoea alba L./ ***I. arveunsis*** L./***I. aquatica*** Fosk

Family — Convolvulaceae

Herb / Shrub / Tree — Shrub (2-3 m height)

Annual / Biennial / Perennial — Perennial

Used as — Ornamental / vegetables but grown as weed in marshy land crops

1. *Climatic requirements* — Warm climate in marshy land
2. *Soil requirements* — Alluvial, loamy, clayey soils
3. *Required land situations* — Medium / low land (marshy land)
4. *Importance of Crop*— White flowers (like moon)
5. *Fertilizer management* — 60 kg N, 30 kg P_2O_5 and 30 kg K_2O/ha, if grown for vegetable / flower
6. *Water management* — 2-3 irrigations in medium land
7. *Duration of crop* — 2-3 years
8. *Parts used* — For vegetable, twig; but for ornamental, flower only
9. *Where grown in India* — All marshy lands in India except too hot and cold areas
10. *Export possibility* — High
11. *Economic yields (q / ha)* — 50-100 (green)
12. *Economic value (Rs. / ha)* — It is a valuable plant for vegetable or for aesthetic value; but as weed it is damaging to some crops (rice/arum) in *kharif*/season)
13. *By-product (If any)* — Stems
14. *Use of by-product* — Fuel / cattle feed / organic matter.

311. *CROP* — MIKANIA / Jallata / *Jhiari lata*

Scientific Name* — *Mikania scandens Hoog. / ***M. cordata*** B.L. Robinson

Family — Compositae

Herb / Shrub / Tree — Climbing creeper - herb

Annual / Biennial / Perennial — Annual (2-3 m height)

Used as — Medicine (snake bite / scorpion sting) but grown as weed in mesta in warm season (even in water)

1. *Climatic requirements* — Warm climate and marshy conditions
2. *Soil requirements* — Loamy, clayey or alluvial soils
3. *Required land situations* — Medium / low land

4. *Importance of Crop*— Left twisted in mesta plant. Useful for checking soil erosion
5. *Fertilizer management* — 60 kg N, 30 kg P_2O_5 and 30 kg K_2O/ha, if grown as medicine
6. *Water management* — 2-3 irrigation, if land is dry
7. *Duration of crop* — 120-160 days
8. *Parts used* — Vine
9. *Where grown in India* — Assam, West Bengal, T.N., Kerala, Orissa etc.
10. *Export possibility* — High
11. *Economic yields (q / ha)* — 60-100 (green)
12. *Economic value (Rs. / ha)* — It is very valuable as medicine but it is very damaging in crops like mesta / sunnhemp etc.
13. *By-product (If any)* — Leaves
14. *Use of by-product* — Cattle feed / fuel / organic matter.

312. *CROP* — CHITRABEEZ / Tikiokra / Thuiak / *Bunu pat*

Scientific Name — Melochia corchorifolia L.

Family — Sterculiaceae

Herb / Shrub / Tree — Shrub (0.5-1.0 m height)

Annual / Biennial / Perennial — Perennial

Used as — Medicine (boiled with oil, this is remedy of bite of water snakes) but grows as weed in hot areas

1. *Climatic requirements* — High hot climatic conditions
2. *Soil requirements* — Loamy, sandy and alluvial soils
3. *Required land situations* — High / Medium lands
4. *Importance of Crop*— Stems (soft) and leaves are used as vegetable / medicine
5. *Fertilizer management* — 60 kg N, 30 kg P_2O_5 and 30 kg K_2O/ha, if grown as medicine
6. *Water management* — 2-3 irrigations
7. *Duration of crop* — 120-140 days
8. *Parts used* — Stem and leaves
9. *Where grown in India* — Gujarat, Rajasthan, Maharashtra, Tamil Nadu, A.P., Orissa, West Bengal etc.
10. *Export possibility* — High
11. *Economic yields (q / ha)* — 50-150 (green)

12. *Economic value (Rs. / ha)* — It is a valuable plant as medicine but as a weed, it damages the crop heavily
13. *By-product (If any)* — Hard stems
14. *Use of by-product* — Fuel / organic matter

313. *CROP* — SCOPARIA

Scientific Name* — *Scoparia dulcis L.

Family — Scrophulariaceae

Herb / Shrub / Tree — Herb

Annual / Biennial / Perennial — Annual

Used as — Medicine (remedy for fever / cough / bronchitis), but grows as weed in warm season

1. *Climatic requirements* — Warm climate
2. *Soil requirements* — Loamy or alluvial soils
3. *Required land situations* — Medium / low lands
4. *Importance of Crop*— It is useful for diabetes
5. *Fertilizer management* — 60 kg N, 30 kg P_2O_5 and 30 kg K_2O/ha, if grown as medicine
6. *Water management* — 2-3 irrigations
7. *Duration of crop* — 120-140 days
8. *Parts used* — Stem and leaves
9. *Where grown in India* — All over India except too low or too high temperature areas
10. *Export possibility* — High
11. *Economic yields (q / ha)* — 60-150 (green)
12. *Economic value (Rs. / ha)* — It is a valuable plant when grown as medicine but it is damaging to crops when grown as weed
13. *By-product (If any)* — Nil
14. *Use of by-product* — Nil

314. *CROP* — MOLLUGO

Scientific Name* — *Mollugo stricta L./ ***M. pentaphylla*** L.

Family — Molluginaceae

Herb / Shrub / Tree — Herb

Annual / Biennial / Perennial — Annual

Used as — Pot herb / vegetable / medicine (poultice for sore legs), but grows as weed in warm season

1. *Climatic requirements* — Warm climate

2. *Soil requirements* — Loamy, sand, loamy soils
3. *Required land situations* — Medium / low lands
4. *Importance of Crop*— Much branched annual
5. *Fertilizer management* — 60 kg N, 30 kg P_2O_5 and 30 kg K_2O/ha, if grown as vegetable and medicine
6. *Water management* — 2-3 irrigations
7. *Duration of crop* — 120-140 days
8. *Parts used* — Whole plant
9. *Where grown in India* — All over India except too low and too hot temperature areas
10. *Export possibility* — High
11. *Economic yields (q / ha)* — 100-120 (green)
12. *Economic value (Rs. / ha)* — It is a valuable crop when grown as medicinal and vegetable crop, but is damaging to crops, as a weed
13. *By-product (If any)* — Nil
14. *Use of by-product* — Nil

315. *CROP* — BEGGAR WEED / *Vikshuk*

Scientific Name* — *Desmodium tortuosum D.C.

Family — *Papilionaceae*

Herb / Shrub / Tree — Herb

Annual / Biennial / Perennial — Annual

Used as — Fodder but grows wild (weed) in warm season

1. *Climatic requirements* — Warm climate
2. *Soil requirements* — Loamy / alluvial / sandy loam soils
3. *Required land situations* — High / Medium lands
4. *Importance of Crop*— Plant with good flowers
5. *Fertilizer management* — 60 kg N, 30 kg P_2O_5 and 30 kg K_2O/ha when grown as fodder
6. *Water management* — 2-3 irrigations
7. *Duration of crop* — 120-140 days
8. *Parts used* — Whole plant
9. *Where grown in India* — In South India, mainly Tamil Nadu, Kerala and Karnataka
10. *Export possibility* — High
11. *Economic yields (q / ha)* — 150-200 (green)
12. *Economic value (Rs. / ha)* — It is a good crop when grown as fodder, but very damaging when it grows as weed

13. *By-product (If any)* — Nil
14. *Use of by-product* — Nil

316. *CROP* — PARTHENIUM / Congress - topi / *Sadatopi* / *Partheneam*

Scientific Name* — *Parthenium hysterophorus L.

Family — Compositae

Herb / Shrub / Tree — Herb (80-90 cm height)

Annual / Biennial / Perennial — Perennial

Used as — Medicine (decoction of root is given in dysentery) but grows as obnoxious weed in warm climate

1. *Climatic requirements* — Warm climate
2. *Soil requirements* — Alluvial, loamy, sandy and clayey soils
3. *Required land situations* — High / medium lands
4. *Importance of Crop*— White huge number of flowers (known as congress *topi*), it spreads quickly
5. *Fertilizer management* — 60 kg N, 30 kg P_2O_5 and 30 kg K_2O/ha, if grown as medicinal plant
6. *Water management* — 2-3 irrigations
7. *Duration of crop* — 2-3 years
8. *Parts used* — Roots (as medicine)
9. *Where grown in India* — Maharashtra, West Bengal, Bihar, Delhi etc.
10. *Export possibility* — High
11. *Economic yields (q / ha)* — 5-7
12. *Economic value (Rs. / ha)* — It is highly valuable when grown as medicinal plant, but obnoxious when grows as a weed on road side, as it is very unhealthy and damaging to human beings and cattle
13. *By-product (If any)* — Stem and leaves
14. *Use of by-product* — Organic matter / fuel (if harvested before flowering)

317. *CROP* — ZIZYPHUS / JHAR BER / *Ban Kul* / Jujube / Badara / Wild jujube / Chinese date / *Barai*

Scientific Name—***Zizyphus rotandifolia*** Lam. / ***Z. nummularia*** (Burm.f.) Wight.

Family — Rhamnaceae

Herb / Shrub / Tree — Shrub / Spiny tree (medium)

***Annual / Biennial / Perennial* —** Perennial (3-4 m height)

***Used as* —** Fruits / leaves for fodder / but grows as weed in warm climate

1. *Climatic requirements* — Jujube is a warm climate crop
2. *Soil requirements* — Loamy, sandy, rocky, alluvial soils
3. *Required land situations* — High / Medium lands
4. *Importance of Crop*— Spines on the stem
5. *Fertilizer management* — 60 kg N, 30 kg P_2O_5 and 30 kg K_2O/ha annually, if grown as fruit tree
6. *Water management* — 2-3 irrigations
7. *Duration of crop* — 10-15 years
8. *Parts used* — Fruits
9. *Where grown in India* — Punjab, U.P., Bihar, Rajasthan, West Bengal, Orissa etc.
10. *Export possibility* — High
11. *Economic yields (q / ha)* — 50-150 (green)
12. *Economic value (Rs. / ha)* — It is a fruit crop tree and highly valuable but when it grows as wild weed, it damages the other crops
13. *By-product (If any)* — Stems and leaves
14. *Use of by-product* — Fuel / sticks / cattle feed / organic manure

318. *CROP* — SOPHERA SENNA / Kasaundi / *Kalkasunde*

Scientific Name* — *Casia sophera L.

Family — Caesalpiniaceae

Herb / Shrub / Tree — Shrubby herbs

***Annual / Biennial / Perennial* —** Annual (0.5-0.8 m height)

***Used as* —** Medicine (asthma), Ornamental plants (hedge), but grows as weed in warm climate

1. *Climatic requirements* — Warm climate
2. *Soil requirements* — Alluvial, loamy, sand and clayey soils
3. *Required land situations* — High / medium lands
4. *Importance of Crop*— Club like capsule in *sophera* but sickle like pod in *tora*
5. *Fertilizer management* — 40 kg N, 50 kg P_2O_5 and 30 kg K_2O/ha when grown as medicine
6. *Water management* — 2-3 irrigations
7. *Duration of crop* — 120-160 days

8. *Parts used* — Whole plant for ornamental use, but seeds for medicine or preparation of sweets
9. *Where grown in India* — All over India, except too low / too high temperature areas
10. *Export possibility* — High
11. *Economic yields (q / ha)* — 15-20
12. *Economic value (Rs. / ha)* — It is a valuable plant when grown as medicine; but it grows as an obnoxious weed on road sides or in crops
13. *By-product (If any)* — Stems and leaves
14. *Use of by-product* — Fuel / organic matter

319. *CROP* — SICKLE SENNA / Pamaar / Barakalkasunde / *Kalkasunde*/

Scientific Name* — *Cassia tora L.

Family — Caesalpiniaceae

Herb / Shrub / Tree — Herb / Shrub

Annual / Biennial / Perennial — Annual (0.8-1.5 m height)

Used as — Seeds for preparation of sweets, but it grows as weed

1. *Climatic requirements* — Warm climate
2. *Soil requirements* — Loamy / alluvial and sandy soils
3. *Required land situations* — High / Medium lands
4. *Importance of Crop*— Sickle like pod
5. *Fertilizer management* — 40 kg N, 50 kg P_2O_5 and 30 kg K_2O/ha, if grown as industrial use
6. *Water management* — 2-3 irrigations annually
7. *Duration of crop* — 2-3 years
8. *Parts used* — Seeds
9. *Where grown in India* — All over India, except too low / too high temperature areas
10. *Export possibility* — High
11. *Economic yields (q / ha)* — 10-15
12. *Economic value (Rs. / ha)* — It is a good crop when grown for manufacturing sweets, but as weed it is a menace
13. *By-product (If any)* — Stems and leaves
14. *Use of by-product* — Fuel / organic matter

320. *CROP* — LANTANA / *Sugandhi ful* / Lantena

Scientific Name* — *Lantana camara L.

Family — Verbenaceae

Herb / Shrub / Tree — Climbing aromatic shrub (0.5-2 m height)

Annual / Biennial / Perennial — Perennial

Used as — Stalk for paper pulp; but grows in *warm* areas. Medicine can be obtained.

1. *Climatic requirements* — Warm climate
2. *Soil requirements* — Alluvial, loamy, sandy soils
3. *Required land situations* — High land
4. *Importance of Crop*— Aromatic flower / quick growing plant
5. *Fertilizer management* — 60 kg N, 30 kg P_2O_5 and 30 kg K_2O/ha annually, if grown for industrial use and medicine
6. *Water management* — 2-3 irrigations
7. *Duration of crop* — 2-3 years
8. *Parts used* — Flowers and leaves (for lowering blood pressure); stalk for paper pulp.
9. *Where grown in India* — All over India, except too low / too high temperature areas
10. *Export possibility* — High
11. *Economic yields (q / ha)* — 50-80 (stalk)
12. *Economic value (Rs. / ha)* — It is a good plant for flower / medicine / industrial use; but as weed it is very obnoxious and damaging to the environment
13. *By-product (If any)* — Leaves
14. *Use of by-product* — Cattle feed / organic matter / fuel.

321. *CROP* — TRIDAX / *Datitara*

Scientific Name* — *Tridax procumbens L.

Family — Compositae

Herb / Shrub / Tree — Herb (50-60 cm height)

Annual / Biennial / Perennial — Annual

Used as — Fodder, but grows as weed

1. *Climatic requirements* — Warm climate
2. *Soil requirements* — Alluvial, loamy and sandy soils
3. *Required land situations* — High / Medium / low lands
4. *Importance of crop* — The stalk of inflorescence is bending and long

5. *Fertilizer management* — 60 kg N, 30 kg P_2O_5 and 30 kg K_2O/ha, if grown as fodder
6. *Water management* — 2-3 irrigations
7. *Duration of crop* — 120-140 days
8. *Parts used* — Entire plant
9. *Where grown in India* — All over India, except too low / too high temperature areas
10. *Export possibility* — High
11. *Economic yields (q / ha)* — 100 - 120
12. *Economic value (Rs. / ha)* — It is a valuable plant when it is grown as fodder but obnoxious when it grows as weed in pre-*kharif* / *kharif* crops
13. *By-product (If any)* — Nil
14. *Use of by-product* — Nil

322. *CROP* — COUNTRY-MALLOW / Kungyi

Scientific Name — Sida cordifolia L.

Family — Malvaceae

Herb / Shrub / Tree — Shrub

Annual / Biennial / Perennial — Annual / Perennial

Used as — Medicine (seeds for bowel complaints) but it grows as weed in hot climate

1. *Climatic requirements* — Hot humid climate
2. *Soil requirements* — Sandy / loamy / alluvial soils
3. *Required land situations* — High / Medium / low lands
4. *Importance of Crop*— Big round leaves / Ephedrine alkaloid present
5. *Fertilizer management* — 60 kg N, 30 kg P_2O_5 and 30 kg K_2O/ha, if grown for medicine
6. *Water management* — Normally rainfed in East India
7. *Duration of crop* — 120-130 days / 2-3 years
8. *Parts used* — Seeds for medicine
9. *Where grown in India* — Bengal, Assam, Tripura, Bihar and Madhya Pradesh, U.P. etc.
10. *Export possibility* — High
11. *Economic yields (q / ha)* — 5-6 (Seed)
12. *Economic value (Rs. / ha)* — 50,000 to 60,000
13. *By-product (If any)* — Stalks and leaves

14. *Use of by-product* — Fuel / organic matter / cattle feed / mulch

323. *CROP* — PURPLE TEPROSIA / Sorphonka / Bannil

Scientific Name* — *Teprosia purpurea (L.) Pers.

Family — Papilionaceae

Herb / Shrub / Tree — Herb

Annual / Biennial / Perennial — Annual

Used as — Green manuring crop / cover crop and medicine, but it grows as weed in warm climate

1. *Climatic requirements* — Warm climate
2. *Soil requirements* — Alluvial, loamy, sandy and clayey soils
3. *Required land situations* — Medium / low land
4. *Importance of crop* — Green manuring crop / cover crop / medicine
5. *Fertilizer management* — 50 kg N, 60 kg P_2O_5 and 30 kg K_2O/ha, if grown for medicinal purpose (powdered leaves smoking is relief for asthma)
6. *Water management* — Normally rainfed, 2-3 irrigations, if irrigated
7. *Duration of crop* — 120-140 days
8. *Parts used* — Whole plant green manure / leaves for medicine
9. *Where grown in India* — All over India, except too low / too high temperature areas
10. *Export possibility* — High
11. *Economic yields (q / ha)* — 5-7 (leaves)
12. *Economic value (Rs. / ha)* — This is very important plant when grown as medicinal or green manuring crop, but as weed, it is damaging to warm crops
13. *By-product (If any)* — Stems
14. *Use of by-product* — Fuel / Organic matter / Cattle feed

324. *CROP* — INDIAN ABUTILON / Kanghi / Country mallow

Scientific Name* — *Abutilon indicum (L.) Sweet

Family — Malvaceae

Herb / Shrub / Tree — Shrub (robust hairy)

Annual / Biennial / Perennial — Perennial

Used as — Fibres (for ropes) but it grows as weed in hotter parts

1. *Climatic requirements* — Warm climate
2. *Soil requirements* — Alluvial, loamy and sandy soils

3. *Required land situations* — High / medium lands
4. *Importance of Crop*— It produces fibre. It also gives oils
5. *Fertilizer management* — 60 kg N, 30 kg P_2O_5 and 30 kg K_2O/ha annually, if grown as fibre crop; but as weed it is obnoxious
6. *Water management* — Rainfed, 1-2 irrigations, when irrigated
7. *Duration of crop* — 2-3 years
8. *Parts used* — Stem
9. *Where grown in India* — All over India, except too low / too high temperature areas
10. *Export possibility* — High
11. *Economic yields (q / ha)* — 15-20
12. *Economic value (Rs. / ha)* — It is a valuable crop, if grown for fibre, but obnoxious when it grows as weed
13. *By-product (If any)* — Leaves
14. *Use of by-product* — Cattle feed / fuel / organic matter

325. *CROP* — MADAR / AK

Scientific Name* — *Calotropis gigantea Air

Family — Asclepiadaceae

Herb / Shrub / Tree — Shrub

Annual / Biennial / Perennial — Perennial

Used as — Medicine (asthma / fibre for fishing net) and twine, but grows as weed in hot climate

1. *Climatic requirements* — Warm climate
2. *Soil requirements* — Alluvial, loamy and sandy soils
3. *Required land situations* — High land
4. *Importance of Crop*— Root-bark is used in dysentery
5. *Fertilizer management* — 60 kg N, 30 kg P_2O_5 and 30 kg K_2O/ha annually, if grown for medicine or fibre
6. *Water management* — 2-3 irrigations annually
7. *Duration of crop* — 3-5 years
8. *Parts used* — Stem or root barks / leaves
9. *Where grown in India* — West Bengal, Assam, Bihar, Orissa, Maharashtra and A.P. etc.
10. *Export possibility* — High
11. *Economic yields (q / ha)* — 15-20 (all stem, root and leaves)
12. *Economic value (Rs. / ha)* — It is a valuable plant when grown as medicine / fibre crop, but is damaging to field crops as weed

13. *By-product (If any)* — Nil
14. *Use of by-product* — Nil

326. *CROP* — BOERHAAVIA/ Punarnava / Hog weed / Sant / Santhi / Horse-purslane

Scientific Name* — *Boerhaavia diffusa L.

Family — Nyctaginaceae

Herb / Shrub / Tree — Herb (20-30 cm height)

Annual / Biennial / Perennial — Annual

Used as — Medicine (in dropsy and asthma), but grows as weed in warm climate

1. *Climatic requirements* — Warm climate
2. *Soil requirements* — Alluvial, loamy and sandy soils
3. *Required land situations* — High / medium lands
4. *Importance of Crop*— Drug '*punarnava*' is manufactured from it
5. *Fertilizer management* — 60 kg N, 30 kg P_2O_5 and 30 kg K_2O/ha, if grown for medicine
6. *Water management* — 2-3 irrigations
7. *Duration of crop* — 120-140 days
8. *Parts used* — Whole plant
9. *Where grown in India* — All over India in warm seasons
10. *Export possibility* — High
11. *Economic yields (q / ha)* — 50-100 (green)
12. *Economic value (Rs. / ha)* — It is highly valuable crop when grown as medicinal plant, but is damaging to crops when it grows as weed
13. *By-product (If any)* — Nil
14. *Use of by-product* — Nil

327. *CROP* — KAKRANDA / Blumea

Scientific Name* — *Blumea lacera DC

Family — Compositae

Herb / Shrub / Tree — Herb

Annual / Biennial / Perennial — Annual

Used as — Medicine : (Juice of leaf anthelmintic / diuretic / stimulant and febrifuge, but grows in warm season as weed

1. *Climatic requirements* — Warm climate
2. *Soil requirements* — Alluvial, sandy or loamy soils

3. *Required land situations* — High / medium / low lands
4. *Importance of Crop*— Blue-clustered flowers; bushy plants
5. *Fertilizer management* — 60 kg N, 30 kg P_2O_5 and 30 kg K_2O/ha, if grown as medicine
6. *Water management* — 2-3 irrigations
7. *Duration of crop* — 120-140 days
8. *Parts used* — Whole plant
9. *Where grown in India* — All over plains of India in warm season
10. *Export possibility* — High
11. *Economic yields (q / ha)* — 100-150 (green)
12. *Economic value (Rs. / ha)* — It is very valuable, when grown as medicine, but it is damaging when it grows as weed
13. *By-product (If any)* — Nil
14. *Use of by-product* — Nil

328. *CROP* — BELLYACHE BUSH / Bherenda / *Reri*

Scientific Name — Jatropha gossypifolia L.

Family — Euphorbiacea

Herb / Shrub / Tree — Shrub

Annual / Biennial / Perennial — Perennial

Used as — Medicine / seed as bio-fuel / antibiotics against *Escherichia coli*, but grows as weed in hot climate

1. *Climatic requirements* — Warm climate
2. *Soil requirements* — Alluvial, loamy or sandy soils
3. *Required land situations* — High / medium lands
4. *Importance of Crop*— A much-branched plant is industrially used for bio-fuel
5. *Fertilizer management* — 60 kg N, 30 kg P_2O_5 and 30 kg K_2O/ha annually, when grown for medicine or other purposes
6. *Water management* — 2-3 irrigations
7. *Duration of crop* — 2-3 years
8. *Parts used* — Extract of shoots
9. *Where grown in India* — All over India except extreme too low / too high temperature areas
10. *Export possibility* — Very high
11. *Economic yields (q / ha)* — 10-15
12. *Economic value (Rs. / ha)* — It is a valuable plant for medicine / bio-oil, but it is obnoxious weed in warm season on roadsides.

13. *By-product (If any)* — Stems and leaves
14. *Use of by-product* — Cattle feed, fuel, organic matter

329. *CROP* — INDIAN ACALYPHA / *Khokali*/ Kuppi

Scientific Name* — *Acalypha indica L.

Family — Euphorbiaceae

Herb / Shrub / Tree — Herb

Annual / Biennial / Perennial — Annual

Used as — Medicine for asthma / pneumonia bronchitis etc. but grows as weed in warm season

1. *Climatic requirements* — Warm climate
2. *Soil requirements* — Loamy, sandy or alluvial soils
3. *Required land situations* — High / medium lands
4. *Importance of Crop*— Whole plant as drug
5. *Fertilizer management* — 60 kg N, 30 kg P_2O_5 and 30 kg K_2O/ha when grown as medicinal plant
6. *Water management* — 2-3 irrigations
7. *Duration of crop* — 120-140 days
8. *Parts used* — Whole plant
9. *Where grown in India* — All over India, except too low / too high temperature areas
10. *Export possibility* — High
11. *Economic yields (q / ha)* — 100-200 (green)
12. *Economic value (Rs. / ha)* — It is a valuable plant, when grown as medicinal plant but as weed, it is damaging to crops
13. *By-product (If any)* — Nil
14. *Use of by-product* — Nil

330. CROP — WILD TOBACCO / Bantamak / Jangli tambaku

Scientific Name* — *Nicotiana plumbaginifolia L.

Family — Solanaceae

Herb / Shrub / Tree — Herb (50-80 cm height)

Annual / Biennial / Perennial — Annual

Used as — Narcotics / mulching materials / fuel but it grows as weed in cool / warm season

1. *Climatic requirements* — Warm / cool climate
2. *Soil requirements* — Alluvial, sandy or loamy soils
3. *Required land situations* — High / medium / low lands

4. *Importance of Crop*— 'Nicotine' type smell in the plant
5. *Fertilizer management* — 60 kg N, 30 kg P_2O_5 and 30 kg K_2O/ha if grown as narcotics / mulching materials
6. *Water management* — 2-3 irrigations
7. *Duration of crop* — 120-140 days
8. *Parts used* — Whole plant
9. *Where grown in India* — West Bengal, Assam, Bihar, Orissa and U.P
10. *Export possibility* — High
11. *Economic yields (q / ha)* — 80-120 (green)
12. *Economic value (Rs. / ha)* — It is a valuable plant when used as narcotics, but it is an obnoxious plant when it is grown as weed, damaging to the crops
13. *By-product (If any)* — Nil
14. *Use of by-product* — Nil

331. *CROP* — COCKLEBUR / Burweed / *Banokra* / *Chhagalguti* / Adhasisi

Scientific Name* — *Xanthium strumarium L.

Family — Compositae

Herb / Shrub / Tree — Herb (30-60 cm height)

Annual / Biennial / Perennial — Annual

Used as — Organic manure / medicine (for urinary troubles) but grows as weed in cool / warm season

1. *Climatic requirements* — Warm climate
2. *Soil requirements* — Alluvial, sandy, loamy and clayey soils
3. *Required land situations* — High / medium / low lands
4. *Importance of Crop*— Hard spiny fruits
5. *Fertilizer management* — 60 kg N, 30 kg P_2O_5 and 30 kg K_2O/ha when grown as medicine
6. *Water management* — 2-3 irrigations
7. *Duration of crop* — 120-140 days
8. *Parts used* — Seeds / whole plant (organic matter)
9. *Where grown in India* — All over India, except too low / too high temperature areas
10. *Export possibility* — High
11. *Economic yields (q / ha)* — 10-15 (seeds)

12. *Economic value (Rs. / ha)* — It is a valuable crop when grown as medicine, but it is damaging, when it grows as weed on the road side
13. *By-product (If any)* — Stems and leaves
14. *Use of by-product* — Fuel / organic matter

332. *CROP* — TRAILING INDIGO / Nil (Latano)

Scientific Name — Indigofera endecaphylla Jacq.

Family — Papilionaceae

Herb / Shrub / Tree — Herb

Annual / Biennial / Perennial — Annual

Used as — Fodder / cover crop / green manuring crop, but it grows as weed in warm seasons

1. *Climatic requirements* — Warm climate
2. *Soil requirements* — Alluvial, loamy and sandy soils
3. *Required land situations* — High / medium lands
4. *Importance of crop* — It is a good fodder crop. It may be used as cover crop.
5. *Fertilizer management* — 40 kg N, 60 kg P_2O_5 and 30 kg K_2O/ha when grown as fodder
6. *Water management* — 2-3 irrigations
7. *Duration of crop* — 120-140 days
8. *Parts used* — Whole plant for fodder
9. *Where grown in India* — All over India, except too low / too high temperature areas
10. *Export possibility* — High
11. *Economic yields (q / ha)* — 150-180
12. *Economic value (Rs. / ha)* — It is a valuable plant, when grown as fodder / green manuring crop, but is damaging to crops when it grows as weed
13. *By-product (If any)* — Nil
14. *Use of by-product* — Nil

333. *CROP* — INDIAN SORREL / Amrul / *Susni* / *Chuka-tripati*

Scientific Name — Oxalis corniculata L.

Family — Oxalidaceae

Herb / Shrub / Tree — Herb

Annual / Biennial / Perennial — Annual / Perennial

Used as — Medicine (for dyspepsia / piles), but grows as weed in warm season

1. *Climatic requirements* — Warm climate
2. *Soil requirements* — Alluvial, sandy or loamy soils
3. *Required land situations* — High / medium / low lands
4. *Importance of Crop*— Mostly grows in marshy lands
5. *Fertilizer management* — 60 kg N, 30 kg P_2O_5 and 30 kg K_2O/ha, if grown as medicine / vegetable
6. *Water management* — 2-3 irrigations
7. *Duration of crop* — 150-180 days / 2 years
8. *Parts used* — Whole plant
9. *Where grown in India* — All over India, except too low / too high temperature areas
10. *Export possibility* — High
11. *Economic yields (q / ha)* — 90-120 (green)
12. *Economic value (Rs. / ha)* — It is a good and valuable plant, when grown as medicine / vegetable, but is damaging to the crops when it grows as weed
13. *By-product (If any)* — Nil
14. *Use of by-product* — Nil

334. *CROP* — HOARY BASIL / Kalitulsi / Mamri / *Bantulsi*

Scientific Name — Ocimum americanum L. / ***O. basilicum*** L.

Family — Labiatae

Herb / Shrub / Tree — Herb (aromatic)

Annual / Biennial / Perennial — Perennial

Used as — Medicine for dysentery / chronic diarrhoea, but grows as weed in warm season

1. *Climatic requirements* — Warm climate
2. *Soil requirements* — Alluvial, loamy, sandy soils
3. *Required land situations* — High / medium lands
4. *Importance of Crop*— Good smell in plant
5. *Fertilizer management* — 60 kg N, 30 kg P_2O_5 and 30 kg K_2O/ha, as medicine
6. *Water management* — 4-5 irrigations, when grown for medicine
7. *Duration of crop* — 2-3 years
8. *Parts used* — Leaves and twigs

9. *Where grown in India* — All over India, except too low / too high temperature areas
10. *Export possibility* — High
11. *Economic yields (q / ha)* — 40-80 (leaves and twigs)
12. *Economic value (Rs. / ha)* — It is highly valuable plant, when grown as medicinal crop, but as weed in wasteland, it is obnoxious plant
13. *By-product (If any)* — Stems
14. *Use of by-product* — Fuel / organic matter / medicine

335. *CROP*— MEXICAN PRICKLY POPPY / *Sialkanta* / Kateli / Kandiari

Scientific Name* — *Argemone mexicana L.

Family — Papaveraceae

Herb / Shrub / Tree — Prickly herb / shrub

Annual / Biennial / Perennial — Annual (30-80 cm height)

Used as — Medicine (roots are used for skin disease), juice for jaundice, but grows in cool / warm season as weed

1. *Climatic requirements* — Warm climate
2. *Soil requirements* — Alluvial, loamy, sandy or clayey soils
3. *Required land situations* — High / medium / low lands
4. *Importance of Crop*— Yellow good flowers in spiny plant
5. *Fertilizer management* — 60 kg N, 30 kg P_2O_5 and 30 kg K_2O/ha, when grown as medicine
6. *Water management* — 2-3 irrigations
7. *Duration of crop* — 120-150 days
8. *Parts used* — Seeds (like mustard seeds)
9. *Where grown in India* — All over India, except too low / too high temperature areas
10. *Export possibility* — High
11. *Economic yields (q / ha)* — 5-10
12. *Economic value (Rs. / ha)* — It is a valuable plant when grown as medicinal/industrial crop, but as weed it is very detrimental for crops and also for field operations
13. *By-product (If any)* — Stems and leaves
14. *Use of by-product* — Fuel / organic matters

336. *CROP* — *MALVA* / MALLOWS / Panirak / Kasbeeze-mradupatra

Scientific Name* — *Malva parviflora L.

Family — Malvaceae

Herb / Shrub / Tree — Herb (60-80 cm height)

Annual / Biennial / Perennial — Perennial

Used as — Medicine / seeds for cough / ulcers in bladder and roots are used for cleansing hair and washing woolen clothes, but grows as weed in warm season

1. *Climatic requirements* — Warm climate
2. *Soil requirements* — Alluvial, loamy and sandy soils
3. *Required land situations* — High / medium lands
4. *Importance of Crop*— Big leaves
5. *Fertilizer management* — 60 kg N, 30 kg P_2O_5 and 30 kg K_2O/ha when grown as medicinal crop
6. *Water management* — 2-3 irrigations
7. *Duration of crop* — 1-2 years
8. *Parts used* — Leaves and seeds, also fibre is obtained from stem
9. *Where grown in India* — Assam, Maharashtra W. Bengal, Bihar, Orissa, U.P., Karnataka etc.
10. *Export possibility* — High
11. *Economic yields (q / ha)* — 10-15 (seed)
12. *Economic value (Rs. / ha)* — It is a valuable plant when grown as medicine; but damaging when grows as weed
13. *By-product (If any)* — Nil
14. *Use of by-product* — Nil

337. CROP — SENSITIVE PLANT / Lajwanti / Lajjabati

Scientific Name — Mimosa pudica L.

Family — Mimosaceae

Herb / Shrub / Tree — Trailing shrub but leaves are very sensitive to droopness

Annual / Biennial / Perennial — Perennial (10-20 cm height)

Used as — Ornamentals, but grows as weed in warm season

1. *Climatic requirements* — Warm climate
2. *Soil requirements* — Alluvial, loamy and sandy soils
3. *Required land situations* — High / medium lands
4. *Importance of Crop*— Small ball – like flowers, pink in colour; plant spiny / sensitive
5. *Fertilizer management* — 40 kg N, 60 kg P_2O_5 and 30 kg K_2O/ha when grown as ornamental

6. *Water management* — 2-3 irrigations
7. *Duration of crop* — 2-3 years
8. *Parts used* — Leaves and flowers
9. *Where grown in India* — All over India, except too low / too high temperature areas
10. *Export possibility* — High
11. *Economic yields (q / ha)* — Asthetic aspects / value
12. *Economic value (Rs. / ha)* — It is valuable plant when grown as ornamental plant, but as weed it is damaging in road side areas
13. *By-product (If any)* — Stem, leaves, flower
14. *Use of by-product* — Fuel / organic matter

338. *CROP* — YELLOW DOCK / Chukhah / Curleddock / *Janglipalak / Ban palang* / Golden dock

Scientific Name* — *Rumex crispus L. / ***R. maritimus*** L.

Family — Polygonaceae

Herb / Shrub / Tree — Herb (20-80 cm height)

Annual / Biennial / Perennial — Perennial / Annual

Used as — Vegetable / Medicine (roots used in homeopathy), but grows as weed in warm season

1. *Climatic requirements* — Warm climate (also in cool climate)
2. *Soil requirements* — Alluvial, loamy and sandy soils
3. *Required land situations* — High / medium / low
4. *Importance of Crop*— Like spinach but with big leaves
5. *Fertilizer management* — 100 kg N, 50 kg P_2O_5 and 30 kg K_2O/ha when grown as vegetable / medicine
6. *Water management* — 3-4 irrigations
7. *Duration of crop* — 2-3 years / 120-140 days
8. *Parts used* — Leaves (vegetable), root (medicine)
9. *Where grown in India* — All over India, except too low / too high temperature areas
10. *Export possibility* — High
11. *Economic yields (q / ha)* — 50-100 (green)
12. *Economic value (Rs. / ha)* — It is a valuable crop when grown as medicinal plant / vegetable, but damaging to crops when grows as weed
13. *By-product (If any)* — Stems
14. *Use of by-product* — Fuel / organic matter / cattle feed.

339. *CROP* — CUDWEED / *Rupajhuri* / Gnaphalium

***Scientific Name* — *Gnaphalium indicum* L.**

Family — Compositae

Herb / Shrub / Tree — Herb (5-15 cm height)

Annual / Biennial / Perennial — Annual

Used as — Mulch / fodder but also grows as weed in cool / warm season

1. *Climatic requirements* — Warm / cool climate
2. *Soil requirements* — Alluvial, loamy and sandy soils
3. *Required land situations* — High / medium / low lands
4. *Importance of Crop*— Snow-white plant, smaller in size
5. *Fertilizer management* — 30 kg N, 30 kg P_2O_5 and 30 kg K_2O/ha, when grown crop for mulching
6. *Water management* — 2-3 irrigations
7. *Duration of crop* — 120-140 days
8. *Parts used* — Entire plant
9. *Where grown in India* — All over India, except too low / too high temperature areas
10. *Export possibility* — Less
11. *Economic yields (q / ha)* — 30-40 (green)
12. *Economic value (Rs. / ha)* — It is a valuable crop when grown for mulching (to conserve moisture of soil), but as weed, it damages the crops
13. *By-product (If any)* — Nil
14. *Use of by-product* — Nil

340. *CROP* — GOAT WEED / Floss flower / Mist flower

***Scientific Name* — *Ageratum conyzoides* L.**

Family — Compositae

Herb / Shrub / Tree — Herb (20-60 cm height)

Annual / Biennial / Perennial — Annual

Used as — Ornamentals / flowers (blue), but grows as weed in warm season crops

1. *Climatic requirements* — Cool climate
2. *Soil requirements* — Alluvial, sandy or loamy soils
3. *Required land situations* — High / medium / low lands
4. *Importance of Crop*— Blue flowers (cluster)
5. *Fertilizer management* — 60 kg N, 30 kg P_2O_5 and 30 kg K_2O/ha, if grown as ornamental

6. *Water management* — 2-3 irrigations
7. *Duration of crop* — 120-140 days
8. *Parts used* — Whole plant
9. *Where grown in India* — All over India, except too hot or too cold areas
10. *Export possibility* — High
11. *Economic yields (q / ha)* — Aesthetic value
12. *Economic value (Rs. / ha)* — It is a valuable plant when grown as ornamental one, but as weed is damaging to the crops
13. *By-product (If any)* — Nil
14. *Use of by-product* — Nil

341. *CROP* — DAUDOTPALA / SAHADEVI / VERNONIA / Kuksima / Ash coloured fleabane / *Kuksima*

Scientific Name* — *Vernonia cinerea (L.) Less

Family — Compositae

Herb / Shrub / Tree — Herb (30-80 cm height)

Annual / Biennial / Perennial — Annual

Used As—Medicinal (leaves for eczema / ring worm, but grows as weed)

1. *Climatic requirements* — Cool climate
2. *Soil requirements* — Alluvial, sandy or loamy soil
3. *Required land situations* — High / medium / low lands
4. *Importance of crop* — Ash coloured flowers are used for rheumatism
5. *Fertilizer management* — 60 kg N, 30 kg P_2O_5 and 30 kg K_2O/ha, when grown as medicinal plant
6. *Water management* — 2-3 irrigations
7. *Duration of crop* — 120-140 days
8. *Parts used* — Leaves / whole plant / flower
9. *Where grown in India* — All over India except too hot or too cold areas
10. *Export possibility* — High
11. *Economic yields (q / ha)* — 100-120 (green)
12. *Economic value (Rs. / ha)* — It is a valuable plant when grown as medicinal one, but as weed, is damaging to the crops
13. *By-product (If any)* — Nil
14. *Use of by-product* — Nil

342. *CROP* — SUFAID MURGHA / *Murgi Jhuti*

***Scientific Name* — *Celosia argentea* L.**

Family — Amaranthaceae

Herb / Shrub / Tree — Herb (50-80 cm height)

Annual / Biennial / Perennial — Annual

Used as — Vegetable (stem / twig), seeds for blood diseases, but grows as weed in warm season

1. *Climatic requirements* — Warm / cool climate
2. *Soil requirements* — Alluvial, sandy or loamy soil
3. *Required land situations* — High / medium / low lands
4. *Importance of Crop*— Inflorescence is like cock's head
5. *Fertilizer management* — 60 kg N, 30 kg P_2O_5 and 30 kg K_2O/ha,if grown as vegetable
6. *Water management* — 3-4 irrigations
7. *Duration of crop* — 120-140 days
8. *Parts used* — Stem and twigs (vegetable) / seeds for medicine
9. *Where grown in India* — All over India, except too hot or too cold areas
10. *Export possibility* — High
11. *Economic yields (q / ha)* — 100-120 (green)
12. *Economic value (Rs. / ha)* — It is a valuable plant when grown as medicine / vegetable but it is damaging, if it grows as weed
13. *By-product (If any)* — Nil
14. *Use of by-product* — Nil.

343. *CROP* — LAMBSQUATERS / Bathua / *Betho* / Pig weed

***Scientific Name* — *Chenopodium album* L.**

Family — Chenopodiaceae

Herb / Shrub / Tree — Herb (50-80 cm length)

Annual / Biennial / Perennial — Annual

Used as — Vegetable / fodder, but grows in winter season with wheat as weed

1. *Climatic requirements* — Cool climate
2. *Soil requirements* — Alluvial, sand or loamy soils
3. *Required land situations* — High / medium / low land
4. *Importance of Crop*— Bushy inflorescence
5. *Fertilizer management* — 60 kg N, 30 kg P_2O_5 and 30 kg K_2O/ha, if grown as vegetable / fodder

6. *Water management* — 2-3 irrigations
7. *Duration of crop* — 120-140 days
8. *Parts used* — Leaves and twigs
9. *Where grown in India* — All over India except too hot / too high temperature areas
10. *Export possibility* — High
11. *Economic yields (q / ha)* — 40-80 (green)
12. *Economic value (Rs. / ha)* — It is a valuable vegetable / fodder crop but as weed it is damaging to other crops (wheat)
13. *By-product (If any)* — Stem
14. *Use of by-product* — Cattle feed / fuel / organic matter

344. *CROP* — MARSILEA

Scientific Name* — *Marsilea quadrifolia L.

Family — Marsileaceae

Herb / Shrub / Tree — Herb

Annual / Biennial / Perennial — Perennial

Used as — Fodder, vegetable and mulch but grows as weed in *Kharif* in paddy fields

1. *Climatic requirements* — Warm climate
2. *Soil requirements* — Alluvial, loamy and clayey soils
3. *Required land situations* — Medium / low land
4. *Importance of Crop*— It grows better in marshy lands
5. *Fertilizer management* — 60 kg N, 30 kg P_2O_5 and 30 kg K_2O/ha when grown as vegetable or mulching materials
6. *Water management* — In dry land 2-3 irrigations
7. *Duration of crop* — 120-140 days
8. *Parts used* — Whole plant, stalks eaten as a pot-herb in time of scarcity
9. *Where grown in India* — All over India in marshy land in wet season
10. *Export possibility* — High
11. *Economic yields (q / ha)* — 50 – 100 (green)
12. *Economic value (Rs. / ha)* — It is a valuable plant when grown as vegetable and fodder but it is the damaging plant when it grows as weed in rice / arum
13. *By-product (If any)* — Nil
14. *Use of by-product* — Nil

NARROW LEAVED WEEDS

345. ***CROP*** **— NUTGRASS / Motha /** ***Mutha*** **/ Earth almond**

Scientific Name* — *Cyperus rotundus / esculentus L.

Family — Cyperaceae

Herb / Shrub / Tree — Sedge (5-20 cm height)

Annual / Biennial / Perennial — Perennial

Used as — Essential oil from rhizomes used in perfumes, but grows as weeds throughout the year

1. *Climatic requirements* — Moderate cool / warm climate
2. *Soil requirements* — Alluvial, sandy, loamy and clayey soils
3. *Required land situations* — High / medium / low lands
4. *Importance of Crop*— Nut with smell is there below the ground and is used in perfumery
5. *Fertilizer management* — 60 kg N, 30 kg P_2O_5 and 30 kg K_2O/ha, when grown as industrial use
6. *Water management* — 2-3 irrigations annually
7. *Duration of crop* — 2-4 years
8. *Parts used* — Nut / rhizomes
9. *Where grown in India* — All over India in all seasons
10. *Export possibility* — High (nut)
11. *Economic yields (q / ha)* — 50-60 (nuts)
12. *Economic value (Rs. / ha)* — It is a valuable crop when grown as industrial crop, but as weed, it is very obnoxious and detrimental to crops of all seasons
13. *By-product (If any)* — 50-100 (green)
14. *Use of by-product* — Cattle feed / fuel / mulch / organic matter

346. ***CROP*****—BARNYARD GRASS / SHYAMA GHASH / Samak / Sanwak**

Scientific Name* — *Echinochloa crus-galli (L.) Beauv

Family — Gramineae

Herb / Shrub / Tree — Herb (30-80 cm height)

Annual / Biennial / Perennial — Annual

Used as — Forage / food, though weed in rice

1. *Climatic requirements* — Warm climate
2. *Soil requirements* — Alluvial, loamy and clayey soils
3. *Required land situations* — Medium / low land

4. *Importance of Crop*— Rice is affected and it gives small grains like rice
5. *Fertilizer management* — 100 kg N, 30 kg P_2O_5 and 30 kg K_2O/ha when grown as fodder / food grain
6. *Water management* — 3-4 irrigations
7. *Duration of crop* — 120-140 days
8. *Parts used* — Whole plant (if used fodder)
9. *Where grown in India* — All over India in marshy land with rice
10. *Export possibility* — High
11. *Economic yields (q / ha)* — 150-180 (green)
12. *Economic value (Rs. / ha)* — It is very valuable as fodder but as weed in rice, it is very damaging
13. *By-product (If any)* — Nil
14. *Use of by-product* — Nil

347. *CROP* — LATMAHURIA / Lesua / Gangatiya

Scientific Name — Digera arvensis Forsk / ***D. muricata*** (L.) Mart

Family — Amaranthaceae

Herb / Shrub / Tree — Herb (30-80 cm height)

Annual / Biennial / Perennial — Annual

Used as — Vegetable (leaves and twigs) / but weed on road sides

1. *Climatic requirements* — Warm climate
2. *Soil requirements* — Alluvial, loamy, and sandy soils
3. *Required land situations* — High / medium lands
4. *Importance of Crop*— Leaves and twigs are used as vegetable
5. *Fertilizer management* — 60 kg N, 30 kg P_2O_5 and 30 kg K_2O/ha, when used as vegetable
6. *Water management* — 2-3 irrigations
7. *Duration of crop* — 120-140 days
8. *Parts used* — Leaves and twigs
9. *Where grown in India* — All over India except too hot / too cold areas
10. *Export possibility* — High
11. *Economic yields (q / ha)* — 50-80 (green)
12. *Economic value (Rs. / ha)* — It is a valuable plant when used as vegetable, but as weed it is damaging, though mostly it grows on roadsides
13. *By-product (If any)* — Stem

14. *Use of by-product* — Cattle feed / mulch / organic matter.

348. *CROP* — DIGITARIA

Scientific Name* — *Digitaria sanguinalis Scope

Family — Gramineae

Herb / Shrub / Tree — Herb (30-80 cm height)

Annual / Biennial / Perennial — Annual

Used as — Fodder / weed in summer

1. *Climatic requirements* — Warm climate
2. *Soil requirements* — Alluvial, sandy, loamy and clayey soil
3. *Required land situations* — Medium / low lands
4. *Importance of Crop*— It is a good fodder (grass)
5. *Fertilizer management* — 100 kg N, 30 kg P_2O_5 and 30 kg K_2O/ha, if grown as fodder
6. *Water management* — 2-3 irrigations
7. *Duration of crop* — 120-140 days
8. *Parts used* — Whole plant
9. *Where grown in India* — All over India except too low or too high temperature areas
10. *Export possibility* — High
11. *Economic yields (q / ha)* — 100-150 (green)
12. *Economic value (Rs. / ha)* — It is very valuable plant, when grown as fodder, but as weed, it is damaging to the crops.
13. *By-product (If any)* — Nil
14. *Use of by-product* — Nil

349. *CROP* — BERMUDA GRASS / Dub / *Durba* / Bahama grass / Doop

Scientific Name* — *Cynodon dactylon (L) Pers.

Family — Gramineae

Herb / Shrub / Tree — Herb (spreading, 15 cm height but 30-80 cm length)

Annual / Biennial / Perennial — Perennial

Used as — Lawn grass / fodder and it grows in all seasons in India as weeds

1. *Climatic requirements* — Both warm and cool climates
2. *Soil requirements* — Alluvial, sandy, loamy, clayey, laterite, black cotton soils etc.
3. *Required land situations* — High / medium / low lands
4. *Importance of Crop*— Drought and flood resistant grass

5. *Fertilizer management* — 60 kg N, 30 kg P_2O_5 and 30 kg K_2O/ha, when grown as fodder / lawn grass
6. *Water management* — 4-5 irrigation annually
7. *Duration of crop* — 2-3 years or more
8. *Parts used* — Entire plant
9. *Where grown in India* — All over in India, except too low / too high temperature areas
10. *Export possibility* — High
11. *Economic yields (q / ha)* — 50-100 (green)
12. *Economic value (Rs. / ha)* — It is a valuable, fodder / lawn grass, but as weed it is damaging to the crops of different seasons
13. *By-product (If any)* — Nil
14. *Use of by-product* — Nil

350. *CROP* — GAMBA GRASS / Gamba / *Chorkanta* ghash / *Batui* / Andropogon

Scientific Name* — *Andropogon gayanus Kunth

Family — Gramineae

Herb / Shrub / Tree — Herb (30-50 cm height)

Annual / Biennial / Perennial — Perennial

Used as — Fodder / used as soil erosion check; but weed on roadsides

1. *Climatic requirements* — Warm climate
2. *Soil requirements* — Alluvial, loamy, sandy, clayey soils
3. *Required land situations* — High / Medium / low lands
4. *Importance of crop* — Seeds with pointed end, get into the clothing, if come in contact / good for pasture land
5. *Fertilizer management* — 60 kg N, 30 kg P_2O_5 and 30 kg K_2O/ha, when grown as fodder
6. *Water management* — 2-3 irrigations annually
7. *Duration of crop* — 2-3 years or more
8. *Parts used* — Whole plant
9. *Where grown in India* — All over India except too low / high temperature areas
10. *Export possibility* — High
11. *Economic yields (q / ha)* — 50-100 (green)
12. *Economic value (Rs. / ha)* — It is a valuable plant as fodder, but as weed it is damaging to crops / human beings
13. *By-product (If any)* — Nil

14. *Use of by-product* — Nil

351. *CROP*— GOOSE GRASS / Balraja / Mandla / Fowl fit grass / *Ban ragi*

Scientific Name* — *Eleusine indica (L) Gaertn

Family — Gramineae

Herb / Shrub / Tree — Herb (30-40 cm height)

Annual / Biennial / Perennial — Annual

Used as — Fodder / soil-binder (but weed on road sides)

1. *Climatic requirements* — Warm climate
2. *Soil requirements* — Alluvial, loamy and sandy soils
3. *Required land situations* — High / medium / low land
4. *Importance of Crop*— Very hard clustered grass nearer to soil
5. *Fertilizer management* — 60 kg N, 30 kg P_2O_5 and 30 kg K_2O/ha, when grown as fodder
6. *Water management* — 2-3 irrigation
7. *Duration of crop* — 120-140 days
8. *Parts used* — Entire plant
9. *Where grown in India* — All over in India except too low / too high temperature areas
10. *Export possibility* — High
11. *Economic yields (q / ha)* — 15-100 (green)
12. *Economic value (Rs. / ha)* — It is a valuable plant when grown as fodder, but when grows as weed, it is damaging to crops
13. *By-product (If any)* — Nil
14. *Use of by-product* — Nil

352. *CROP* — SUGAR SORGHUM / *Deodhan* / Sorgho / *Misti sorghum*

Scientific Name* — *Sorghum vulgare Pers

Family — Gramineae

Herb / Shrub / Tree — Herb (70-80 cm length)

Annual / Biennial / Perennial — Annual

Used as — Food / fodder; weed in fields

1. *Climatic requirements* — Warm climate
2. *Soil requirements* — Alluvial, loamy, sandy soils
3. *Required land situations* — Medium / high / low lands
4. *Importance of Crop*— It is like rice, but small sweet juice comes out of stem

5. *Fertilizer management* — 100 kg N, 50 kg P_2O_5 and 30 kg K_2O/ha, when grown as food / fodder
6. *Water management* — 3-4 irrigations
7. *Duration of crop* — 120-150 days
8. *Parts used* — Leaves stem (for fodder) / grains (for food)
9. *Where grown in India* — All over India except too low / too high temperature areas
10. *Export possibility* — High
11. *Economic yields (q / ha)* — 100-120 (green)
12. *Economic value (Rs. / ha)* — It is a valuable plant when grown as fodder / food, but as weed it is damaging to crops
13. *By-product (If any)* — Nil
14. *Use of by-product* — Nil

353. *CROP* — MAKRAGHASH / Brachiaria / Dalghash / *Banpara ghash*

Scientific Name — Brachiaria ramosa (L) Stapf.

Family — Gramineae

Herb / Shrub / Tree — Herb

Annual / Biennial / Perennial — Annual

Used as — Fodder but weed in marshy field

1. *Climatic requirements* — Warm climate
2. *Soil requirements* — Clayey, loamy and alluvial soils
3. *Required land situations* — Medium / low lands
4. *Importance of Crop*— Spreading type, even in water
5. *Fertilizer management* — 60 kg N, 30 kg P_2O_5 and 30 kg K_2O/ha, when grown as fodder
6. *Water management* — 2-3 irrigations, when sown dry
7. *Duration of crop* — 140-160 days
8. *Parts used* — Whole plant
9. *Where grown in India* — All over India except too low / too high temperature areas
10. *Export possibility* — High
11. *Economic yields (q / ha)* — 100-120 (green)
12. *Economic value (Rs. / ha)* — It is a valuable plant when grown as fodder, but as weed it is damaging to the crops
13. *By-product (If any)* — Nil
14. *Use of by-product* — Nil

354. *CROP* — HILOGRASS / Paspalum

Scientific Name — Paspalum conjugatum Berg

Family — Gramineae

Herb / Shrub / Tree — Herb (80-90 cm height)

Annual / Biennial / Perennial — Perennial

Used as — Fodder / grows as weed

1. *Climatic requirements* — Warm season
2. *Soil requirements* — Loamy, sandy and alluvial soils
3. *Required land situations* — High / medium lands
4. *Importance of Crop*— Tall grass with strong stem
5. *Fertilizer management* — 100 kg N, 50 kg P_2O_5 and 30 kg K_2O/ha, if grown as fodder
6. *Water management* — 3-4 irrigations
7. *Duration of crop* — 2-3 years
8. *Parts used* — Entire plant
9. *Where grown in India* — All over India in high lands mostly, except too low / too high temperature areas
10. *Export possibility* — High
11. *Economic yields (q / ha)* — 100-150 (green)
12. *Economic value (Rs. / ha)* — It is a valuable plant as fodder, but when it grows as weed, it is detrimental to crops
13. *By-product (If any)* — Nil
14. *Use of by-product* — Nil

355. *CROP* — CHIRIA-KA-DANA / Sporobolus

Scientific Name — Sporobolus diander (Retz) Beauv.

Family — Gramineae

Herb / Shrub / Tree — Herb (30-80 cm height)

Annual / Biennial / Perennial — Perennial

Used as — Fodder / pasture land weeds in waste lands

1. *Climatic requirements* — Warm season
2. *Soil requirements* — Loamy / sandy / alluvial soils
3. *Required land situations* — High / medium / low lands
4. *Importance of Crop*— Long inflorescence, though plant height less
5. *Fertilizer management* — 60 kg N, 30 kg P_2O_5 and 30 kg K_2O/ha, when grown as fodder
6. *Water management* — 3-4 irrigations

7. *Duration of crop* — 2-3 years
8. *Parts used* — Whole plants
9. *Where grown in India* — All over India except too low / too high temperature areas
10. *Export possibility* — High
11. *Economic yields (q / ha)* — 100-150 (green)
12. *Economic value (Rs. / ha)* — It is a valuable plant when grown as fodder, but damaging to crops, if grown as weed
13. *By-product (If any)* — Nil
14. *Use of by-product* — Nil

356. *CROP* — CAT TAIL MILLET / Bandra

Scientific Name* — *Setaria glauca (L) Beauv

Family — Gramineae

Herb / Shrub / Tree — Herb (30-80 cm height)

Annual / Biennial / Perennial — Annual

Used as — Fodder / food; weed on road sides

1. *Climatic requirements* — Warm seasons
2. *Soil requirements* — Loamy, sandy, alluvial, laterite soils
3. *Required land situations* — High / medium lands
4. *Importance of Crop*— Small grains are also obtained from it and consume5
5. *Fertilizer management* — 60 kg N, 30 kg P_2O_5 and 30 kg K_2O/, when grown as fodder / food
6. *Water management* — 3-4 irrigations
7. *Duration of crop* — 120-140 days
8. *Parts used* — Whole plant
9. *Where grown in India* — All over India except too low / too high temperature areas
10. *Export possibility* — High
11. *Economic yields (q / ha)* — 100-120 (green)
12. *Economic value (Rs. / ha)* — It is a valuable plant when grown as food / fodder, but it is damaging when it grows as weed
13. *By-product (If any)* — Nil
14. *Use of by-product* — Nil

357. *CROP* — COGON GRASS/Imperata/Ulu/Imperita/*Ulukhar* / *Dabh*

Scientific Name* — *Imperata cylindrical (L) Beauv var. ***major*** Nees.

Family — Gramineae

Herb / Shrub / Tree — Herb (50-89 cm height)

Annual / Biennial / Perennial — Perennial

Used as — Paper making / thatching in home or *boroj* / erosion checking but a serious weed in waste land

1. *Climatic requirements* — Warm climate, survives in cool condition also
2. *Soil requirements* — Alluvial, sandy, loamy soils
3. *Required land situations* — High / medium lands
4. *Importance of Crop*— Very hardy thatching material for house or *boroj*, very difficult to control as weed. It can withstand flood water.
5. *Fertilizer management* — 100 kg N, 50 kg P_2O_5 and 50 kg K_2O/ha, when grown as paper making / thatching material
6. *Water management* — 3-4 irrigations
7. *Duration of crop* — 2-5 years
8. *Parts used* — Whole plants
9. *Where grown in India* — All over India except too low / too high temperature areas
10. *Export possibility* — High
11. *Economic yields (q / ha)* — 100-200 (green)
12. *Economic value (Rs. / ha)* — It is highly valuable, when grown for roofing / paper or board making, but as weed it is damaging to the land
13. *By-product (If any)* — Nil
14. *Use of by-product* — Nil

358. *CROP* — RICE GRASS / Dal ghash / Jalghash

Scientific Name* — *Leersia hexandra Sw.

Family — Gramineae

Herb / Shrub / Tree — Herb (Creeping / spreading)

Annual / Biennial / Perennial — Perennial

Used as — Fodder; weed on water surface

1. *Climatic requirements* — Warm climate
2. *Soil requirements* — Alluvial, clayey and loamy soils

3. *Required land situations* — Low marshy lands
4. *Importance of Crop*— Rhizomatous aquatic plant
5. *Fertilizer management* — 100 kg N, 50 kg P_2O_5 and 30 kg K_2O/ha, when grown as fodder
6. *Water management* — No irrigation necessary (as it grows in water / marshy land)
7. *Duration of crop* — 2-3 years
8. *Parts used* — Entire plant
9. *Where grown in India* — All over India except too low / too high temperature areas
10. *Export possibility* — High
11. *Economic yields (q / ha)* — 100-200 (green)
12. *Economic value (Rs. / ha)* — It is a valuable plant when grown as fodder, but it is damaging to marshy land crops, as weed
13. *By-product (If any)* — Nil
14. *Use of by-product* — Nil

359. *CROP* — SCIRPUS / *Rasuighash*

Scientific Name* — *Scirpus erectus L.

Family — Cyperaceae

Herb / Shrub / Tree — Herb (in water, weed mainly in rice)

Annual / Biennial / Perennial — Annual

Used as — Fodder / mulch material but grows as weed in rice field in warm season

1. *Climatic requirements* — Warm season
2. *Soil requirements* — Clayey, loamy and alluvial soils
3. *Required land situations* — Medium / low lands
4. *Importance of Crop*— Grows just over the soil surface but below water (in rice)
5. *Fertilizer management* — 50 kg N, 30 kg P_2O_5 and 30 kg K_2O/ha, when grown as fodder
6. *Water management* — No water management as it grows in water
7. *Duration of crop* — 120-160 days
8. *Parts used* — Whole plant
9. *Where grown in India* — All over India except too low / too high temperature areas
10. *Export possibility* — High
11. *Economic yields (q / ha)* — 50-80 (green)

12. *Economic value (Rs. / ha)* — It is a valuable plant when grown as fodder; but it is very damaging to crops (rice), when grows as weed in water (just above soil surface)
13. *By-product (If any)* — Nil
14. *Use of by-product* — Nil

360. CROP — BRACHIARIA / Paragrass / Dalghash / Mauritius grass

Scientific Name — Brachiaria mutica (Forsk) Stapf

Family — Gramineae

Herb / Shrub / Tree — Herb

Annual / Biennial / Perennial — Perennial

Used as — Fodder, weed in marshy lands or deep water

1. *Climatic requirements* — Wet climatic conditions
2. *Soil requirements* — Clayey, loamy, alluvial soils
3. *Required land situations* — Low / medium lands
4. *Importance of Crop*— Creeping / branching type / with hair on the body
5. *Fertilizer management* — 80 kg N, 30 kg P_2O_5 and 30 kg K_2O/ha, if grown as fodder
6. *Water management* — No need (as it grows in marshy / waterlogged condition)
7. *Duration of crop* — 2-3 years
8. *Parts used* — Whole plant
9. *Where grown in India* — All over India except too low / too high temperature areas
10. *Export possibility* — High
11. *Economic yields (q / ha)* — 100-200 (green)
12. *Economic value (Rs. / ha)* — It is a valuable plant as fodder, but when it grows as weed, it is detrimental
13. *By-product (If any)* — Nil
14. *Use of by-product* — Nil

3. RAINY / WET / *KHARIF* SEASON

BROAD LEAVED WEEDS (as in pre-*kharif* season) PLUS

361. *CROP* — HEMP / Ganja / Bhang / *Charas*

Scientific Name — Cannabis sativa L.

Family — Cannabinaceae

Herb / Shrub / Tree — Herb / Shrub (smelling)

Annual / Biennial / Perennial — Annual / Perennial

Used as — Fibre / narcotics; weed in waste land / irrigation channels

1. *Climatic requirements* — Warm season
2. *Soil requirements* — Alluvial, loamy, sandy soils
3. *Required land situations* — High / medium lands
4. *Importance of Crop*— Strong smelling plant
5. *Fertilizer management* — 60 kg N, 30 kg P_2O_5 and 30 kg K_2O/ha, when grown as narcotics
6. *Water management* — 2-4 irrigations
7. *Duration of crop* — 120-160 days / 2-3 years
8. *Parts used* — Stem (fibre) / oil (seed) / leaves (narcotics)
9. *Where grown in India* — All over India except too low / too high temperature areas
10. *Export possibility* — Very high
11. *Economic yields (q / ha)* — 10-15 (green)
12. *Economic value (Rs. / ha)* — It is highly valuable when grown as narcotics, but it is obnoxious weed, damages the crops / places
13. *By-product (If any)* — Stems
14. *Use of by-product* — Cattle feed / mulch / organic matter / fuel

362. *CROP — KALMEGH* /KING-OF-BITTERS / Kiryat

Scientific Name — Andrographis paniculata Wall

Family — Acanthaceae

Herb / Shrub / Tree — Herb (30-80 cm height)

Annual / Biennial / Perennial — Annual

Used as — Medicine (worms / dysentery)

1. *Climatic requirements* — Warm seasons
2. *Soil requirements* — Alluvial, loamy, clayey and sandy soils
3. *Required land situations* — High / medium lands
4. *Importance of Crop*— Branched black / ash coloured plant
5. *Fertilizer management* — 60 kg N, 50 kg P_2O_5 and 30 kg K_2O/ha, when grown as medicine
6. *Water management* — 3-4 irrigations
7. *Duration of crop* — 120-140 days
8. *Parts used* — All parts (stem, flower and leaves)

9. *Where grown in India* — All over India except too low / too high temperature areas
10. *Export possibility* — High
11. *Economic yields (q / ha)* — 50-80 (green)
12. *Economic value (Rs. / ha)* — It is a valuable plant as fodder / medicine but when it grows as weed, it is detrimental
13. *By-product (If any)* — Nil
14. *Use of by-product* — Nil

363. *CROP* — RASNA / Vanda / Banda

Scientific Name* — *Vanda tessellata (Roxb) Hook

Family — Orchidaceae

Herb / Shrub / Tree — Herb

Annual / Biennial / Perennial — Annual

Used as — Medicine (rheumatism / bronchitis); weed on road side

1. *Climatic requirements* — Warm climatic condition
2. *Soil requirements* — Alluvial, loamy and sandy soils
3. *Required land situations* — High / medium
4. *Importance of Crop*— Roots are used for curing rheumatism
5. *Fertilizer management* — 60 kg N, 30 kg P_2O_5 and 30 kg K_2O/ha, when grown as medicinal crop
6. *Water management* — 3-4 irrigations
7. *Duration of crop* — 120-140 days
8. *Parts used* — Roots / leaves
9. *Where grown in India* — All over India (W.B. / Bihar / M.P. / T.N.)
10. *Export possibility* — High
11. *Economic yields (q / ha)* — 15-20
12. *Economic value (Rs. / ha)* — It is a valuable plant when grown as medicinal plant, but as weed, it is damaging to the crops.
13. *By-product (If any)* — Stem
14. *Use of by-product* — Fuel / organic matter

NARROW LEAVED WEED (as in pre-*kharif* season) PLUS

364. *CROP* — BRACHIARIA / *KHUDEDHAN*

Scientific Name* — *Brachiaria deflexa Hubb

Family — Gramineae

Herb / Shrub / Tree — Herb

Annual / Biennial / Perennial — Perennial

Used as — Fodder / food; weed in marshy land

1. *Climatic requirements* — Warm climate
2. *Soil requirements* — Alluvial, clayey, loamy soils
3. *Required land situations* — Low / medium lands
4. *Importance of Crop*— Spreading type with aerial tillers
5. *Fertilizer management* — 60 kg N, 30 kg P_2O_5 and 30 kg K_2O/ha, when grown as fodder / food
6. *Water management* — No irrigation required as it is grown in marshy land
7. *Duration of crop* — 2-3 years
8. *Parts used* — Whole plant
9. *Where grown in India* — All over India except too low / too high temperature areas
10. *Export possibility* — High
11. *Economic yields (q / ha)* — 100-150 (green)
12. *Economic value (Rs. / ha)* — It is very valuable plant as fodder, but as weed, it is very damaging to crops
13. *By-product (If any)* — Nil
14. *Use of by-product* — Nil

365. *CROP* — GOOSE GRASS / Fowl-foot grass / Balraja / *Buno Ragi / Mandla*

Scientific Name — Eleusine indica (L) Gaertn

Family — Gramineae

Herb / Shrub / Tree — Herb (30-50 cm height)

Annual / Biennial / Perennial — Annual

Used as — Fodder / soil-binder; weed on road sides

1. *Climatic requirements* — Warm climate
2. *Soil requirements* — Loamy, sandy and loamy soils
3. *Required land situations* — High / medium lands
4. *Importance of Crop*— Long stalk of star like inflorescence
5. *Fertilizer management* — 60 kg N, 30 kg P_2O_5 and 30 kg K_2O/ha, when grown as fodder
6. *Water management* — 3-4 irrigations
7. *Duration of crop* — 120-150 days
8. *Parts used* — Whole plants

9. *Where grown in India* — All over India except too low / too high temperature areas
10. *Export possibility* — High
11. *Economic yields (q / ha)* — 100-150 (green)
12. *Economic value (Rs. / ha)* — It is a very valuable crop when grown as fodder, but it is damaging to crops / places, when grows as weed
13. *By-product (If any)* — Nil
14. *Use of by-product* — Nil

366. *CROP* — BARNYARD GRASS / Samak / Sanwak / *Sada shyama*

Scientific Name* — *Echinochloa hispidula Retz

Family — Gramineae

Herb / Shrub / Tree — Herb (50-80 cm height)

Annual / Biennial / Perennial — Annual

Used as — Fodder / food; weed in rice fields

1. *Climatic requirements* — Warm season
2. *Soil requirements* — Clayey, loamy, alluvial soils
3. *Required land situations* — Low / medium land
4. *Importance of Crop*— Plant like *shyamaghash*
5. *Fertilizer management* — 60 kg N, 30 kg P_2O_5 and 30 kg K_2O/ha, when grown as fodder / food
6. *Water management* — 3-4 irrigations
7. *Duration of crop* — 120-14 days
8. *Parts used* — Whole plant for fodder
9. *Where grown in India* — Maharashtra, West Bengal, M.P., Karnataka, Assam, Bihar, Orissa, T.N., A.P., etc.
10. *Export possibility* — High
11. *Economic yields (q / ha)* — 100-150 (green)
12. *Economic value (Rs. / ha)* — It is a valuable plant when grown as fodder / food crop but as weed, it is damaging to crops
13. *By-product (If any)* — Nil
14. *Use of by-product* — Nil

Part 2 :
PLANTATION CROPS

Chapter - 5

FRUIT CROPS

INTRODUCTION

Fruits are actually the reproductive portions of the plants. These fruits are with different tastes and are grown in different seasons either in plains or in hills, having different mineral contents. The fruit maintains the requirements of minerals, vitamins, carbohydrate, fats, protein etc. in the human body and as a result the man becomes healthy, showing resistance to diseases. It also increases the inner strength of the human body.

Every man and woman should get some fruits in his/her every-day-diet, to get rid of some diseases and also for maintaining proper health, in terms of minerals and others.

The man should consume the seasonal fruits, for meeting with the body requirements in those particular seasons. It is the gift of nature. Some informations on these fruit crops are stated as follows :

367. *CROP*—COCONUT / Nariyal / Nirikel

Scientific Name — Cocos nucifera L.

Family — Palmae

Herb / Shrub / Tree — Tall palm tree

Annual / Biennial / Perennial — Perennial

Used as — Food / fruit / drink / oil / fibre / fuel / timber

1. *Climatic requirements* — Warm climate
2. *Soil requirements* — Alluvial, loamy, sandy soils with slight salinity
3. *Required land situations* — High / medium lands
4. *Importance of crop* — 'Miracle Crop' for all purposes
5. *Fertilizer management* — 200 kg N, 100 kg P_2O_5 and 100 kg K_2O/ha annually

6. *Water management* — 3-4 irrigation annually
7. *Duration of crop* — 10-15 years
8. *Parts used* — Nut / trunk / leave / shell
9. *Where grown in India* — Kerala, T.N., Karnataka, W. Bengal, Orissa, A.P., Maharashtra etc.
10. *Export possibility* — High
11. *Economic yields (q / ha)* — 180-220 (green) annually
12. *Economic value (Rs. / ha)* — 80,000 to 1,20,000 annually
13. *By-product (If any)* — Trunk / shell / leaves
14. *Use of by-product* — Timber / mattress / utensil / broom / cattle feed / organic matter / fuel / mulch / roofing materials for *boroj*.

368. ***CROP* — MANGO / Am**

Scientific Name* — *Mangifera indica L.

Family — Anacardiaceae

Herb / Shrub / Tree — Tree

Annual / Biennial / Perennial — Perennial

Used as — Fruit / fuel / timber

1. *Climatic requirements* — Moderate warm climate
2. *Soil requirements* — Alluvial, laterite, black cotton and loamy soils
3. *Required land situations* — High / medium land
4. *Importance of crop* — Strong smell while flowering
5. *Fertilizer management* — 200 kg N, 100 kg P_2O_5 and 50 kg K_2O/ha annually
6. *Water management* — 4-5 irrigations annually
7. *Duration of crop* — 10-25 years
8. *Parts used* — Fruits
9. *Where grown in India* — U.P., Punjab, Maharashtra, A.P., W. Bengal, T.N., Karnataka, Bihar, Orissa etc.
10. *Export possibility* — Very high
11. *Economic yields (q / ha)* — 100-200 (green) annually (average)
12. *Economic value (Rs. / ha)* — 70,000 to 90,000 annually
13. *By-product (If any)* — Trunk / leaves
14. *Use of by-product* — Wood for fuel / furniture / fodder

369. ***CROP* — GUAVA / AMRUD / JAMPHAL / *Peyara***

Scientific Name* — *Psidium guajava L.

Family — Myrtaceae

Herb / Shrub / Tree — Tree

Annual / Biennial / Perennial — Perennial

Used as — Fruit / agricultural implements from branches

1. *Climatic requirements* — Moderate temperature with moderate rainfall
2. *Soil requirements* — Loamy, sandy and alluvial soils
3. *Required land situations* — High / medium lands
4. *Importance of crop* — Pectin is present in it and it is good for gelly preparation
5. *Fertilizer management* — 120 kg N2, 60 kg P_2O_5 and 60 kg K_2O/ha annually
6. *Water management* — 1-2 times irrigations, if irrigated
7. *Duration of crop* — 5-10 years
8. *Parts used* — Fruit / branches / leaves
9. *Where grown in India* — Maharashtra U.P., Bihar, Punjab, W.B., A.P., Orissa, Karnataka etc.
10. *Export possibility* — High
11. *Economic yields (q / ha)* — 200-300 (green) annually
12. *Economic value (Rs. / ha)* — 60,000 to 80,000
13. *By-product (If any)* — Stems and leaves
14. *Use of by-product* — Fuel / tanning and dyeing / stem made into spear / handle / leaves for cattle feeds.

370. *CROP* — JACK FRUIT / Kathal / Kanthal / Ichar / Jack tree / Panas

Scientific Name — Artocarpus heterophyllus Lamk

Family — Moraceae

Herb / Shrub / Tree — Tree

Annual / Biennial / Perennial — Perennial

Used as — Fruits / leaf as fodder / timber

1. *Climatic requirements* — High rainfall with moderate temperature
2. *Soil requirements* — Loamy / alluvial / sandy soils
3. *Required land situations* — High
4. *Importance of crop* — Fruit tubes come from the stems
5. *Fertilizer management* — 120 kg N, 60 kg P_2O_5 and 60 kg K_2O/ha annually
6. *Water management* — Normally rainfed; 1-2 times irrigations annually, if irrigated

7. *Duration of crop* — 10-20 years
8. *Parts used* — Fruits
9. *Where grown in India* — Assam, Tripura, West Bengal, Bihar, Orissa, Karnataka, and Maharashtra
10. *Export possibility* — High
11. *Economic yields (q / ha)* — 230-260 (green) annually
12. *Economic value (Rs. / ha)* — 60,000 to 80,000
13. *By-product (If any)* — Stem and leaves
14. *Use of by-product* — Stem for furniture / implements and leaves for goat-cattle feeds / organic matter / fuel

371. *CROP*—SAPODILLA / Sapota / Chiku / *Sapeda*

Scientific Name* — *Manilkara achras (Mill) Fosberg.

Family — Sapotaceae

Herb / Shrub / Tree — Tree (ever green)

Annual / Biennial / Perennial — Perennial

Used as — Fruit / fuel

1. *Climatic requirements* — Moderate temperature with moderate rain fall
2. *Soil requirements* — Loamy / clayey / sandy / alluvial soils
3. *Required land situations* — High land
4. *Importance of crop* — Round brownish / yellowish fruits, sweet when ripe. Chicle – milky juice, is used in manufacture of chewing gum
5. *Fertilizer management* — 120 kg N, 60 kg P_2O_5 and 60 kg K_2O/ha annually
6. *Water management* — Normally rainfed, 1-2 times irrigation, if irrigated
7. *Duration of crop* — 10-15 years
8. *Parts used* — Fruits
9. *Where grown in India* — Gujarat, Maharashtra, T.N., A.P., Karnataka, West Bengal, Bihar, Orissa, U.P. etc.
10. *Export possibility* — High
11. *Economic yields (q / ha)* — 100-150 (green) annually
12. *Economic value (Rs. / ha)* — 30,000 to 60,000
13. *By-product (If any)* — Stem and leaves
14. *Use of by-product* — Stem as fuel / furniture and leaves as cattle feed / organic matter

372. *CROP* — PAPAYA / PAPAW / *Papita / Penpe*)

Scientific Name* — *Carica papaya L.

Family — Caricaceae

Herb / Shrub / Tree — Shrub

Annual / Biennial / Perennial — Perennial

Used as — Fruit (both green and ripe) / vegetable

1. *Climatic requirements* — Moderate temperature with moderate rainfall
2. *Soil requirements* — Loamy / alluvial / sandy soils
3. *Required land situations* — High lands
4. *Importance of crop* — Latex comes from green fruits / leaves
5. *Fertilizer management* — 100 kg N, 50 kg P_2O_5 and 50 kg K_2O/ha annually
6. *Water management* — 5-6 times irrigations annually
7. *Duration of crop* — 2-3 years
8. *Parts used* — Fruits and 'Papain'-proteolytic enzyme present in latex
9. *Where grown in India* — Assam, W.B., Bihar, U.P. and Gujarat
10. *Export possibility* — High (particularly green fruit)
11. *Economic yields (q / ha)* — 260-280 (green) annually
12. *Economic value (Rs. / ha)* — 50,000 to 80,000
13. *By-product (If any)* — Soft stem and leaves
14. *Use of by-product* — Organic matter / fuel / cattle feed / roofing

373. *CROP* — PINEAPPLE / Ananas /*Anaras*

Scientific Name* — *Ananas comosus (L.) Merr.

Family — Bromeliaceae

Herb / Shrub / Tree — Herb

Annual / Biennial / Perennial — Biennial

Used as — Fruits / leaves (pointed) for fibre

1. *Climatic Requirements*— High rainfall with moderate temperature
2. *Soil requirements* — Loamy / laterite / alluvial soils
3. *Required land situations* — High and medium land
4. *Importance of crop* — Leaves and fruits are spiny; fruits first green then yellow
5. *Fertilizer management* — 120 kg N, 60 kg P_2O_5 and 60 kg K_2O/ha annually

6. *Water management* — 5-6 irrigations annually
7. *Duration of crop* — 2 years
8. *Parts used* — Fruits and leaves
9. *Where grown in India* — West Bengal, T.N., A.P. (Coastal), Assam, Tripura, Kerala, Karnataka, Orissa, etc.
10. *Export possibility* — Very High
11. *Economic yields (q / ha)* — 180-190 (green)
12. *Economic value (Rs. / ha)* — 60,000 to 80,000
13. *By-product (If any)* — Stalk (sometimes leaves)
14. *Use of by-product* — Strong, white, silky fibre for pina cloth / strong cords / fishing nets and organic matter / fuel

374. *CROP* — APPLE / Seb / Apel

Scientific Name* — *Malus pumila Mill

Family — Rosaceae

Herb / Shrub / Tree — Tree

Annual / Biennial / Perennial — Perennial

Used as — Fruit / fuel

1. *Climatic requirements* — Low temperature with less rainfall in high altitude
2. *Soil requirements* — Stony / gravel / loamy soils
3. *Required land situations* — High land
4. *Importance of crop* — Fruits with no stones
5. *Fertilizer management* — 100 kg N, 50 kg P_2O_5 and 50 kg K_2O/ha annually
6. *Water management* — Normally rainfed; 1-2 times irrigations, if irrigated
7. *Duration of crop* — 10-15 years
8. *Parts used* — Fruits
9. *Where grown in India* — H.P., Kashmir, Kulu, Kumayun, Assam, Nilgiri's, U.P. etc.
10. *Export possibility* — High
11. *Economic yields (q / ha)* — 140-190 (green) annually
12. *Economic value (Rs. / ha)* — 80,000 to 1,80,000
13. *By-product (If any)* — Stem and leaves
14. *Use of by-product* — Leaves : cattle feeds / organic matter / fuel and different uses in the house.

375. *CROP* — ARECANUT / Betelnut palm / *Supari* / *Arekanut*

Scientific Name* — *Areca catechu L.

Family — Palmae

Herb / Shrub / Tree — Palm (erect)

Annual / Biennial / Perennial — Perennial

Used as — Fruit along with paan / beverage / medicine

1. *Climatic requirements* — Moderately warm climate
2. *Soil requirements* — Loamy and alluvial soils
3. *Required land situations* — High land
4. *Importance of crop* — Small nut fruits in cluster, masticated and chewed with '*paan*'
5. *Fertilizer management* — 120 kg N, 60 kg P_2O_5 and 60 kg K_2O/ha annually
6. *Water management* — 4-5 irrigations annually
7. *Duration of crop* — 10-12 years
8. *Parts used* — Nuts
9. *Where grown in India* — Karnataka, Kerala, T.N., West Bengal, Assam, Maharashtra, Orissa, Bihar etc.
10. *Export possibility* — High
11. *Economic yields (q / ha)* — 80-90 (with husk)
12. *Economic value (Rs. / ha)* — 80,000 to 90,000
13. *By-product (If any)* — Stem / leaves / husk
14. *Use of by-product* — Stem for pillars / leaves for covering *boroj* / husk for medicine

376. *CROP* — BANANA / Kela / Plantain / Kayla / *Kala* / *Astarambha*

Scientific Name* — *Musa paradisiaca (L.) / ***M. sapientum*** L.

Family — Musaceae

Herb / Shrub / Tree — Shrub (with pseudostem)

Annual / Biennial / Perennial — Perennial

Used as — Fruit / medicine / vegetable

1. *Climatic requirements* — Warm climate
2. *Soil requirements* — Alluvial / loamy / clays soils
3. *Required land situations* — High / medium lands
4. *Importance of crop* — Green banana also eaten as vegetable
5. *Fertilizer management* — 120 kg N, 60 kg P_2O_5 and 60 kg K_2O/ha annually

6. *Water management* — 5-6 irrigations annually
7. *Duration of crop* — 2-3 years
8. *Parts used* — Fruits / leaves / pseudostem
9. *Where grown in India* — Assam, W. Bengal, Orissa, Bihar, M.P., A.P., Kerala, Karnataka, Maharashtra, T.N. etc.
10. *Export possibility* — High
11. *Economic yields (q / ha)* — 170-180 (green) annually
12. *Economic value (Rs. / ha)* — 90,000 to 1,20,000
13. *By-product (If any)* — Leaves / stem
14. *Use of by-product* — Leaf : Plates / inner portion for medicine / cattle feed/fuel

377. *CROP* — LIME / Nimbu / Lebu / Kaghzi nimbu / (*Kagaji lebu*)

Scientific Name — Citrus aurantifolia (Christm) Swingle

Family — Rutaceae

Herb / Shrub / Tree — Tree / shrub

Annual / Biennial / Perennial — Perennial

Used as — Fruits / medicine / industrial uses

1. *Climatic requirements* — Warm cool climate with rainfall
2. *Soil requirements* — Alluvial / loamy / sandy and clays soil
3. *Required land situations* — High / medium lands
4. *Importance of crop* — Some smell from fruits / leaves of thorny plants
5. *Fertilizer management* — 100 kg N, 50 kg P_2O_5 and 50 kg K_2O/ha annually
6. *Water management* — 3-4 irrigation annually
7. *Duration of crop* — 5-7 years
8. *Parts used* — Fruits
9. *Where grown in India* — Assam, West Bengal, Orissa, Maharashtra (Nagpur), A.P., Karnataka, Bihar, U.P., Punjab, M.P. etc.
10. *Export possibility* — Less / High / very high
11. *Economic yields (q / ha)* — 100-150 (green)
12. *Economic value (Rs. / ha)* — 60,000 to 80,000
13. *By-product (If any)* — Stem / leaves
14. *Use of by-product* — Stem : Fuel / vining; Leaves : Oil extraction .

278. *CROP* — LEMON / Baranimbu / Pahari nimbu / Pat nimbu / Hill lemon / *Pati lebu*

Scientific Name* — *Citrus limon (L) Burm.f.

Family — Rutaceae

Herb / Shrub / Tree — Tree

Annual / Biennial / Perennial — Perennial

Used as — Food / medicine (stomachic / carminative) / Industries (lemonades / squashes etc.)

1. *Climatic requirements* — Warm climate with high rainfall
2. *Soil requirements* — Loamy, clays or alluvial soils
3. *Required land situations* — High / medium lands
4. *Importance of crop* — Heavy thorns are on the stems
5. *Fertilizer management* — 100 kg N, 50 kg P_2O_5 and 50 kg K_2O/ha
6. *Water management* — 3-4 irrigations
7. *Duration of crop* — 3-4 years
8. *Parts used* — Fruits (for using with major food)
9. *Where grown in India* — Punjab, Assam, W. Bengal, U.P., Maharashtra, Orissa, T.N. etc.
10. *Export possibility* — High
11. *Economic yields (q / ha)* — 60-80 (green)
12. *Economic value (Rs. / ha)* — 60,000 to 80,000
13. *By-product (If any)* — Branches / leaves
14. *Use of by-product* — Fuel / lemon decoction / organic matter

379. *CROP* — MANDARIN / Tangerine / Santra / Nagpur orange (Khasi orange / Coorg orange) / Desi (Punjab), Sikkim orange and Darjeeling orange

Scientific Name* — *Citrus reticulata Blanco

Family — Rutaceae

Herb / Shrub / Tree — Tree (small)

Annual / Biennial / Perennial — Perennial

Used as — Food / fruit / industries (confectioneries)

1. *Climatic requirements* — Warm cool climate with less rainfall
2. *Soil requirements* — Loamy / alluvial / sandy soils
3. *Required land situations* — High / medium land
4. *Importance of crop* — Peel (coating) is very thin and used in *marmalades*

5. *Fertilizer management* — 100 kg N, 50 kg P_2O_5 and 50 kg K_2O/ha annually
6. *Water management* — 3-4 irrigations annually
7. *Duration of crop* — 5-10 years
8. *Parts used* — Fruits
9. *Where grown in India* — Assam, Sikkim, Maharashtra, W.B., Karnataka, Punjab, Kashmir, etc.
10. *Export possibility* — High
11. *Economic yields (q / ha)* — 160-180 (green)
12. *Economic value (Rs. / ha)* — 60,000 to 80,000
13. *By-product (If any)* — Branches and leaves
14. *Use of by-product* — Fuel / crop trailing / organic matter / furniture/ stick

380. *CROP* — SWEET ORANGE / Sweet lime / Mitha Nimbu / *Misti Kamala / Kamala*

Scientific Name* — *Citrus aurantifolia var. ***limetta*** (Christm)

Family — Rutaceae

Herb / Shrub / Tree — Shrub

Annual / Biennial / Perennial — Perennial

Used as — Food / fruit / confectionery

1. *Climatic requirements* — Warm climate with less rainfall
2. *Soil requirements* — Loamy, stony, sandy, alluvial soils
3. *Required land situations* — High / medium land
4. *Importance of crop* — The fruit is sweet. Essential oil – 'Petitgrain oil' is obtained
5. *Fertilizer management* — 100 kg N, 50 kg P_2O_5 and 50 kg K_2O/ha annually
6. *Water management* — 3-4 irrigations annually
7. *Duration of crop* — 3-4 years
8. *Parts used* — Fruits (also for root stock)
9. *Where Grown in India*—U.P., Sikkim, Assam, W.B., M.P. etc.
10. *Export possibility* — Very high
11. *Economic yields (q / ha)* — 60-80 (Green)
12. *Economic value (Rs. / ha)* — 60,000 to 80,000
13. *By-product (If any)* — Branches and leaves
14. *Use of by-product* — Fuel / organic matter / for utilizing vine crops/ leaves for essential oil.

381. *CROP* — TANGERINE / Santara

Scientific Name* — *Citrus reticulata var. ***nobilis*** Blanco

Family — Rutaceae

Herb / Shrub / Tree — Tree (small)

Annual / Biennial / Perennial — Perennial

Used as — Food / fruit / industrial use (confectionery)

1. *Climatic requirements* — Moderate warm climate with less rainfall
2. *Soil requirements* — Loamy, alluvial and sandy soils
3. *Required land situations* — High / medium land
4. *Importance of crop* — Edible fruits
5. *Fertilizer management* — 100 kg N, 50 kg P_2O_5 and 50 kg K_2O/ha annually
6. *Water management* — 4-5 irrigations annually
7. *Duration of crop* — 5-10 years
8. *Parts used* — Fruits
9. *Where grown in India* — Maharashtra, West Bengal, Kashmir, Sikkim, Assam, Karnataka, Punjab, etc.
10. *Export possibility* — High
11. *Economic yields (q / ha)* — 160-180 (green)
12. *Economic value (Rs. / ha)* — 60,000 to 60,000
13. *By-product (If any)* — Branches and leaves
14. *Use of by-product* — Fuel and organic matter / cattle feed

382. *CROP* — GRAPE/ Wine gape / *Angur*

Scientific Name* — *Vitis vinifera L.

Family — Vitaceae

Herb / Shrub / Tree — A woody but shrubby vine obtained

Annual / Biennial / Perennial — Perennial

Used as — Fruit / industrial uses (Brandy is distilled from fermented fruit juice)

1. *Climatic requirements* — Warm climate with less rainfall
2. *Soil requirements* — Loamy / late rite / sandy / alluvial soils
3. *Required land situations* — High / medium land
4. *Importance of crop* — A branch of fruits on the trailed vine obtained
5. *Fertilizer management* — 100 kg N, 60 kg P_2O_5 and 50 kg K_2O/ha annually

6. *Water management* — 5-6 irrigations annually
7. *Duration of crop* — 5-10 years
8. *Parts used* — Fruits
9. *Where grown in India* — Maharashtra, A.P., Karnataka, Punjab, Rajasthan, Uttar Pradesh, Uttaranchal, Tamil Nadu, etc.
10. *Export possibility* — Very high
11. *Economic yields (q / ha)* — 160-180 (green)
12. *Economic value (Rs. / ha)* — 60,000 to 80,000
13. *By-product (If any)* — Vine and leaves
14. *Use of by-product* — Fuel / organic matter / cattle feed

383. *CROP* — POMEGRANATE / Anar / *Dalim / Bedana*

Scientific Name* — *Punica granatum L.

Family — Punicaceae

Herb / Shrub / Tree — Shrub / tree

Annual / Biennial / Perennial — Perennial

Used as — Fruits / medicine

1. *Climatic requirements* — Warm and dry climate
2. *Soil requirements* — Loamy / sandy / alluvial / clays soils
3. *Required land situations* — High / medium land
4. *Importance of crop* — Seeds are edible; round big and red fruits obtained
5. *Fertilizer management* — 60 kg N, 40 kg P_2O_5 and 30 kg K_2O/ha annually
6. *Water management* — 3-4 irrigations annually
7. *Duration of crop* — 3-4 years
8. *Parts used* — Fruits (Peel from fruits is used as medicine)
9. *Where grown in India* — All over India
10. *Export possibility* — High
11. *Economic yields (q / ha)* — 140-180 (green)
12. *Economic value (Rs. / ha)* — 50,000 to 80,000
13. *By-product (If any)* — Branches and leaves
14. *Use of by-product* — Fuel / cattle feed / pegs (sticks) / organic matter

384. *CROP* — AVOCADO / Alligator pear / *Naspati*

Scientific Name* — *Persea americana Mill

Family — Lauraceae

***Herb / Shrub / Tree* —** Tree (small)

***Annual / Biennial / Perennial* —** Perennial

***Used as* —** Fruit

1. *Climatic requirements* — Moderately warm climate
2. *Soil requirements* — Loamy, alluvial, stony and laterite soils
3. *Required land situations* — High (Hills) land
4. *Importance of crop* — Small fruits
5. *Fertilizer management* — 100 kg N, 50 kg P_2O_5 and 30 kg K_2O/ha annually
6. *Water management* — 3-4 irrigations annually
7. *Duration of crop* — 4-5 years
8. *Parts used* — Fruits
9. *Where grown in India* — Bangalore, Palni Hill, Nandi Hill, Nilgiris etc.
10. *Export possibility* — High
11. *Economic yields (q / ha)* — 160-180 (green)
12. *Economic value (Rs. / ha)* — 60,000 to 80,000
13. *By-product (If any)* — Branches and leaves
14. *Use of by-product* — Fuel / cattle feeds / organic matter

385. *CROP* — CASHEWNUT / Kaju / Kajubadam

Scientific Name* — *Anacardium occidentale L

***Family* —** Anacardiaceae

***Herb / Shrub / Tree* —** Tree (small)

***Annual / Biennial / Perennial* —** Perennial

***Used as* —** Fruits / industrial use (confectionary)

1. *Climatic requirements* — Warm climate of coastal (saline) areas
2. *Soil requirements* — Sandy, loamy, alluvial and saline soils
3. *Required land situations* — High / medium land
4. *Importance of crop* — 'Cashew apple' (Fleshy swollen pedicel) present in cashew fruits
5. *Fertilizer management* — 120 kg N, 50 kg P_2O_5 and 50 kg K_2O/ha annually
6. *Water management* — 4-5 irrigations, annually
7. *Duration of crop* — 10-15 years
8. *Parts used* — Seeds
9. *Where grown in India* — Kerala, Karnataka, T.N., A.P., Goa, Orissa and West Bengal

10. *Export possibility* — Very high
11. *Economic yields (q / ha)* — 10-20 (dry) annually
12. *Economic value (Rs. / ha)* — 80,000 to 1,00,000 (or more)
13. *By-product (If any)* — Trees / branch / leaves
14. *Use of by-product* — Timber / fuel / cattle feed / organic matter

386. *CROP* — LOQUAT / Japan medlar / Lokat

Scientific Name* — *Eriobotrya japonica (Thunb) Lind L.

Family — Rosaceae

Herb / Shrub / Tree — Tree (small)

Annual / Biennial / Perennial — Perennial (ever green)

Used as — Fruits / ornamental

1. *Climatic requirements* — Moderately warm climate
2. *Soil requirements* — Loamy, stony and alluvial soils
3. *Required land situations* — High
4. *Importance of crop* — Yellow orange fruits
5. *Fertilizer management* — 60 kg N, 40 kg P_2O_5 and 30 kg K_2O/ha annually
6. *Water management* — 3-4 irrigations annually
7. *Duration of crop* — 3-4 years
8. *Parts used* — Fruits
9. *Where grown in India* — Dehradun, Muzaffarnagar, Meerut, Kanpur, Bareilly (U.P.) Amritsar, Gurudaspur, Hoshiarpur (Punjab), Assam and Maharashtra
10. *Export possibility* — High
11. *Economic yields (q / ha)* — 140-160 (green)
12. *Economic value (Rs. / ha)* — 40,000 to 60,000
13. *By-product (If any)* — Branches and leaves
14. *Use of by-product* — Fuel / cattle feed / organic matter

387. *CROP* — MANGOSTEEN / Mangustan / *Vilaitigab*

Scientific Name* — *Garcinia mangostana L.

Family — Guttiferae

Herb / Shrub / Tree — Tree

Annual / Biennial /Perennial — Perennial

Used as — Fruits / medicine (diarrhoea / dysentery)

1. *Climatic requirements* — Warm dry climate

2. *Soil requirements* — Loamy, sandy and alluvial soils
3. *Required land situations* — High land
4. *Importance of crop* — Ever fresh with red fruits and hairs on them
5. *Fertilizer management* — 100 kg N, 50 kg P_2O_5 and 50 kg K_2O/ha annually
6. *Water management* — 3-4 irrigations
7. *Duration of crop* — 5-10 years
8. *Parts used* — Fruits
9. *Where grown in India* — Lower hills of Nilgiri, T.N. and other areas in a small amount
10. *Export possibility* — High
11. *Economic yields (q / ha)* — 160-180 (green)
12. *Economic value (Rs. / ha)* — 50,000 to 90,000
13. *By-product (If any)* — Branches / trunk / leaves
14. *Use of by-product* — Timber / fuel / cattle feed / organic matter

388. *CROP* — LITCHI / Leechee / Lichi / *Lichu*

Scientific Name* — *Litchi chinensis (Gaertn) Sonn

Family — Sapindaceae

Herb / Shrub / Tree — Tree (small)

Annual / Biennial / Perennial — Perennial

Used as — Fruit / fuel

1. *Climatic requirements* — Warm and humid climate
2. *Soil requirements* — Loamy and sandy loam soils
3. *Required land situations* — High land
4. *Importance of crop* — Fleshy aril (just below peel but before stone) is white, for eating)
5. *Fertilizer management* — 100 kg N, 50 kg P_2O_5 and 30 kg K_2O/ha annually
6. *Water management* — 4-5 irrigations annually
7. *Duration of crop* — 5-10 years
8. *Parts used* — Fruits
9. *Where grown in India* — Bihar, Uttar Pradesh, West Bengal, H.P. and Karnataka
10. *Export possibility* — High
11. *Economic yields (q / ha)* — 80-100 (green) annually
12. *Economic value (Rs. / ha)* — 60,000 to 80,000

13. *By-product (If any)* — Branches and leaves
14. *Use of by-product* — Fuel / cattle feed / organic matter / timber

389. *CROP* — DATE PALM / Pind *khajur* / *Khejur*

Scientific Name — Phoenix dactylifera L.

Family — Palmae

Herb / Shrub / Tree — Tree (tall, but single stem)

Annual / Biennial / Perennial — Perennial

Used as — Fruits / fibre (from leaves)

1. *Climatic requirements* — Hot dry climate
2. *Soil requirements* — Sandy and sandy loam soils
3. *Required land situations* — High land
4. *Importance of crop* — A tall tree having spiny pointed big leaves (juice obtained from stem) grown in dry condition
5. *Fertilizer management* — 50 kg N, 30 kg P_2O_5 and 30 kg K_2O/ha annually
6. *Water management* — 2-3 irrigations annually
7. *Duration of crop* — 10-15 years
8. *Parts used* — Fruits / Leaves
9. *Where grown in India* — Punjab, Rajasthan, Gujarat, West Bengal, Madhya Pradesh, Haryana etc.
10. *Export possibility* — High
11. *Economic yields (q / ha)* — 20-30
12. *Economic value (Rs. / ha)* — 20,000 to 30,000
13. *By-product (If any)* — Trunk / leaves
14. *Use of by-product* — Pillars / ropes / stick / fuel / organic matter

390. *CROP* — PHALSA / Phalsai / *Falsha*

Scientific Name — Grewla subinaequalis DC

Family — Tiliaceae

Herb / Shrub / Tree — Tree (small)

Annual / Biennial / Perennial — Perennial

Used as — Fruit

1. *Climatic requirements* — Warm climate
2. *Soil requirements* — Loamy, sandy loam, alluvial and sandy soils
3. *Required land situations* — High land
4. *Importance of crop* — Spreading crown

5. *Fertilizer management* — 50 kg N, 30 kg P_2O_5 and 30 kg K_2O/ha annually
6. *Water management* — 2-3 irrigations annually
7. *Duration of crop* — 5-7 years
8. *Parts used* — Fruits
9. *Where grown in India* — All over India except too low / too high temperature areas
10. *Export possibility* — High
11. *Economic yields (q / ha)* — 80-160 (green)
12. *Economic value (Rs. / ha)* — 40,000 to 50,000
13. *By-product (If any)* — Branches / leaves
14. *Use of by-product* — Wood (branches) used for golf shafts / cattle feed / organic matter / fuel.

391. *CROP* — WOODAPPLE / Elephant-apple / Bale / *Kath Bel* / Kavitha / Kut bel

Scientific Name — Limonia acidissima L.

Family — Rutaceae

Herb / Shrub / Tree — Tree

Annual / Biennial / Perennial — Perennial

Used as — Fruit / fuel

1. *Climatic requirements* — Warm climate
2. *Soil requirements* — Loamy, sandy loam and alluvial soils
3. *Required land situations* — High land
4. *Importance of crop* — Spines on the branches / Smell on fruits
5. *Fertilizer Management*— 80 kg N, 50 kg P_2O_5 and 30 kg K_2O/ha annually
6. *Water management* — 4-5 irrigations annually
7. *Duration of crop* — 10-15 years
8. *Parts used* — Fruits
9. *Where grown in India* — All over India
10. *Export possibility* — High
11. *Economic yields (q / ha)* — 30-50 (ripen green)
12. *Economic value (Rs. / ha)* — 20,000 to 30,000
13. *By-product (If any)* — Wood (branches) and leaves
14. *Use of by-product* — Fuel (branches) and leaves (used in puja of Hindu family) stick / organic matter

392. ***CROP* — BULLOCK'S HEART / Common custard-apple / Ramphal / Lavnianona / *Nona***

Scientific Name — Annona reticulata L.

Family — Annonaceae

Herb / Shrub / Tree —Shrub / tree (small)

Annual / Biennial / Perennial — Perennial

Used as — Fruits / insecticides

1. *Climatic requirements* — Warm humid climate
2. *Soil requirements* — Loamy and sandy loam soils
3. *Required land situations* — High lands
4. *Importance of crop* — Peel wavy in nature
5. *Fertilizer management* — 50 kg N, 30 kg P_2O_5 and 30 kg K_2O/ha annually
6. *Water management* — 2-3 irrigations annually
7. *Duration of crop* — 2-4 years
8. *Parts used* — Fruits
9. *Where grown in India* — W. Bengal, Assam, Karnaaka, A.P., Tamil Nadu and Kerala etc.
10. *Export possibility* — High
11. *Economic yields (q / ha)* — 50-60 (green)
12. *Economic value (Rs. / ha)* — 20,000 to 30,000
13. *By-product (If any)* — Branches / leaves
14. *Use of by-product* — Fuel / cattle feed / organic matter

393. ***CROP* — EMBLIC / Aonla / Amla / *Amlaki* / Indian goosebery**

Scientific Name — Emblica officinalis Gaertn

Family — Euphorbiaceae

Herb / Shrub / Tree — Tree (deciduous)

Annual / Biennial / Perennial — Perennial

Used as — Fruit / medicinal (antibiotic effects)

1. *Climatic requirements* — Warm climate with dry conditions
2. *Soil requirements* — Loamy and sandy loam soils
3. *Required land situations* — High lands
4. *Importance of crop* — Moderate sized deciduous plant with fine small leaves
5. *Fertilizer management* — 50 kg N, 30 kg P_2O_5 and 30 kg K_2O/ha annually
6. *Water management* — 3-4 irrigations annually

7. *Duration of crop* — 10-15 years
8. *Parts used* — Fruits
9. *Where grown in India* — All over India except too low / too high temperature areas
10. *Export possibility* — High
11. *Economic yields (q / ha)* — 40-50 (green)
12. *Economic value (Rs. / ha)* — 80,000 to 90,000 annually
13. *By-product (If any)* — Branches / trunk / leaves
14. *Use of by-product* — Fuel / tanning / *hookah* (produced from this wood)

394. *CROP* — FIG / Anjir / *Damur* / Dumur

Scientific Name* — *Ficus carica L.

Family — Moraceae

Herb / Shrub / Tree — Tree (small)

Annual / Biennial / Perennial — Perennial

Used as — Fruits / medicine (warts)

1. *Climatic requirements* — Warm climate
2. *Soil requirements* — Loamy and sandy loam soils
3. *Required land situations* — High / Medium / low lands
4. *Importance of crop* — Latex (fruits come from stems) is present
5. *Fertilizer Management*— 50 kg N, 30 kg P_2O_5 and 30 kg K_2O/ha annually
6. *Water management* — 2-3 irrigations annually
7. *Duration of crop* — 5-7 years
8. *Parts used* — Fruit
9. *Where grown in India* — Maharashtra, W.B., U.P., Rajasthan, Punjab, A.P., Bihar, Orissa etc.
10. *Export possibility* — High
11. *Economic yields (q / ha)* — 10-30
12. *Economic value (Rs. / ha)* — 10,000 to 30,000
13. *By-product (If any)* — Trunk / branches / leaves
14. *Use of by-product*— Fuel / cattle feed / organic matter / pegs.

395. *CROP* — CHINESE DATE / Ber / JUJUBE / Vadara / Kul / Badara / Karkandhu Kuvala / Pemdiber

Scientific Name* — *Ziziphus mauritiana Lamk

Family — Rhamnaceae

Herb / Shrub / Tree — Tree (small)

Annual / Biennial / Perennial — Perennial

Used as — Fruits / fuel

1. *Climatic requirements* — Warm climate
2. *Soil requirements* — Loamy, sandy loam and stony soils
3. *Required land situations* — High / medium lands
4. *Importance of crop* — Spiny branched tree
5. *Fertilizer Management*— 80 kg N, 50 kg P_2O_5 and 30 kg K_2O/ha annually
6. *Water management* — 2-3 irrigations annually
7. *Duration of crop* — 10-15 years
8. *Parts used* — Fruits / leaves / branches
9. *Where grown in India* — Punjab, Bihar, Rajasthan, W. Bengal, Orissa, Assam, A.P., M.P. etc.
10. *Export possibility* — High
11. *Economic yields (q / ha)* — 15-100 (green)
12. *Economic value (Rs. / ha)* — 10,000 to 40,000
13. *By-product (If any)* — Trunk / leaves
14. *Use of by-product* — Fuel / fodder (tree) / pegs / timber / organic matter

396. *CROP* — APRICOT / Khubani / Zardalu

Scientific Name* — *Prunus armeniaca L.

Family — Rosaceae

Herb / Shrub / Tree — Tree

Annual / Biennial / Perennial — Perennial

Used as — Fruits / Industry (hair oil)

1. *Climatic requirements* — Cool climate
2. *Soil requirements* — Stony, loamy and sandy loam soils
3. *Required land situations* — High land
4. *Importance of crop* — Sweet fruits
5. *Fertilizer Management*— 50 kg N, 30 kg P_2O_5 and 30 kg K_2O/ha annually
6. *Water management* — 2-3 irrigations annually
7. *Duration of crop* — 2-4 years
8. *Parts used* — Fruit
9. *Where grown in India* — Kashmir, H.P., U.P. and Kumayun

10. *Export possibility* — High
11. *Economic yields (q / ha)* — 40-80 (green)
12. *Economic value (Rs. / ha)* — 30,000 to 80,000
13. *By-product (If any)* — Trunk / branch / leaves
14. *Use of by-product* — Fuel / cattle feed / organic matter

397. *CROP* — CHERRY / Sour Cherry / Alubalu / Dwarf cherry / *Cheri / Gilas*

Scientific Name* — *Prunus cerasus L.

Family — Rosaceae

Herb / Shrub / Tree — Tree / shrub

Annual / Biennial / Perennial — Perennial

Used as — Fruits / confectionery

1. *Climatic requirements* — Cool climate
2. *Soil requirements* — Loamy, stony and sandy loam soils
3. *Required land situations* — High land
4. *Importance of crop* — Small red fruits for cakes etc.
5. *Fertilizer Management*— 50 kg N, 50 kg P_2O_5 and 50 kg K_2O/ha annually
6. *Water management* — 2-3 irrigations annually
7. *Duration of crop* — 3-4 years
8. *Parts used* — Fruits
9. *Where grown in India* — In the hilly regions (Kashmir, Himachal Pradesh and Uttarakhand (Kumayun)
10. *Export possibility* — High
11. *Economic yields (q / ha)* — 40-80 (green)
12. *Economic value (Rs. / ha)* — 40,000 to 80,000
13. *By-product (If any)* — Branches and leaves
14. *Use of by-product* — Musical instruments / cattle feed / organic matter / pegs / fuel

398. *CROP* — PEACH / Aru / Pis

Scientific Name* — *Prunus persica (L) Batsch

Family — Rosaceae

Herb / Shrub / Tree — Tree

Annual / Biennial / Perennial — Perennial

Used as — Fruit / Oil (from seed)

1. *Climatic requirements* — Cool climate in hills

2. *Soil requirements* — Stony, loamy and sandy loam soils
3. *Required land situations* — High land
4. *Importance of crop* — Pitcher like fruits
5. *Fertilizer Management*— 80 kg N, 50 kg P_2O_5 and 50 kg K_2O/ha annually
6. *Water management* — 4-5 irrigations annually
7. *Duration of crop* — 10-15 years
8. *Parts used* — Fruits
9. *Where grown in India* — In the hilly regions (Kashmir, Himachal Pradesh and Uttarakhand (Kumayun)
10. *Export possibility* — High
11. *Economic yields (q / ha)* — 40-80 (green)
12. *Economic value (Rs. / ha)* — 50,000 to 60,000
13. *By-product (If any)* — Trunk / branches / leaves
14. *Use of by-product* — Fuel / cattle feed / organic matter / timber

399. *CROP* — PEAR / Nakh / *Naspati*

Scientific Name* — *Pyrus communis L.

Family — Rosaceae

Herb / Shrub / Tree — Tree

Annual / Biennial / Perennial — Perennial

Used as — Fruits / fuel

1. *Climatic requirements* — Cool climate
2. *Soil requirements* — Loamy and sandy loam soils
3. *Required land situations* — High lands
4. *Importance of crop* — Fruit pitcher like structure
5. *Fertilizer Management*— 80 kg N, 50 kg P_2O_5 and 50 kg K_2O/ha annually
6. *Water management* — 3-4 irrigations annually
7. *Duration of crop* — 10-15 years
8. *Parts used* — Fruits
9. *Where grown in India* — In hilly regions (Kashmir, Himachal Pradesh and U.P. (Kumayun)
10. *Export possibility* — High
11. *Economic yields (q / ha)* — 40-80 (green)
12. *Economic value (Rs. / ha)* — 40,000 to 50,000
13. *By-product (If any)* — Branches / leaves

14. *Use of by-product* — Fuel / cattle feed / organic matter

400. *CROP* — PLUM / BULACE / Alucha / *Alubukhara / Alubakra*

Scientific Name* — *Prunus domestica L.

Family — Rosaceae

Herb / Shrub / Tree — Tree

Annual / Biennial / Perennial — Perennial

Used as — Fuits

1. *Climatic requirements* — Cool climate
2. *Soil requirements* — Loamy and sandy loam soils
3. *Required land situations* — High
4. *Importance of crop* — Used in preparation of pickle (*chatni*)
5. *Fertilizer Management*— 80 kg N, 50 kg P_2O_5 and 30 kg K_2O/ha annually
6. *Water management* — 2-3 irrigations annually
7. *Duration of crop* — 5-7 years
8. *Parts used* — Fruit
9. *Where grown in India* — All high hilly regions (Kashmir, Himachal Pradesh and U.P.
10. *Export possibility* — High
11. *Economic yields (q / ha)* — 40-80 (green)
12. *Economic value (Rs. / ha)* — 30,000 to 50,000
13. *By-product (If any)* — Branches and leaves
14. *Use of by-product* — Fuel and cattle feed / organic matter

401. *CROP* — STRAW BERRY / Strawberi / Yellow straw berry / *Kiphaliya*

Scientific Name* — *Fragaria indica Andr. / ***Duchesnea indica*** (Andr.) Focke

Family — Rosaceae

Herb / Shrub / Tree — Herb

Annual / Biennial / Perennial — Perennial

Used as — Fruits

1. *Climatic requirements* — Cool climate
2. *Soil requirements* — Loamy and sandy loam soils
3. *Required land situations* — High land
4. *Importance of crop* — Small beautiful berry
5. *Fertilizer Management*—100 kg N, 50 kg P_2O_5 and 30 kg K_2O/ha annually.

6. *Water management* — 4-5 irrigations annually
7. *Duration of crop* — 2-3 years
8. *Parts used* — Fruits
9. *Where grown in India* — All hilly areas (Punjab, Himachal Pradesh, Assam, Nilgiri etc.)
10. *Export possibility* — High
11. *Economic yields (q / ha)* — 40-80 (green)
12. *Economic value (Rs. / ha)* — 50,000 to 80,000
13. *By-product (If any)* — Vine and leaves
14. *Use of by-product* — Fuel / cattle feed / organic matter

402. *CROP* — JALPAI / Jalpoi / Jalpay

Scientific Name* — *Elaeocarpus floribundus Blume

Family — Elaeocarpaceae

Herb / Shrub / Tree — Tree

Annual / Biennial / Perennial — Perennial

Used as — Fruits (sour)

1. *Climatic requirements* — Warm humid climate
2. *Soil requirements* — Loamy, clays and sandy loam soils
3. *Required land situations* — High / Medium lands
4. *Importance of Crop*— Fruits are used as 'sour pickle'
5. *Fertilizer Management*—100 kg N, 50 kg P_2O_5 and 30 kg K_2O/ha annually
6. *Water management* — 2-3 irrigations annually
7. *Duration of crop* — 10-15 years
8. *Parts used* — Fruits
9. *Where grown in India* — Assam, West Bengal, Orissa etc.
10. *Export possibility* — High
11. *Economic yields (q / ha)* — 40-80 (green)
12. *Economic value (Rs. / ha)* — 20,000 to 40,000
13. *By-product (If any)* — Trunk / branch / leaves
14. *Use of by-product* — Fuel / cattle feed / organic matter / pegs

403. *CROP* — JAVAPLUM/Jamun/Jambolana/Jambu/ *Jam / Jam bava*

Scientific Name* — *Syzygium cumini (L) Skeels

Family — Myrtaceae

Herb / Shrub / Tree — Tree

Annual / Biennial / Perennial — Perennial

Used as — Fruits / medicine (seeds reduce blood sugar of diabetic patients)

1. *Climatic requirements* — Warm climate
2. *Soil requirements* — Loamy, sandy loam and sandy soils
3. *Required land situations* — High / medium
4. *Importance of Crop*— Large tree with black / pink fruits. Seeds are good for diabetic patients
5. *Fertilizer Management*—100 kg N, 50 kg P_2O_5 and 30 kg K_2O/ha annually
6. *Water management* — 2-3 irrigations annually
7. *Duration of crop* — 20-20 years
8. *Parts used* — Fruits / Seeds
9. *Where grown in India* — All over India
10. *Export possibility* — High
11. *Economic yields (q / ha)* — 50-100 (green)
12. *Economic value (Rs. / ha)* — 40,000 to 50,000
13. *By-product (If any)* — Trunk / branches / leaves
14. *Use of by-product* — Furniture / fuel / cattle feed / organic matter / pegs

404. *CROP* — TAMARIND / Tamarindo / Imli / *Tentul*

Scientific Name* — *Tamarindus indica L.

Family — Caesalpiniaceae

Herb / Shrub / Tree — Tree (big)

Annual / Biennial / Perennial — Perennial

Used as — Fruits / avenue / plant / medicine (poultice) / industrial uses

1. *Climatic requirements* — Warm climate conditions
2. *Soil requirements* — Loamy, sandy loam and sandy soils
3. *Required land situations* — High land
4. *Importance of Crop*— Fruits are sour, used in food / pickle / curry
5. *Fertilizer Management*—100 kg N, 50 kg P_2O_5 and 30 kg K_2O/ha annually
6. *Water management* — 2-3 irrigations annually
7. *Duration of crop* — 10-20 years
8. *Parts used* — Fruits / trunk / leaves
9. *Where grown in India* — Plains all over India

10. *Export possibility* — High
11. *Economic yields (q / ha)* — 40-80 (green)
12. *Economic value (Rs. / ha)* — 40,000 to 60,000
13. *By-product (If any)* — Trunk / branch / leaves
14. *Use of by-product* — Furniture / fuel / cattle feed / organic matter / drinks

405. *CROP* — JAMRUL / Jamrool / Lal jamrul / Watery rose apple

Scientific Name* — *Eugenia aquea Burm f. / ***E. formosa*** Wall

Family — Myrtaceae

Herb / Shrub / Tree — Tree

Annual / Biennial / Perennial — Perennial

Used as — Fruits

1. *Climatic requirements* — Warm humid climate
2. *Soil requirements* — Loamy, sandy loam and sandy soils
3. *Required land situations* — High land
4. *Importance of Crop*— Fleshy watery fruits
5. *Fertilizer Management*—80 kg N, 50 kg P_2O_5 and 30 kg K_2O/ha annually
6. *Water management* — 2-3 irrigations annually
7. *Duration of crop* — 10-15 years
8. *Parts used* — Fruits
9. *Where grown in India* — Assam, West Bengal, Orissa etc.
10. *Export possibility* — High
11. *Economic yields (q / ha)* — 50-100 (green)
12. *Economic value (Rs. / ha)* — 30,000 to 40,000
13. *By-product (If any)* — Trunk / branch / leaves
14. *Use of by-product* — Furniture / fuel / cattle feed / organic matter

406. *CROP* — OLEA / *Kau* / Indian olive

Scientific Name* — *Olea ferruginea Royle

Family — Oleaceae

Herb / Shrub / Tree — Tree

Annual / Biennial / Perennial — Perennial

Used as — Fruits / timber / medicine (gonorrhea)

1. *Climatic requirements* — Moderate warm climate
2. *Soil requirements* — Alluvial, sandy loam and loamy soils

3. *Required land situations* — High / medium
4. *Importance of Crop*— Sour fruit size like orange
5. *Fertilizer Management*—80 kg N, 50 kg P_2O_5 and 30 kg K_2O/ha annually
6. *Water management* — 2-3 irrigations annually
7. *Duration of crop* — 5-10 years
8. *Parts used* — Fruits / Timber
9. *Where grown in India* — Assam, W. Bengal, Orissa etc.
10. *Export possibility* — High
11. *Economic yields (q / ha)* — 40-80 (green)
12. *Economic value (Rs. / ha)* — 10,000 to 40,000
13. *By-product (If any)* — Trunk / branches / leaves
14. *Use of by-product* — Agricultural implements / fuel / cattle feed / pegs / furniture

407. *CROP* — NASPATI / CHINESE PEAR / Sand pear / Chinese pear

Scientific Name* — *Pyrus pyrifolia (Burm.f.) Nakai

Family — Rosaceae

Herb / Shrub / Tree — Tree

Annual / Biennial / Perennial — Perennial

Used as — Fruits

1. *Climatic requirements* — Cool climate
2. *Soil requirements* — Loamy, sandy loam and alluvial soils
3. *Required land situations* — High land
4. *Importance of Crop*— Hard fruits
5. *Fertilizer Management*—80 kg N, 50 kg P_2O_5 and 30 kg K_2O/ha annually
6. *Water management* — 2-3 irrigations annually
7. *Duration of crop* — 5-10 years
8. *Parts used* — Fruits
9. *Where grown in India* — Hilly regions like Darjeeling (W.B.), Assam, Kashmir, Himachal Pradesh, U.P. and Tamil Nadu
10. *Export possibility* — High
11. *Economic yields (q / ha)* — 40-80 (green)
12. *Economic value (Rs. / ha)* — 50,000 to 1,00,000
13. *By-product (If any)* — Trunk / branch / leaves
14. *Use of by-product* — Timber / fuel / cattle feed / organic matter.

Chapter - 6

MEDICINAL PLANTS

INTRODUCTION

The plants, which give some relief to the disease problems of the human beings, or, to other pet animals, are called the medicinal plants.

The medicinal plants may be perennial or annual to any season, either in plains or hills. Some of them may be used on a mass scale, through factories or may be used by some people locally. The cinchona (*Cinchona calisaya*) is a big perennial tree, grown in hilly and cool region, is used for manufacture of 'quinine' in the treatment of malaria disease of human beings, whereas, kalmegh (*Andrographis paniculate*), the small (40-60 cm height) plant is used for the treatment of worms, fevers and dysentery in rural areas, in the form of 'alui' (a home-made small ball like structure from the leaves).

The salient points regarding the identifications, management and crop economics of some medicinal plants are reported as follows :-

408. *CROP*—LEMON GRASS / Sugandh rohisha / *Aghyaghas / Sugandh ghas*

Scientific Name* — *Cymbopogon citratus (DC) Stapf

Family — Gramineae

Herb / Shrub / Tree — Herb (grass)

Annual / Biennial / Perennial — Perennial

Used as — Medicinal purposes (aromatic oil)

1. *Climatic requirements* — Moderate warm climate
2. *Soil requirements* — Alluvial, sandy loam and loamy soils
3. *Required land situations* — High / medium land
4. *Importance of crop* — Aromatic oil obtained from it

5. *Fertilizer management* — 100 kg N, 50 kg P_2O_5 and 30 kg K_2O/ha annually
6. *Water management* — 3-4 irrigation annually
7. *Duration of crop* — 2-3 years
8. *Parts used* — Whole plant
9. *Where grown in India* — Maharashtra, Karnataka, Kerala, West Bengal, U.P., Bihar, Orissa etc.
10. *Export possibility* — Very High
11. *Economic yields (q / ha)* — 5-10 (oil)
12. *Economic value (Rs. / ha)* — 60,000 to 80,000 annually
13. *By-product (If any)* — Extracted grass residues
14. *Use of by-product* — Printing paper / writing paper / cattle feed / fuel

409. *CROP* — CITRONELLA GRASS / Ganjini / *Citronella ghash*

Scientific Name — *Cymbopogon nardus* Rendle

Family — Gramineae

Herb / Shrub / Tree — Herb (grass)

Annual / Biennial / Perennial — Perennial

Used as — Medicine / perfume (mosquito repellant)

1. *Climatic requirements* — Warm climate
2. *Soil requirements* — Loamy, sandy loam and alluvial soils
3. *Required land situations* — High / medium lands
4. *Importance of crop* — Robust aromatic grass
5. *Fertilizer management* — 100 kg N, 50 kg P_2O_5 and 30 kg K_2O/ha annually
6. *Water management* — 3-4 irrigations, annually
7. *Duration of crop* — 2-3 years
8. *Parts used* — Entire grass plant
9. *Where grown in India* — Maharashtra, A.P., W.B., Karnataka, Kerala, U.P., Bihar, Uttaranchal etc.
10. *Export possibility* — High
11. *Economic yields (q / ha)* — 5-10 (oil)
12. *Economic value (Rs. / ha)* — 60,000 to 1,20,000
13. *By-product (If any)* — Extracted grass refuse
14. *Use of by-product* — Cattle feed / writing pad / printing materials / fuel / organic matter.

410. *CROP* — ANISE / Saonf / Saurif / Joan

Scientific Name — Pimpinella anisum L.

Family — Umbelliferae

Herb / Shrub / Tree — Herb

Annual / Biennial / Perennial — Annual

Used as — Medicine (carminative) / perfumery

1. *Climatic requirements* — Moderate cool climate
2. *Soil requirements* — Loamy, sandy loam and alluvial soils
3. *Required land situations* — High / medium lands
4. *Importance of crop* — Aromatic plant
5. *Fertilizer management* — 80 kg N2, 50 kg P_2O_5 and 30 kg K_2O/ha
6. *Water management* — 3-4 irrigations
7. *Duration of crop* — 140-150 days
8. *Parts used* — Whole plant
9. *Where grown in India* — U.P., Punjab, Assam, W. Bengal, Orissa etc.
10. *Export possibility* — High
11. *Economic yields (q / ha)* — 50-100 (green)
12. *Economic value (Rs. / ha)* — 15,000 to 25,000
13. *By-product (If any)* — Plant extracted residues
14. *Use of by-product* — Cattle feed / printing paper / fuel

411. *CROP* — FOX GLOVE / Digitalis

Scientific Name — Digitalis purpurea L.

Family — Scorphulariaceae

Herb / Shrub / Tree — Herb

Annual / Biennial / Perennial — Annual

Used as — Medicine (for weak-heart ointment, for burn and wound)

1. *Climatic requirements* — Cool climate
2. *Soil requirements* — Loamy, sandy loam and sandy soils
3. *Required land situations* — High land
4. *Importance of crop* — Therapeutic activity of dried leaves
5. *Fertilizer management* — 80 kg N, 50 kg P_2O_5 and 30 kg K_2O/ha
6. *Water management* — 2-3 irrigations
7. *Duration of crop* — 140-160 days
8. *Parts used* — Entire plant

9. *Where grown in India* — Hilly regions of North India
10. *Export possibility* — High
11. *Economic yields (q / ha)* — 50-100 (green)
12. *Economic value (Rs. / ha)* — 20,000 to 40,000
13. *By-product (If any)* — Refuse materials after extractions
14. *Use of by-product* — Cattle feed / fuel / paper /organic matter

412. *CROP*—LIQUORICE / Mulhatti / Licorice

Scientific Name — Glycyrrhiza glabra L

Family — Papilionaceae

Herb / Shrub / Tree — Herb (sweet)

Annual / Biennial / Perennial — Perennial

Used as — Medicine (gastric ulcer) / confectionery (root extraction)

1. *Climatic requirements* — Moderate cool climate
2. *Soil requirements* — Loamy, sandy loam and sandy soils
3. *Required land situations* — High land
4. *Importance of crop* — Fifty times sweeter than sugar
5. *Fertilizer management* — 40 kg N, 60 kg P_2O_5 and 30 kg K_2O/ha annually
6. *Water management* — 2-3 irrigations annually
7. *Duration of crop* — 2-3 years
8. *Parts used* — Whole plant (including roots)
9. *Where grown in India* — Punjab, J. & K., Maharashtra, T.N., A.P., Karnataka, Kerala etc.
10. *Export possibility* — High
11. *Economic yields (q / ha)* — 100-150 (green)
12. *Economic value (Rs. / ha)* — 50,000 to 80,000
13. *By-product (If any)* — Extracted refused materials
14. *Use of by-product* — Fuel / cattle feed / organic matter / paper board

413. *CROP* — ROSHA GRASS / PALMAROSA (oil grass) / Gandhejghas / Gandhaghas

Scientific Name — Cympopogon martinii (Roxb) Wats

Family — Gramineae

Herb / Shrub / Tree — Herb (grass)

Annual / Biennial / Perennial — Perennial

Used as — Medicine (oil for skin disease) / confectionery / perfume

1. *Climatic requirements* — Warm climate
2. *Soil requirements* — Loamy, sandy loam and sandy soils
3. *Required land situations* — High land
4. *Importance of crop* — Sweet scented grass
5. *Fertilizer management* — 80 kg N, 50 kg P_2O_5 and 30 kg K_2O/ha annually
6. *Water management* — 3-4 irrigations annually
7. *Duration of crop* — 2-3 years
8. *Parts used* — Whole plant
9. *Where grown in India* — Maharashtra, A.P., Karnataka, Madhya Pradesh and Uttar Pradesh etc.
10. *Export possibility* — Very high
11. *Economic yields (q / ha)* — 150-200 (green)
12. *Economic value (Rs. / ha)* — 80,000 to 1,00,000
13. *By-product (If any)* — Materials after extraction
14. *Use of by-product* — Fuel / organic matter / cattle feed / boards

414. *CROP* — JAPANESE MINT / *Mentha* / Peppermint / *Piparmint*

Scientific Name* — *Mentha arvensis DC. var. ***piperascens*** Holmes.

Family — Labiatae

Herb / Shrub / Tree — Herb

Annual / Biennial / Perennial — Perennial

Used as — Medicine (cold)

1. *Climatic requirements* — Cool climate
2. *Soil requirements* — Loamy, sandy loam and alluvial soils
3. *Required land situations* — High land
4. *Importance of crop* — Menthol-an essential oil from it and this is used in cold
5. *Fertilizer management* — 80 kg N, 50 kg P_2O_5 and 30 kg K_2O/ha annually
6. *Water management* — 2-3 irrigations annually
7. *Duration of crop* — 2-3 years
8. *Parts used* — Leaves
9. *Where grown in India* — Hilly regions particularly in J.& K.
10. *Export possibility* — High
11. *Economic yields (q / ha)* — 10-12 (green)
12. *Economic value (Rs. / ha)* — 50,000 to 80,000

13. *By-product (If any)* — Extract refuse materials
14. *Use of by-product* — Cattle feed / fuel / organic matter

415. CROP — SPEAR MINT / Pahari pudina / Pahare pudina

Scientific Name — Mentha spicata (L) Huds

Family — Labiatae

Herb / Shrub / Tree — Herb

Annual / Biennial / Perennial — Perennial

Used as — Medicine (cough) / Industrial use (flavouring food product)

1. *Climatic requirements* — Moderate warm climate
2. *Soil requirements* — Loamy, sandy loam and alluvial soils
3. *Required land situations* — High / medium lands
4. *Importance of crop* — Leaves are the source of spear mint oil
5. *Fertilizer management* — 80 kg N, 50 kg P_2O_5 and 30 kg K_2O/ha annually
6. *Water management* — 2-3 irrigations annually
7. *Duration of crop* — 2-3 years
8. *Parts used* — Leaves
9. *Where grown in India* — Punjab, U.P., Maharashtra, Karnataka etc.
10. *Export possibility* — High
11. *Economic yields (q / ha)* — 10-15 (green)
12. *Economic value (Rs. / ha)* — 60,000 to 80,000
13. *By-product (If any)* — After extraction of leaves, materials
14. *Use of by-product* — Cattle feed / fuel / organic matter.

416. *CROP* — LEMON MINT / FIELD MINT / Pudina / *Pudin*

Scientific Name — Mentha arvensis L.

Family — Labiatae

Herb / Shrub / Tree — Herb

Annual / Biennial / Perennial — Perennial

Used as — Medicine (carminative / stimulant)

1. *Climatic requirements* — Cool climate
2. *Soil requirements* — Loamy, alluvial and sandy soils
3. *Required land situations* — High land
4. *Importance of crop* — Presence of essential oils needed for cigarettes

5. *Fertilizer management* — 80 kg N, 50 kg P_2O_5 and 30 kg K_2O/ha, annually
6. *Water management* — 3-4 irrigations annually
7. *Duration of crop* — 2-3 years
8. *Parts used* — Leaves
9. *Where grown in India* — Western Himalayas, J.& K., Punjab, Uttaranchal etc.
10. *Export possibility* — High
11. *Economic yields (q / ha)* — 10-15 (green)
12. *Economic value (Rs. / ha)* — 60,000 to 80,000
13. *By-product (If any)* — Materials after leaves extraction
14. *Use of by-product* — Cattle feed / fuel / organic matter

417. CROP — KHUS / vetiver / Khas-khas /Gandar / Khas

Scientific Name — Vetiveria zizaniodes (L) Nash

Family — Gramineae

Herb / Shrub / Tree — Herb

Annual / Biennial / Perennial — Perennial

Used as — Medicine (stimulant) / Industrial uses

1. *Climatic requirements* — Warm climate
2. *Soil requirements* — Loamy, sandy and sandy loam soils
3. *Required land situations* — High land
4. *Importance of crop* — Dried roots are used to make aromatic scented mats / fans / ornamentals
5. *Fertilizer management* — 100 kg N, 50 kg P_2O_5 and 30 kg K_2O/ha annually
6. *Water management* — 4-5 irrigations annually
7. *Duration of crop* — 2-3 years
8. *Parts used* — Whole plant
9. *Where grown in India* — Rajasthan, U.P., Punjab, West Coast
10. *Export possibility* — High
11. *Economic yields (q / ha)* — 100-150 (green)
12. *Economic value (Rs. / ha)* — 60,000 to 80,000
13. *By-product (If any)* — Nil
14. *Use of by-product* — Nil

418. *CROP* — HENNA / *Mehndi* /Egyptian privet / *Mehendi*

Scientific Name — Lawsonia inermis L.

Family — Lythraceae

Herb / Shrub / Tree — Shrub / tree

Annual / Biennial / Perennial — Perennial

Used as — Medicine (antibiotic for boils) / Industrial use (for staining skin).

1. *Climatic requirements* — Warm climate
2. *Soil requirements* — Sandy loam, sandy and loamy soils
3. *Required land situations* — High land
4. *Importance of crop* — Much branched small tree
5. *Fertilizer management* — 80 kg N, 50 kg P_2O_5 and 50 kg K_2O/ha annually
6. *Water management* — 3-4 irrigations annually
7. *Duration of crop* — 4-5 years
8. *Parts used* — Leaves / flower
9. *Where grown in India* — Gujarat, Punjab, Haryana, M.P. and Rajasthan
10. *Export possibility* — High
11. *Economic yields (q / ha)* — 10-15 (green)
12. *Economic value (Rs. / ha)* — 70,000 to 90,000
13. *By-product (If any)* — Stem / used leaves
14. *Use of by-product* — Fuel / Fuel / cattle feed / organic matter

419. *CROP* — SWEET FLAG / Safed Bach / Ghorabach / *Sadakachu*

Scientific Name* — *Acorus calamus L.

Family — Araceae

Herb / Shrub / Tree — Herb

Annual / Biennial / Perennial — Annual

Used as — Medicine (carminative / stimulant of nervous system / insecticides)

1. *Climatic requirements* — Moderate cool climate
2. *Soil requirements* — Loamy, sandy loam and sandy soils
3. *Required land situations* — High land
4. *Importance of crop* — Much branched aromatic rhizomes, grow in marshy lands
5. *Fertilizer management* — 80 kg N, 50 kg P_2O_5 and 30 kg K_2O/ha
6. *Water Management*— No need for irrigation (give when necessary)
7. *Duration of crop* — 130-150 days

8. *Parts used* — Rhizomes / leaves
9. *Where grown in India* — J.&K., Manipur, Karnataka and other places
10. *Export possibility* — High
11. *Economic yields (q / ha)* — 150-200 (green)
12. *Economic value (Rs. / ha)* — 50,000 to 70,000
13. *By-product (If any)* — Petiole
14. *Use of by-product* — Fuel / Organic matter / vegetable

420. *CROP* — PEPPERMINT / Vilaiti Pudina / *Barapudina*

Scientific Name* — *Mentha piperita L.

Family — Labiatae

Herb / Shrub / Tree — Herb

Annual / Biennial / Perennial — Perennial

Used as — Medicine (Vomiting / headache) / flavouring purpose

1. *Climatic requirements* — Cool climate
2. *Soil requirements* — Loamy and sandy loam soils
3. *Required land situations* — High lands
4. *Importance of crop* — Leaves used for flavouring curries
5. *Fertilizer management* — 80 kg N, 50 kg P_2O_5 and 30 kg K_2O/ha annually
6. *Water management* — 3-4 irrigations annually
7. *Duration of crop* — 2-4 years
8. *Parts used* — Leaves
9. *Where grown in India* — Maharashtra, J. & K. Punjab etc.
10. *Export possibility* — High
11. *Economic yields (q / ha)* — 10-15 (green)
12. *Economic value (Rs. / ha)* — 60,000 to 80,000
13. *By-product (If any)* — Used leaves / branches
14. *Use of by-product* — Cattle feed / organic matter / fuel

421. *CROP* — ARJUN / ARJUNA / Terminalia / Maruthu / Kahu

Scientific Name* — *Terminalia arjuna (Roxb) Wight & Arn.

Family — Combretaceae

Herb / Shrub / Tree — Tree

Annual / Biennial / Perennial — Perennial

Used as — Medicine (dry powder of bark is for hypertension / cirrhosis of liver)

1. *Climatic requirements* — Moderate warm climate
2. *Soil requirements* — Loamy, sandy loam and sandy soils
3. *Required land situations* — High land
4. *Importance of crop* — Bark is the source of medicine. It is the avenue tree also
5. *Fertilizer management* — 100 kg N, 50 kg P_2O_5 and 30 kg K_2O/ha annually
6. *Water management* — 5-6 irrigations annually
7. *Duration of crop* — 10-15 years
8. *Parts used* — Bark / leaves
9. *Where grown in India* — All over India except too low / too high temperature areas
10. *Export possibility* — High
11. *Economic yields (q / ha)* — 5-10 (Green)
12. *Economic value (Rs. / ha)* — 60,000 to 80,000
13. *By-product (If any)* — Trunk / branches / leaves
14. *Use of by-product* — Timber / fuel / cattle feed / organic matter

422. *CROP* — QUININE / Cinchona / Kuinine

Scientific Name* — *Cinchona calisaya Wedd

Family — Rubiaceae

Herb / Shrub / Tree — Tree

Annual / Biennial / Perennial — Perennial

Used as — Medicine [material for quinine (fcr malaria) from this plant]

1. *Climatic requirements* — Cool climate
2. *Soil requirements* — Loamy and sandy loam soils
3. *Required land situations* — High land
4. *Importance of crop* — Bark of the plant is the source of quinine
5. *Fertilizer management* — 100 kg N, 50 kg P_2O_5 and 50 kg K_2O/ha annually
6. *Water management* — 4-5 irrigations annually
7. *Duration of crop* — 10-15 years
8. *Parts used* — Bark
9. *Where grown in India* — Hills of India particularly Nilgiris / Sikkim / Darjeeling (W.B.), J. & K. etc.
10. *Export possibility* — High
11. *Economic yields (q / ha)* — 5-10 (green)

12. *Economic value (Rs. / ha)* — 60,000 to 80,000
13. *By-product (If any)* — Trunk / branch / leaves
14. *Use of by-product* — Timber / fuel / cattle feed / organic matter

423. *CROP* — CAMPHOR / Muskapur / *Karpur*

Scientific Name* — *Cinnamomum camphora (L) Nees & Eberm

Family — Lauraceae

Herb / Shrub / Tree — Tree

Annual / Biennial / Perennial — Perennial

Used as — Medicine (for inflammation / pains / sprains)

1. *Climatic requirements* — Moderate cool temperature
2. *Soil requirements* — Loamy and sandy soils
3. *Required land situations* — High lands
4. *Importance of crop* — 'Safrole' high value constituent, used in preparation of perfumes
5. *Fertilizer management* — 60 kg N, 50 kg P_2O_5 and 30 kg K_2O/ha annually
6. *Water management* — 2-3 irrigations annually
7. *Duration of crop* — 3-5 years
8. *Parts used* — Wood (for camphor production)
9. *Where grown in India* — Hilly regions (Ootacamund etc.) and also in plains of India
10. *Export possibility* — Very high
11. *Economic yields (q / ha)* — 5-6 (oil)
12. *Economic value (Rs. / ha)* — 60,000 to 80,000
13. *By-product (If any)* — Used wood / branch / leaves
14. *Use of by-product* — Fuel / cattle feed / organic matter

424. *CROP* — HARJORI / FALSE PAREIRA ROOT / Akanadi / *Harjora*/ Patat ki bel

Scientific Name* — *Cissampelos pareira L.

Family — Menispermaceae

Herb / Shrub / Tree — Shrub (climbing)

Annual / Biennial / Perennial — Perennial

Used as — Medicine (colic pains / urinary troubles diarrhorea)

1. *Climatic requirements* — Warm climate
2. *Soil requirements* — Loamy and sandy loam soils
3. *Required land situations* — High land

4. *Importance of crop* — Twining vine / supported by trees but leafless
5. *Fertilizer management* — 50 kg N, 30 kg P_2O_5 and 30 kg K_2O/ha annually
6. *Water management* — 3-4 irrigations annually
7. *Duration of crop* — 5-6 years
8. *Parts used* — Root / Vine
9. *Where grown in India* — All over India under humid conditions
10. *Export possibility* — High
11. *Economic yields (q / ha)* — Root 5-6 (green) and vine 10-20 (green)
12. *Economic value (Rs. / ha)* — 50,000 to 60,000
13. *By-product (If any)* — Nil
14. *Use of by-product* — Nil

425. *CROP* — CUBAN BAST / YELLOW MALLOW TREE / Mahoe / Chelwa

Scientific Name — Hibiscus tiliaceus L

Family — Malvaceae

Herb / Shrub / Tree — Shrub / tree (small)

Annual / Biennial / Perennial — Perennial

Used as — Medicine (leaves for ulcers / wounds)

1. *Climatic requirements* — Warm climate
2. *Soil requirements* — Loamy, clayey and sandy loam and saline soils
3. *Required land situations* — Medium / low lands
4. *Importance of crop* — Growth in coastal areas is good
5. *Fertilizer management* — 80 kg N, 50 kg P_2O_5 and 30 kg K_2O/ha annually
6. *Water management* — 2-3 irrigations annually
7. *Duration of crop* — 2-3 years
8. *Parts used* — Leaves
9. *Where grown in India* — W. Bengal (Sundarban / Andaman), Assam etc.
10. *Export possibility* — High
11. *Economic yields (q / ha)* — 10-15 (green)
12. *Economic value (Rs. / ha)* — 50,000 to 60,000
13. *By-product (If any)* — Leaves after extraction and branches
14. *Use of by-product* — Cattle feed / fuel / organic matter

426. *CROP* — GUM-ARABIC ACACIA / Kumta / *Babul* / Babool

Scientific Name* — *Acacia senegal (L) Willd

Family — Mimosaceae

Herb / Shrub / Tree — Tree (small)

Annual / Biennial / Perennial — Perennial

Used as — Medicine (demulcent) / confectionery / industrial uses

1. *Climatic requirements* — Hot and warm climate
2. *Soil requirements* — Sandy / sandy loam and loamy soils
3. *Required land situations* — High
4. *Importance of crop* — Branched from near the base with hooked / straight spines
5. *Fertilizer management* — 40 kg N, 30 kg P_2O_5 and 30 kg K_2O/ha annually
6. *Water management* — 2-3 irrigations, annually
7. *Duration of crop* — 10-15 years
8. *Parts used* — Stem (for gum)
9. *Where grown in India* — Punjab, Haryana and Rajasthan, Gujarat, Maharashtra etc.
10. *Export possibility* — High
11. *Economic yields (q / ha)* — 100-200 (green stem)
12. *Economic value (Rs. / ha)* — 50,000 to 60,000
13. *By-product (If any)* — Used stems / branches / leaves
14. *Use of by-product* — Cattle feed / furniture / fuel / organic matter

427. *CROP* — INDIAN ACALYPHA / Khokali / Kuppi

Scientific Name* — *Acalypha indica L.

Family — Euphorbiaceae

Herb / Shrub / Tree — Herb

Annual / Biennial / Perennial — Annual

Used as — Medicine (asthma / pneumonia / bronchitis / rheumatism)

1. *Climatic requirements* — Warm climate
2. *Soil requirements* — Loamy, sandy loam and alluvial soils
3. *Required land situations* — High land
4. *Importance of crop* — Bronchitis / rheumatism cured with this plant
5. *Fertilizer management* — 50 kg N, 30 kg P_2O_5 and 30 kg K_2O/ha
6. *Water management* — 3-4 irrigations

7. *Duration of crop* — 120-160 days
8. *Parts used* — Whole plants
9. *Where grown in India* — All over the country except too low or too high temperature areas
10. *Export possibility* — High
11. *Economic yields (q / ha)* — 80-100 (green)
12. *Economic value (Rs. / ha)* — 40,000 to 50,000
13. *By-product (If any)* — Used material
14. *Use of by-product* — Fuel / cattle feed / organic matter

428. *CROP* — BABUL ACACIA / *babul* / Kikar / Babool

Scientific Name* — *Acacia nilotica (L) Willd

Family — Mimosaceae

Herb / Shrub / Tree — Tree

Annual / Biennial / Perennial — Perennial

Used as — Medicine (demulcent effect) / Industrial uses (gum)

1. *Climatic requirements* — Hot climate
2. *Soil requirements* — Sandy, sand loam, loamy soils
3. *Required land situations* — High land
4. *Importance of crop* — Gum from stem / leaves or seed for fodder
5. *Fertilizer management* — 30 kg N, 30 kg P_2O_5 and 30 kg K_2O/ha annually
6. *Water management* — 2-3 irrigations annually
7. *Duration of crop* — 10-15 years
8. *Parts used* — Stems / leaves and seeds
9. *Where grown in India* — Drier regions of India (Gujarat, M.P. Punjab, Rajasthan, Haryana, etc.)
10. *Export possibility* — High
11. *Economic yields (q / ha)* — 100-150 (green gum)
12. *Economic value (Rs. / ha)* — 40,000 to 60,000
13. *By-product (If any)* — Used materials
14. *Use of by-product* — Fuel / organic matter / cattle feed

429. *CROP* — CINNAMON / Dalchini / *Darchini*

Scientific Name* — *Cinnamomum zeylanicum Blume

Family — Lauraceae

Herb / Shrub / Tree — Tree

Annual / Biennial / Perennial — Perennial

Used as — Spice / condiment / medicine (stimulant / carminative / nausea/ vomiting)

1. *Climatic requirements* — High rainfall areas (Kerala)
2. *Soil requirements* — Loamy and sandy loam soils
3. *Required land situations* — High / medium lands
4. *Importance of crop* — Evergreen tree / delicate fragrance and warm sweet taste
5. *Fertilizer management* — 50 kg N, 30 kg P_2O_5 and 30 kg K_2O/ha annually
6. *Water management* — 4-5 irrigations annually
7. *Duration of crop* — 10-15 years
8. *Parts used* — Bark (dried)
9. *Where grown in India* — Western Ghats, Kerala etc.
10. *Export possibility* — Very High
11. *Economic yields (q / ha)* — 5-10 (dried barks)
12. *Economic value (Rs. / ha)* — 60,000 to 80,000
13. *By-product (If any)* — Used materials / leaves
14. *Use of by-product* — Fuel / organic matter / cattle feed

430. *CROP* — EUCALYPTUS / Lemon scented eucalyptus / *Ecaliptus gandhi*

Scientific Name — Eucalyptus citriodora Hook

Family — Myrtaceae

Herb / Shrub / Tree — Tree

Annual / Biennial / Perennial — Perennial

Used as — Medicine (oil for cough) / wood / perfumery / ornamental

1. *Climatic requirements* — Moderate temperature
2. *Soil requirements* — Loamy, sandy loam and alluvial soils
3. *Required land situations* — High / medium / low lands
4. *Importance of crop* — A tall handsome tree with good odour. Birds avoid this plant
5. *Fertilizer management* — 100 kg N, 50 kg P_2O_5 and 30 kg K_2O/ha annually
6. *Water management* — 2-3 irrigations annually
7. *Duration of crop* — 10-15 years
8. *Parts used* — Wood / branches / leaves

9. *Where grown in India* — Punjab, A.P., West Bengal, Karnataka, Haryana, U.P., etc.
10. *Export possibility* — Very High
11. *Economic yields (q / ha)* — 100-200 (green)
12. *Economic value (Rs. / ha)* — 70,000 to 90,000
13. *By-product (If any)* — Used wood / branch / leaves
14. *Use of by-product* — Fuel / paper / cattle feed / organic matter

431. *CROP* — ASOKA / *Asok* / Ampich / *Asoke*

Scientific Name* — *Saraca asoca (Roxb) de Wilde

Family — Caesalpiniaceae

Herb / Shrub / Tree — Tree

Annual / Biennial / Perennial — Perennial

Used as — Medicine (for excessive menstruation and as a uterine sedative)

1. *Climatic requirements* — Warm climate
2. *Soil requirements* — Loamy, sandy loam and alluvial soils
3. *Required land situations* — High / medium land
4. *Importance of crop* — Garden tree for scarlet orange flower
5. *Fertilizer management* — 80 kg N, 50 kg P_2O_5 and 30 kg K_2O/ha annually
6. *Water management* — 2-3 irrigations annually
7. *Duration of crop* — 5-10 years
8. *Parts used* — Stem bark as medicine
9. *Where grown in India* — West Bengal, Assam, Orissa, Bihar, T.N., Karnataka, A.P., Kerala, etc.
10. *Export possibility* — Very High
11. *Economic yields (q / ha)* — 5-10 (dry bark)
12. *Economic value (Rs. / ha)* — 60,000 to 70,000
13. *By-product (If any)* — Used bark / trunk / leaves
14. *Use of by-product* — Fuel / timber / cattle fed / organic matter

432. *CROP* — *BASAKA* / MALABARNUT / Arusa / Adulasa / *Basak*

Scientific Name* — *Adhatoda vasica Nees

Family — Acanthaceae

Herb / Shrub / Tree — Shrub (3-4 m height)

Annual / Biennial / Perennial — Perennial

Used as — Medicine (syrup / decoction for cough from leaves)

1. *Climatic requirements* — Warm climate
2. *Soil requirements* — Loamy, sandy loam and alluvial soils
3. *Required land situations* — High / medium lands
4. *Importance of crop* — Alkaloid 'vasicine' and essential oil are obtained
5. *Fertilizer Management*— 80 kg N, 50 kg P_2O_5 and 30 kg K_2O/ha annually
6. *Water management* — 2-3 irrigations annually
7. *Duration of crop* — 5-6 years
8. *Parts used* — Leaves
9. *Where grown in India* — All over India (plains submontane regions)
10. *Export possibility* — High
11. *Economic yields (q / ha)* — 5-10 (dry leaves)
12. *Economic value (Rs. / ha)* — 50,000 to 80,000
13. *By-product (If any)* — Branches
14. *Use of by-product* — Fuel / sticks / organic matter

433. *CROP* — KALMEGH / KING-OF-BITTERS / Kiryat / *Kalamegh*

Scientific Name* — *Andrographis paniculata (Burm.f.) Wall.

Family — *Acanthaceae*

Herb / Shrub / Tree —Herb (30-60 cm height)

Annual / Biennial / Perennial — Annual / Perennial

Used as — Medicine (worms / liver / fevers / dysentery), ornamental plant

1. *Climatic requirements* — Warm climate
2. *Soil requirements* — Alluvial, loamy and sandy loam soils
3. *Required land situations* — High / medium lands
4. *Importance of crop* — Leaves are very bitter and 'alui' (ball) is produced
5. *Fertilizer management* — 50 kg N, 30 kg P_2O_5 and 30 kg K_2O/ha
6. *Water management* — 3-4 irrigations
7. *Duration of crop* — 120-160 days
8. *Parts used* — Stems / inflorescence / leaves
9. *Where grown in India* — All over India (mainly in plains)
10. *Export possibility* — Very high
11. *Economic yields (q / ha)* — 20-25 (dry)

12. *Economic value (Rs. / ha)* — 50,000 to 60,000
13. *By-product (If any)* — Nil
14. *Use of by-product* — Nil

Chapter - 7

TIMBERS AND PLANTS FOR FURNITURE

INTRODUCTION

Timbers are the big plants, mostly obtained as trunks / logs mainly in the hilly / forest regions. In some plains also some of the timbers may be obtained. Some plants other than timbers, are mostly used for furniture, can also be obtained either from forests / hilly regions or in the plains. Cane (*Calamus guruba* Buch) is used for preparations of furnitures such as tables, chairs, baskets, sofa sets, cradles etc. from timbers, many furnitures such as : tables, chairs, dinning sets, doors, almirahs, stools, boxes, cots, sofa sets are also prepared, other than poles / railway sleeper etc.

Salient points, for identification, managements and crop economics of some of the timbers and other furniture crops, are also given as follows -

434. *CROP*—TEAK / Sagwan / *Tik* / *Segun*

Scientific Name* — *Tectona grandis Lf

Family — Verbenaceae

Herb / Shrub / Tree — Tree

Annual / Biennial / Perennial — Perennial

Used as — Timber / furniture's / instruments

1. *Climatic requirements* — Warm climate
2. *Soil requirements* — Loamy, sandy loam and alluvial soils
3. *Required land situations* — High l and
4. *Importance of crop* — Tree with big trunk and big leaf
5. *Fertilizer management* — 80 kg N, 50 kg P_2O_5 and 30 kg K_2O/ha
6. *Water management* — 4-5 times irrigations annually

7. *Duration of crop* — 10-15 years
8. *Parts used* — Wood / branches
9. *Where grown in India* — West Peninsular / Central India / Bihar / West Bengal etc.
10. *Export possibility* — Very High
11. *Economic yields (q / ha)* — 100-200 yearly
12. *Economic value (Rs. / ha)* — 11,00,000 to 12,00,000 yearly
13. *By-product (If any)* — Wood waste / leaves
14. *Use of by-product* — Paper / boards / cattle feed / fuel / organic matter

N.B. : Most of the timbers are felled after 10-16 years' of growth, then felling continues yearly, gradually and raw timber is obtained.

435. *CROP — Arjun* / ARJOON / Terminalia / Maruthu / Kahu

Scientific Name — Terminalia arjuna (Roxb) Wight & Arn.

Family — Combretaceae

Herb / Shrub / Tree — Tree

Annual / Biennial / Perennial — Perennial

Used as — Timber / medicine (relieves of hypertension / cirrhosis of liver)

1. *Climatic requirements* — Warm climate
2. *Soil requirements* — Loamy, sandy loam and alluvial soils
3. *Required land situations* — High land
4. *Importance of crop* — Tasar silk worms are reared with the leaves
5. *Fertilizer management* — 80 kg N, 50 kg P_2O_5 and 30 kg K_2O/ha annually
6. *Water management* — 4-5 irrigations, annually
7. *Duration of crop* — 10-15 years
8. *Parts used* — Bark (for dyeing / tanning / medicine) / leaves (for rearing larva of silk worm)
9. *Where grown in India* — All over India, except too cold and too hot areas
10. *Export possibility* — High
11. *Economic yields (q / ha)* — 100-200 (green) yearly
12. *Economic value (Rs. / ha)* — 1,50,000 to 2,60,000 yearly
13. *By-product (If any)* — Trunk / branch / leaves
14. *Use of by-product* — Furniture / fuel / cattle feed / organic matter

436. ***CROP* — TAMARIND / Tamarindo / Imli / *Tentul***

Scientific Name — Tamarindus indica L.

Family — Caesalpiniaceae

Herb / Shrub / Tree — Trees (big)

Annual / Biennial / Perennial — Perennial

Used as — Timber (agricultural implements) / medicine (carminative / luxative) / fruit

1. *Climatic requirements* — Warm weather
2. *Soil requirements* — Loamy, sandy loam and alluvial soils
3. *Required land situations* — High land
4. *Importance of crop* — Seed mass is soury in taste / timber is hard
5. *Fertilizer management* — 50 kg N, 60 kg P_2O_5 and 30 kg K_2O/ha, annually
6. *Water management* — 3-4 irrigations, annually
7. *Duration of crop* — 10-12 years
8. *Parts used* — Seed pulp / leaves / green pod / trunk
9. *Where grown in India* — All over India (plains)
10. *Export possibility* — High
11. *Economic yields (q / ha)* — Timber – 100-200 yearly
12. *Economic value (Rs. / ha)* — 1,50,000 to 2,80,000 yearly
13. *By-product (If any)* — Trunk / leaves
14. *Use of by-product* — Furniture / implements / cattle feed

437. ***CROP* — CHINESE DATE / Jujube / Ber / Badara / *Kul* / Pemdiber / Kharkandhu / Kuvala**

Scientific Name — Ziziphus jujuba (L.) Lam

Family — Rhamnaceae

Herb / Shrub / Tree — Tree

Annual / Biennial / Perennial — Perennial

Used as — Timber / fodder / tanning

1. *Climatic requirements* — Warm climate
2. *Soil requirements* — Loamy, sandy loam and alluvial soils
3. *Required land situations* — High land
4. *Importance of crop* — Spiny branches, can withstand heavy pruning
5. *Fertilizer management* — 80 kg N, 50 kg P_2O_5 and 30 kg K_2O/ha, annually
6. *Water management* — 3-4 irrigations annually

7. *Duration of crop* — 10-15 years
8. *Parts used* — Bark for tanning / leaves for fodder / laxative (for throat troubles) / host for lac insect) / fruits
9. *Where grown in India* — Punjab, Bihar, West Bengal, Orissa, Assam, M.P., A.P., U.P. etc.
10. *Export possibility* — High
11. *Economic yields (q / ha)* — 100-150 (green) yearly
12. *Economic value (Rs. / ha)* — 1,50,000 to 1,60,000 yearly
13. *By-product (If any)* — Trunk / leaves
14. *Use of by-product* — Timber / fuel / cattle feed / organic matter

438. *CROP*— EUCALYPTUS (lemon scented) / *Ucaliptus*

Scientific Name — Eucalyptus citriodora Hook

Family — Myrtaceae

Herb / Shrub / Tree — Tree (long white tall)

Annual / Biennial / Perennial — Perennial

Used as — Timber / oil (for perfumes) / medicine (cough)

1. *Climatic requirements* — Warm climate / cool climate
2. *Soil requirements* — Stony, loamy and sandy loam soils
3. *Required land situations* — High land
4. *Importance of crop* — Tall light silvery white and handsome plant
5. *Fertilizer management* — 80 kg N, 50 kg P_2O_5 and 30 kg K_2O/ha annually
6. *Water management* — 2-3 irrigations annually
7. *Duration of crop* — 10-15 years
8. *Parts used* — Trunk for wood / furniture
9. *Where grown in India* — Punjab, U.P., T.N., A.P., West Bengal etc.
10. *Export possibility* — High
11. *Economic yields (q / ha)* — 100-150 (green) yearly
12. *Economic value (Rs. / ha)* — 1,60,000 to 2,80,000 yearly
13. *By-product (If any)* — Used wood / branches / leaves
14. *Use of by-product* — Paper / board / fuel / cattle feed / organic matter

439. *CROP* — KOKKO / *Siris* / Lebbeck / *Kalosiris*

Scientific Name — Albizia lebbeck (L.) Benth

Family — Mimosaceae

Herb / Shrub / Tree — Tree (spreading type)

Annual / Biennial / Perennial — Perennial

Used as — Timber (for furniture) / Shade tree for tea / coffee

1. *Climatic requirements* — Warm climate (with high rainfall)
2. *Soil requirements* — Loamy, sandy loam and alluvial soils
3. *Required land situations* — High lands
4. *Importance of crop* — Big spreading type tree with big width trunk
5. *Fertilizer management* — 40 kg N, 50 kg P_2O_5 and 50 kg K_2O/ha, annually
6. *Water management* — 2-3 irrigations annually (if needed)
7. *Duration of crop* — 10-15 years
8. *Parts used* — Trunk / branches
9. *Where grown in India* — West Bengal, Andaman, Assam, T.N., M.P., Punjab, U.P., Bihar, Orissa
10. *Export possibility* — High
11. *Economic yields (q / ha)* — 100-200 yearly
12. *Economic value (Rs. / ha)* — 1,80,000 to 2,00,000 yearly
13. *By-product (If any)* — Branches / leaves
14. *Use of by-product* — Fuel / cattle feed / organic matter

440. *CROP* — SISSOO / Shisham / *Sisu*

Scientific Name — Dalbergia sisso Roxb.

Family — Papilionaceae

Herb / Shrub / Tree — Tree

Annual / Biennial / Perennial — Perennial

Used as — Timber (furniture) / wood pulp (for paper)

1. *Climatic requirements* — Warm climate
2. *Soil requirements* — Loamy, sandy loam and alluvial soils
3. *Required land situations* — High land
4. *Importance of crop* — Road side ornamental plant also
5. *Fertilizer management* — 80 kg N, 50 kg P_2O_5 and 30 kg K_2O/ha, annually
6. *Water management* — 3-3 irrigations annually
7. *Duration of crop* — 10-15 years
8. *Parts used* — Trunk / branches
9. *Where grown in India* — In most of the states of India
10. *Export possibility* — High

11. *Economic yields (q / ha)* — 100-150 (Green) yearly
12. *Economic value (Rs. / ha)* — 1,60,000 to 2,80,000 yearly
13. *By-product (If any)* — Branches / leaves
14. *Use of by-product* — Fuel / cattle feed / organic matter

441. *CROP* — CASUARINA / She-oak / Jangli jhau / *Jhau / Janglisaru*

Scientific Name* — *Casuarina equisetifolia JR. & G. Forst.

Family — Casuarinaceae

Herb / Shrub / Tree — Tree

Annual / Biennial / Perennial — Perennial

Used as — Timber / wind break / avenue / sand binding

1. *Climatic requirements* — Warm climate
2. *Soil requirements* — Sandy, sandy loam and loamy soils
3. *Required land situations* — High / medium land
4. *Importance of crop* — A narrow tall evergreen tree
5. *Fertilizer management* — 80 kg N, 50 kg P_2O_5 and 50 kg K_2O/ha annually
6. *Water management* — 2-3 irrigations annually
7. *Duration of crop* — 5-10 years
8. *Parts used* — Trunk
9. *Where grown in India* — Coastal areas (near sea / river) of India
10. *Export possibility* — High
11. *Economic yields (q / ha)* — 100-150 (Green) yearly
12. *Economic value (Rs. / ha)* — 1,50,000 to 2,60,000 yearly
13. *By-product (If any)* — Branches / leaves
14. *Use of by-product* — Fuel / cattle feed / organic matter.

442. *CROP* — NEEM / Margosa / *Nim*

Scientific Name* — *Azadirachta indica A. Juss

Family — Meliaceae

Herb / Shrub / Tree — Tree

Annual / Biennial / Perennial — Perennial

Used as — Timber / medicine / pollution control / ornamental plant / antiseptic / avenue plant

1. *Climatic requirements* — Warm climate
2. *Soil requirements* — Loamy, sandy loam and sandy soils

3. *Required land situations* — High lands
4. *Importance of crop* — Green twigs for tooth brushes / seeds used for fertilizer (for insect repelling) / leaf bitter but germ killer
5. *Fertilizer management* — 80 kg N, 50 kg P_2O_5 and 50 kg K_2O/ha, annually
6. *Water management* — 3-4 irrigations annually
7. *Duration of crop* — 10-15 years
8. *Parts used* — Trunk / branches / leaves
9. *Where grown in India* — All over India (plains)
10. *Export possibility* — High
11. *Economic yields (q / ha)* — 100-250 (green) yearly
12. *Economic value (Rs. / ha)* — 1,60,000 to 2,80,000 yearly
13. *By-product (If any)* — Branch / leaves
14. *Use of by-product* — Fuel / cattle feed / organic matter / medicine

443. *CROP*—MAHOGANI / Mahagani / *Mehagini* / West Indian mahogany

Scientific Name* — *Swietenia mahogoni (L) Jacq

Family — Meliaceae

Herb / Shrub / Tree — Tree

Annual / Biennial / Perennial — Perennial

Used as — Timber (costly) / fuel

1. *Climatic requirements* — Warm climate
2. *Soil requirements* — Loamy, sandy loam, alluvial and saline soils
3. *Required land situations* — High land
4. *Importance of crop* — Good looking, road side tree also
5. *Fertilizer management* — 80 kg N, 50 kg P_2O_5 and 30 kg K_2O/ha annually
6. *Water management* — 2-3 irrigations annually
7. *Duration of crop* — 10-15 years
8. *Parts used* — Trunk / branches
9. *Where grown in India* — All over India (mostly east India)
10. *Export possibility* — High
11. *Economic yields (q / ha)* — 100-300 yearly
12. *Economic value (Rs. / ha)* — 3,60,000 to 4,80,000 yearly
13. *By-product (If any)* — Branches / leaves
14. *Use of by-product* — Fuel / cattle feed / organic matter

444. *CROP* — SUNDRI / *Sundari*

Scientific Name* — *Heritiera fomes Buch-Hum

Family — Sterculiaceae

Herb / Shrub / Tree — Tree

Annual / Biennial / Perennial - - Perennial

Used as — Timber (furniture's) / Industrial use (tan)

1. *Climatic requirements* — Warm saline conditions
2. *Soil requirements* — Clayey and sandy loam soils
3. *Required land situations* — Low / medium lands
4. *Importance of crop* — Wood is hard twice as teak and used in boat / ship buildings / furnitures and it grows in Sundarbans (W.B.).
5. *Fertilizer management* — 80 kg N, 50 kg P_2O_5 and 50 kg K_2O/ha annually
6. *Water management* — 2-3 irrigations annually
7. *Duration of crop* — 10-12 years
8. *Parts used* — Trunk / branches
9. *Where grown in India* — Mostly in Sundarbans (West Bengal)
10. *Export possibility* — High
11. *Economic yields (q / ha)* — 100-150 (Green) yearly
12. *Economic value (Rs. / ha)* — 1,50,000 to 2,50,000 yearly
13. *By-product (If any)* — Branch / leaves
14. *Use of by-product* — Fuel / cattle feed / organic matter

445. *CROP* — BABOOL / Babul / Babul acacia / Babli / *Kikar*

Scientific Name* — *Acacia nilotica (L) Willd ex Del.

Family — Mimosaceae

Herb / Shrub / Tree — Tree

Annual / Biennial / Perennial — Perennial

Used as — Timber / fodder / industrial uses

1. *Climatic requirements* — Hot and warm climate
2. *Soil requirements* — Sandy, sandy loam and loamy soils
3. *Required land situations* — High lands
4. *Importance of crop* — Long serrated pods with small leaves
5. *Fertilizer Management*—40 kg N, 50 kg P_2O_5 and 30 kg K_2O/ha, annually
6. *Water management* — 2-3 irrigations (normally rainfed) annually
7. *Duration of crop* — 10-15 years

8. *Parts used* — Trunk / branches / leaves
9. *Where grown in India* — All drier regions of North, South and Central India
10. *Export possibility* — High
11. *Economic yields (q / ha)* — 80-90 yearly
12. *Economic value (Rs. / ha)* — 1,20,000 to 1,40,000 yearly
13. *By-product (If any)* — Branches / leaves / pods
14. *Use of by-product* — Fuel / cattle feed / organic matter

446. *CROP* — NEEL CHAMELI / *Akashneem* / Indian cork tree

Scientific Name* — *Millingtonia hortensis Lf

Family — Bignoniaceae

Herb / Shrub / Tree — Tall

Annual / Biennial / Perennial — Perennial

Used as — Timber / medicine (antipyretive)

1. *Climatic requirements* — Warm climate
2. *Soil requirements* — Loamy, sandy loam and alluvial soils
3. *Required land situations* — High lands
4. *Importance of crop* — A long / tall tree with very soft branches / Avenue tree
5. *Fertilizer management* — 50 kg N, 40 kg P_2O_5 and 30 kg K_2O/ha, annually
6. *Water management* — 2-3 irrigations annually
7. *Duration of crop* — 5-10 years
8. *Parts used* — Trunk
9. *Where grown in India* — All over India (as avenue tree)
10. *Export possibility* — High
11. *Economic yields (q / ha)* — 100-120 yearly
12. *Economic value (Rs. / ha)* — 1,10,000 to 1,50,000 yearly
13. *By-product (If any)* — Branches / leaves
14. *Use of by-product* — Fuel / cattle feed / organic matter

447. *CROP* — FLAME-OF-FOREST / SILK COTTON / Kapok / Safed simal / Simul

Scientific Name* — *Ceiba pentandra (L) Gaertn / ***Bombax pentandrum*** L.

Family — Bombacaceae

Herb / Shrub / Tree — Tree (spiny trunk)

Annual / Biennial / Perennial — Perennial

Used as — Timber (for life belt / life jackets) / cotton fibre (pillows / quilt)

1. *Climatic requirements* — Warm climate
2. *Soil requirements* — Sandy, sandy loam, loamy and alluvial soils
3. *Required land situations* — High / medium / low lands. Big red flower as if like fire (i.e. flame of the forest)
4. *Importance of crop* — Mature stems are used for house construction, paper manufacture etc.
5. *Fertilizer management* — 80 kg N, 50 kg P_2O_5 and 30 kg K_2O/ha, annually
6. *Water management* — 2-3 irrigations (normally rainfed), annually
7. *Duration of crop* — 10-20 years
8. *Parts used* — Trunk / pod
9. *Where grown in India* — All over India
10. *Export possibility* — High
11. *Economic yields (q / ha)* — 100-150 yearly
12. *Economic value (Rs. / ha)* — 1,20,000 to 1,40,000 yearly
13. *By-product (If any)* — Branches / leaves
14. *Use of by-product* — Fuel / cattle feed / organic matter

448. *CROP* — CASTANOPSIS

Scientific Name — Castanopsis hystrix A.DC.

Family — Fagaceae

Herb / Shrub / Tree — Tree

Annual / Biennial / Perennial — Perennial

Used as — Timber (pole / cart shaft / fruit)

1. *Climatic requirements* — Cool climate
2. *Soil requirements* — Loamy, sandy loam and alluvial soils
3. *Required land situations* — High lands
4. *Importance of crop* — Large tree
5. *Fertilizer management* — 100 kg N, 50 kg P_2O_5 and 50 kg K_2O/ha, annually
6. *Water management* — 2-3 irrigations, if necessary, annually
7. *Duration of crop* — 10-15 years
8. *Parts used* — Trunk
9. *Where grown in India* — Eastern Himalayas, Khasi Hills etc.

10. *Export possibility* — High
11. *Economic yields (q / ha)* — 100-200 yearly
12. *Economic value (Rs. / ha)* — 1,40,000 to 1,80,000
13. *By-product (If any)* — Branches / leaves
14. *Use of by-product* — Fuel / cattle feed / organic matter

449. *CROP* — DEODAR / *Debdaru* / Himalayan cedar / Diar / Paludar

Scientific Name* — *Cedrus deodara (Roxb. ex Lamb.) G. Don.

Family — Pinaceae

Herb / Shrub / Tree — Tree

Annual / Biennial / Perennial — Perennial

Used as — Timber [railway sleeper (construction work / furniture) / avenue plant / medicine (skin disease)]

1. *Climatic requirements* — Moderate cool / warm climate
2. *Soil requirements* — Loamy, rocky, alluvial soils
3. *Required land situations* — High lands
4. *Importance of crop* — Ever green tall decorative avenue plant
5. *Fertilizer management* — 100 kg N, 50 kg P_2O_5 and 50 kg K_2O/ha annually
6. *Water management* — 2-3 irrigations annually
7. *Duration of crop* — 20-30 years
8. *Parts used* — Triml
9. *Where grown in India* — All over India specially N-W Himalayas (Kashmir to Garhwal), Kulu, Chamba Garhwal, Almora, Bushahar, Shimla etc.
10. *Export possibility* — High
11. *Economic yields (q / ha)* — 150-250 yearly
12. *Economic value (Rs. / ha)* — 2,00,000 to 3,00,000 yearly
13. *By-product (If any)* — Branches / leaves
14. *Use of by-product* — Fuel / cattle feed/ decoration / organic matter

450. *CROP* — SILVER WATTLE / *Sada khair* / Khair

Scientific Name* — *Acacia dealbata Link

Family — Mimosaceae

Herb / Shrub / Tree — Tree

Annual / Biennial / Perennial — Perennial

Used as — Timbers and bark (for tanning / dyeing)

1. *Climatic requirements* — Cool climate

2. *Soil requirements* — Rocky / sandy / sandy loam
3. *Required land situations* — High land
4. *Importance of crop* — Quality perfumes are obtained from its flowers
5. *Fertilizer management* — 80 kg N, 50 kg P_2O_5 and 30 kg K_2O/ha annually
6. *Water management* — 2-3 irrigations (if possible) annually
7. *Duration of crop* — 10-15 years
8. *Parts used* — Wood pulp
9. *Where grown in India* — Nilgiris / Palni Hills of T.N., Hilly regions of Uttaranchal, H.P., U.P.
10. *Export possibility* — High
11. *Economic yields (q / ha)* — 150-200 yearly
12. *Economic value (Rs. / ha)* — 1,20,000 to 1,60,000 yearly
13. *By-product (If any)* — Branches / leaves
14. *Use of by-product* — Fuel / cattle feed / paper / dye

451. *CROP* — ACACIA (TORTILIS) / Tortilis (acacia) *Babla*

Scientific Name* — *Acacia tortilis (Forsk) Hayne

Family — Mimosaceae

Herb / Shrub / Tree — Tree

Annual / Biennial / Perennial — Perennial

Used as — Timber (agricultural implements / posts)

1. *Climatic requirements* — Moderate warm climate
2. *Soil requirements* — Loamy, sandy loam
3. *Required land situations* — High land
4. *Importance of crop* — Drought, flood and frost resistant
5. *Fertilizer management* — 40 kg N, 50 kg P_2O_5 and 30 kg K_2O/ha, annually
6. *Water management* — 2-3 irrigations annually
7. *Duration of crop* — 5-10 years
8. *Parts used* — Trunk
9. *Where grown in India* — Rajasthan, Gujarat, Punjab, Madhya Pradesh, Haryana, A.P. etc.
10. *Export possibility* — High
11. *Economic yields (q / ha)* — 150-200 yearly
12. *Economic value (Rs. / ha)* — 1,00,000 to 1,50,000 yearly

13. *By-product (If any)* — Branches / leaves
14. *Use of by-product* — Fuel / cattle feed / paper / dye

452. *CROP* — KADAMBA / Kadam / *Kadom*

Scientific Name* — *Anthocephalus cadamba (Roxb) Miq.

Family — Naucleaceae

Herb / Shrub / Tree — Tree

Annual / Biennial / Perennial — Perennial

Used as — Timber (match boxes / ceiling boards/paper)

1. *Climatic requirements* — Warm climate
2. *Soil requirements* — Loamy, sandy loam and alluvial soils
3. *Required land situations* — High lands
4. *Importance of crop* — A graceful deciduous good tree having yellow ball-like flowers
5. *Fertilizer management* — 80 kg N, 50 kg P_2O_5 and 30 kg K_2O/ha, annually
6. *Water management* — 2-3 irrigations annually
7. *Duration of crop* — 5-8 years
8. *Parts used* — Trunk (for packing boxes)
9. *Where grown in India* — Assam, West Bengal, Andaman, Bihar, Orissa etc.
10. *Export possibility* — High
11. *Economic yields (q / ha)* — 100-150 yearly
12. *Economic value (Rs. / ha)* — 1,00,000 to 1,50,000 yearly
13. *By-product (If any)* — Branches / leaves
14. *Use of by-product* — Fuel / cattle feed / organic matter

453. *CROP*—CANE / *Bet* / East Indian cane / Dangribet / Sarubet / *Motabet*

Scientific Name* — *Calamus guruba Buch – Ham ex Kunth / ***C. latifolius*** Roxb.

Family — Palmae

Herb / Shrub / Tree — Shrub (spiny, climbing)

Annual / Biennial / Perennial — Perennial

Used as — Furniture materials for chair, table, basket medicine

1. *Climatic requirements* — Warm climate and marshy / waterlogged condition
2. *Soil requirements* — Loamy, clayey, alluvial soils

3. *Required land situations* — Low/medium lands (mostly marshy areas)
4. *Importance of crop* — Spiny stem and leaf / climbing cane with white cluster of fruits
5. *Fertilizer management* — 80 kg N, 50 kg P_2O_5 and 30 kg K_2O/ha, annually
6. *Water management* — 2-3 irrigations annually
7. *Duration of crop* — 3-4 years
8. *Parts used* — Cane the (stem) / inner top soft portion (for dysentery)
9. *Where grown in India* — Assam, West Bengal, Orissa, Meghalaya, Kerala, T.N., Sikkim, Tripura etc.
10. *Export possibility* — High
11. *Economic yields (q / ha)* — 150-200 yearly
12. *Economic value (Rs. / ha)* — 1,50,000 to 2,00,000 yearly
13. *By-product (If any)* — Leaf / leaf sheath / fruits
14. *Use of by-product* — Fuel (careful for spines) / organic matter / fruits

454. *CROP* — PSEUDOSTACHYUM / *Bamboo* / Shrubby bamboo / Jharbansh / *Bajali*

Scientific Name* — *Pseudoshachyum polymorphum Munro

Family — Gramineae

Herb / Shrub / Tree — Hollow shrub (long)

Annual / Biennial /Perennial — Perennial

Used as — Furniture (basket / umbrella handles / walking stick) / supporting the betelvine pandel (*boroj*)

1. *Climatic requirements* — Hot and warm climate
2. *Soil requirements* — Clayey, loamy and alluvial soils
3. *Required land situations* — Low / medium lands
4. *Importance of crop* — Long hollow shrubby bamboo in hill / hillocks and resistant to flood.
5. *Fertilizer management* — 100 kg N, 50 kg P_2O_5 and 30 kg K_2O/ha annually
6. *Water management* — 2-3 irrigations (if possible) annually
7. *Duration of crop* — 2-4 years
8. *Parts used* — Bamboo culm (stem) / branches
9. *Where grown in India* — Terai regions, low hills of Sikkim, Assam, Garo Hills, West Bengal, Tripura etc.

10. *Export possibility* — High
11. *Economic yields (q / ha)* — 450-600 yearly
12. *Economic value (Rs. / ha)* — 80,000 to 1,50,000
13. *By-product (If any)* — Branches / leaves
14. *Use of by-product*—Fencing/cattle feed/organic matter/fuel/roofing

455. *CROP* — PLAINS BAMBOO / BANSH / Bhaluka / Bambo / Peka / Chau / Bangsa

Scientific Name* — *Bambusa balcooa Roxb / ***B. tulda*** Roxb

Family — Bambusaceae

Herb / Shrub / Tree — Shrub (long)

Annual / Biennial / Perennial — Perennial

Used as — For manufacturing basket / table / chair / paper and also for propping

1. *Climatic requirements* — Warm climate
2. *Soil requirements* — Loamy, clayey and alluvial soils
3. *Required land situations* — High / medium / low
4. *Importance of crop* — Tall hard plant, resistant to flood
5. *Fertilizer management* — 100 kg N, 50 kg P_2O_5 and 30 kg K_2O/ha, annually
6. *Water management* — 2-3 irrigations annually
7. *Duration of crop* — 4-5 years
8. *Parts used* — Culm / stems
9. *Where grown in India* — Assam, West Bengal, Mizoram, Sikkim, Orissa, Meghalaya etc.
10. *Export possibility* — High
11. *Economic yields (q / ha)* — 450-600 yearly
12. *Economic value (Rs. / ha)* — 80,000 to 1,50,000
13. *By-product (If any)* — Branches / leaves
14. *Use of by-product* — Fencing / cattle feed / fuel / organic matter / roofing.

Chapter - 8

TREE-FODDERS

INTRODUCTION

Normally fodders are obtained from annual plant, but in some special cases fodder (leaves and soft twigs) are obtained from the perennial trees either in plains or hills. This tree fodder is fed to the cattle / goat / camel / buffalo / horse / ass etc. in some special circumstances (dearth of fodder / likeliness of the animal etc.) when and where normal fodders are not available.

It has been observed that particular fodders are liked by particular animals. Goats are very much fond of jackfruit leaves; whereas camels are fond of *babul* plants.

Hence, the salient point of identification, management, crop economics and exportability of some tree fodders are given as follows –

456. *CROP*—VILAITIBAVAL / SUBABUL / Horse tamarind / White popinac / *Subhabul*

Scientific Name — Leucaena leucocephala (Lamk) de Wit

Family — Mimosaceae

Herb / Shrub / Tree — Tree (small) / shrub

Annual / Biennial / Perennial — Perennial

Used as — Fodder (leaves, twigs and pods are fed to cattle, sheep and goats) / fuel / organic matters

1. *Climatic requirements* — Warm climate
2. *Soil requirements* — Sandy loam, sandy and loamy soils
3. *Required land situations* — High lands
4. *Importance of crop* — It is also a shade tree in tea, coffee and rubber plantations, as it fixes nitrogen. It is resistant to flood water

5. *Fertilizer Management*— 40 kg N, 50 kg P_2O_5 and 30 kg K_2O/ha, annually
6. *Water management* — 2-3 irrigations, annually
7. *Duration of crop* — 3-4 years
8. *Parts used* — Branches, twigs, leaves and pods
9. *Where grown in India* — All over India except too hot / too cold areas
10. *Export possibility* — Less
11. *Economic yields (q / ha)* — 100-150 (green annually)
12. *Economic value (Rs. / ha)* — 25,000 to 30,000 annually
13. *By-product (If any)* — Stem
14. *Use of by-product* — Poles / fuels / pegs / organic matter

457. *CROP* — BABOOL / Babul / Kikar / Babul acacia / *Babla*

Scientific Name* — *Acacia nilotica (L) Willd ex. Del

Family — Mimosaceae

Herb / Shrub / Tree — Tree (small)

Annual / Biennial / Perennial — Perennial

Used as — Fodder (leaves, soft twigs and pods)

1. *Climatic requirements* — Very hot climate
2. *Soil requirements* — Sandy, sandy loam and loamy soils
3. *Required land situations* — High lands
4. *Importance of crop* — Gum is produced from bark (for calico and printing) / for afforestation of dry areas
5. *Fertilizer management* — 40 kg N, 50 kg P_2O_5 and 30 kg K_2O/ha, annually
6. *Water management* — 2-3 irrigations annually
7. *Duration of crop* — 2-4 years
8. *Parts used* — Leaves, twigs and pods
9. *Where grown in India* — Drier parts of the country
10. *Export possibility* — High
11. *Economic yields (q / ha)* — 100-150 (green) annually
12. *Economic value (Rs. / ha)* — 30,000 to 40,000 annually
13. *By-product (If any)* — Branches
14. *Use of by-product* — Fuel, poles, organic matter / pegs

458. *CROP* — DHAULA-KHEJRA / *Dholbabla*

Scientific Name* — *Acacia polyacantha Willd

Family — Mimosaceae

Herb / Shrub / Tree — Tree (small)

Annual / Biennial / Perennial — Perennial

Used as — Fodder (leaves, soft twigs and pods)

1. *Climatic requirements* — Warm climate
2. *Soil requirements* — Sandy, sandy loam and loamy soils
3. *Required land situations* — High
4. *Importance of crop* — Bark as tan and heart wood is source of cutch (catechu)
5. *Fertilizer management* — 40 kg N, 50 kg P_2O_5 and 30 kg K_2O/ha, annually
6. *Water management* — 2-3 irrigations, annually
7. *Duration of crop* — 2-3 years
8. *Parts used* — Leaves, soft twig and pods
9. *Where grown in India* — Bihar, Tamil Nadu, Andhra Pradesh, West Bengal, Madhya Pradesh etc.
10. *Export possibility* — High
11. *Economic yields (q / ha)* — 100-150 (green) annually
12. *Economic value (Rs. / ha)* — 30,000 to 40,000 annually
13. *By-product (If any)* — Branches and trunk
14. *Use of by-product* — Poles / fuel / organic matter / implements / timber

459. *CROP* — ACACIA TORTILIS / *Tortilis babla*

Scientific Name* — *Acacia tortilis (Forsk) Hayne

Family — Mimosaceae

Herb / Shrub / Tree — Tree

Annual / Biennial / Perennial — Perennial

Used as — Fodder (leaves, soft twigs and pods)

1. *Climatic requirements* — Hot weather / climate
2. *Soil requirements* — Sandy, sandy loam and loamy soils
3. *Required land situations* — High land
4. *Importance of crop* — Woods for implements in agriculture / afforestation of dry areas
5. *Fertilizer management* — 40 kg N, 50 kg P_2O_5 and 30 kg K_2O/ha, annually
6. *Water management* — 2-3 irrigations annually
7. *Duration of crop* — 2-3 years

8. *Parts used* — Leaves, twigs and pods
9. *Where grown in India* — Very hot areas of India (Rajasthan, Gujarat, Haryana, Punjab etc.)
10. *Export possibility* — High
11. *Economic yields (q / ha)* — 100-150 (green) annually
12. *Economic value (Rs. / ha)* — 30,000 to 40,000 annually
13. *By-product (If any)* — Trunk / branches
14. *Use of by-product* — Fuel / poles / implements / organic matter

460. *CROP*— ISRAIL BABOOL / Kinkar / Bouli / *Kanta babla*

Scientific Name* — *Acacia jacque montii Benth

Family — Mimosaceae

Herb / Shrub / Tree — Tree (small)

Annual / Biennial / Perennial — Perennial

Used as — Fodder (leaves, soft twigs and pods) / timber / furniture

1. *Climatic requirements* — Warm weather
2. *Soil requirements* — Sandy, sandy loam and loamy soils
3. *Required land situations* — High lands
4. *Importance of crop* — Full of thorns with yellow fragrant flowers
5. *Fertilizer management* — 40 kg N, 50 kg P_2O_5 and 30 kg K_2O/ha, annually
6. *Water management* — 2-3 irrigations annually
7. *Duration of crop* — 2-3 years
8. *Parts used* — Leaves, soft twig and pods
9. *Where grown in India* — In hot areas of Rajasthan, Gujarat, Haryana, Punjab, Maharashtra etc.
10. *Export possibility* — High
11. *Economic yields (q / ha)* — 100-150 (green) annually
12. *Economic value (Rs. / ha)* — 30,000 to 40,000 annually
13. *By-product (If any)* — Trunks / branches
14. *Use of by-product* — Poles / fuel / implements / timber / gum

461. *CROP* — Rain tree / KOKKO / Lebbeck tree / Sirish / *Siris*

Scientific Name* — *Albizia lebbeck (L.) Benth

Family — Mimosaceae

Herb / Shrub / Tree — Tree (big)

Annual / Biennial / Perennial — Perennial

Used as — Fodder (twigs, leaves and flowers)

1. *Climatic requirements* — Warm weather
2. *Soil requirements* — Loamy, sandy loam and alluvial soils
3. *Required land situations* — High land
4. *Importance of crop* — Avenue tree / wind break tree / shelter belt/ shade plant
5. *Fertilizer management* — 40 kg N, 50 kg P_2O_5 and 30 kg K_2O/ha, annually
6. *Water management* — 2-3 irrigations annually
7. *Duration of crop* — 20-25 years
8. *Parts used* — Leaves, twigs and flowers
9. *Where grown in India* — West Bengal, Assam, Tamil Nadu, M.P., Punjab, U.P., etc.
10. *Export possibility* — High
11. *Economic yields (q / ha)* — 100-200 (green) annually
12. *Economic value (Rs. / ha)* — 50,000 to 60,000 annually
13. *By-product (If any)* — Trunk / branches
14. *Use of by-product* — Furniture / poles / timber / implements / organic matter

462. *CROP* — INDIAN CORK TREE / Nilchameli / *Akash neem*

Scientific Name — Millingtonia hortensis L.f.

Family — Bignoniaceae

Herb / Shrub / Tree — Tree

Annual / Biennial / Perennial — Perennial

Used as — Fodder (soft twigs, leaves and flowers)/ medicine (antipyretic)

1. *Climatic requirements* — Warm climate
2. *Soil requirements* — Loamy, sandy loam and alluvial soils
3. *Required land situations* — High land
4. *Importance of crop* — Tall avenue tree with soft twigs, good leaves and white flowers
5. *Fertilizer management* — 80 kg N, 50 kg P_2O_5 and 30 kg K_2O/ha, annually
6. *Water management* — 2-3 irrigations annually
7. *Duration of crop* — 5-10 years
8. *Parts used* — Soft twigs / leaves / flowers
9. *Where grown in India* — All over India

10. *Export possibility* — High
11. *Economic yields (q / ha)* — 100-150 (green) annually
12. *Economic value (Rs. / ha)* — 50,000 to 60,000 annually
13. *By-product (If any)* — Trunk / branches
14. *Use of by-product* — Furniture / boxes / fuel / timber / organic matter / medicine

463. *CROP* — JACK FRUIT / Jack tree / *Kanthal* / Kathal / *Ichar* / Panas

Scientific Name* — *Artocarpus heterophyllus Lamk

Family — Moraceae

Herb / Shrub / Tree — Tree (evergreen)

Annual / Biennial / Perennial — Perennial

Used as — Fodder (leaves and soft twig for goats) / fruits

1. *Climatic requirements* — Warm and humid climate
2. *Soil requirements* — Loamy, sandy loam, alluvial soils
3. *Required land situations* — High lands
4. *Importance of crop* — Fruits (big) come from the stems profusely and consumed as vegetable, susceptible to flood water
5. *Fertilizer Management*—80 kg N, 50 kg P_2O_5 and 30 kg K_2O/ha, annually
6. *Water management* — 2-3 irrigations annually
7. *Duration of crop* — 10-15 years
8. *Parts used* — Twig and leaves
9. *Where grown in India* — All over India particularly Assam, West Bengal, Orissa, Meghalaya, Karnataka, Maharashtra, Tripura, Western Ghats etc.
10. *Export possibility* — High
11. *Economic yields (q / ha)* — 100-150 (green) annually
12. *Economic value (Rs. / ha)* — 40,000 to 60,000 annually
13. *By-product (If any)* — Trunk / branches / fruits
14. *Use of by-product* — Furniture / poles / fuel / timber / dye / instruments / vegetable

464. *CROP* — ANJAL / PONGAM / *Karanja*

Scientific Name* — *Pongamia pinnata (L.) Pierre / ***P. glabra*** Vent

Family — Papilionaceae

Herb / Shrub / Tree — Tree

Annual / Biennial / Perennial — Perennial

Used as — Fodder (fruits / leaves) / seeds (for oil) / medicine (skin diseases)

1. *Climatic requirements* — Warm climate near the bank of the river
2. *Soil requirements* — Loamy, sandy loam and alluvial soils
3. *Required land situations* — High
4. *Importance of crop* — Seed-oil is used in soap making / oil / fuel
5. *Fertilizer management* — 40 kg N, 50 kg P_2O_5 and 30 kg K_2O/ha, annually
6. *Water management* — 2-3 irrigations annually
7. *Duration of crop* — 15-20 years
8. *Parts used* — Fruits and seeds and leaves
9. *Where grown in India* — All over India, mostly near the bank of streams (S. India, West Bengal, Orissa, Kerala etc.)
10. *Export possibility* — High
11. *Economic yields (q / ha)* — 150-180, annually and 10-15 (seed) annually
12. *Economic value (Rs. / ha)* — 40,000 to 80,000 yearly
13. *By-product (If any)* — Trunk / branchs
14. *Use of by-product* — Furniture / fuel / timber / organic matter

465. *CROP*— BAMBOO / *Bansa* / *Bansh* / Peka / Chau

Scientific Name — Bambusa tulda (Roxb)

Family — Bambusaceae

Herb / Shrub / Tree — Culms (Shrub)

Annual / Biennial / Perennial — Perennial

Used as — Fodder (leaves and small twigs)

1. *Climatic requirements* — Warm climate with humid conditions
2. *Soil requirements* — Loamy, sandy loam, alluvial and clayey soils
3. *Required land situations* — Medium / high / low lands
4. *Importance of crop* — A long culms with long nodes
5. *Fertilizer management* — 100 kg N, 50 kg P_2O_5 and 30 kg K_2O/ha annually
6. *Water management* — 2-3 irrigations, if irrigated
7. *Duration of crop* — 2-3 years
8. *Parts used* — Leaves and small twigs
9. *Where grown in India* — All over India specially Assam, West Bengal, Sikkim, Bihar, Orissa etc.

10. *Export possibility* — High
11. *Economic yields (q / ha)* — 300-600 (green) annually
12. *Economic value (Rs. / ha)* — 40,000 to 80,000 annually
13. *By-product (If any)* — Straight culm and branches
14. *Use of by-product* — Furniture / pole / fencing / organic matter / use in *boroj* (betelvine pandal)

466. CROP — FIG / Anjir / Dumur / Jagga dumur

Scientific Name — Ficus carica L.

Family — Moraceae

Herb / Shrub / Tree — Tree (small)

Annual / Biennial / Perennial — Perennial

Used as — Fodder (leaves and soft twigs) / fruits (vegetable)

1. *Climatic requirements* — Warm humid climate with high humidity
2. *Soil requirements* — Loamy, sandy loam and alluvial soils
3. *Required land situations* — High / medium / low lands
4. *Importance of crop* — Fruits come from stems; resistant to flood water
5. *Fertilizer management* — 80 kg N, 50 kg P_2O_5 and 30 kg K_2O/ha annually
6. *Water management* — 2-3 irrigations annually
7. *Duration of crop* — 10-12 years
8. *Parts used* — Leaves and twigs / fruits
9. *Where grown in India* — All over India, under humid conditions
10. *Export possibility* — High
11. *Economic yields (q / ha)* — 50-60 (green) annually
12. *Economic value (Rs. / ha)* — 15,000 to 20,000 annually
13. *By-product (If any)* — Trunk / branches
14. *Use of by-product* — Furniture / fuel / poles / organic matter

467. *CROP* — GLIRICIDIA / Madre tree / Spotted gliricidia

Scientific Name — Gliricidia sepium (Jacq) Walp / ***G. maculata*** (H.B. & K.) Steud

Family — Papilionaceae

Herb / Shrub / Tree — Tree

Annual / Biennial / Perennial — Perennial

Used as — Fodder (leaves and soft twig) / green manure / ornamentals

1. *Climatic requirements* — Warm and humid climate

2. *Soil requirements* — Sandy loam, loamy and alluvial soils
3. *Required land situations* — High land
4. *Importance of crop* — Good looking flowers in a row of inflorescence
5. *Fertilizer Management* –40 kg N, 80 kg P_2O_5 and 30 kg K_2O/ha, annually
6. *Water management* — 2-3 irrigations annually
7. *Duration of crop* — 5-7 years
8. *Parts used* — Leaves and small twigs
9. *Where grown in India* — All over India under humid conditions
10. *Export possibility* — High
11. *Economic yields (q / ha)* — 140-190 (green) annually
12. *Economic value (Rs. / ha)* — 40,000 to 60,000 annually
13. *By-product (If any)* — Trunk / branches
14. *Use of by-product* — Poles / fencing / fuel / organic matter.

468. *CROP* — UMBRELLA THORN / *Kanta acacia*

Scientific Name — Acacia planifrons Wight & Arn

Family — Mimosaceae

Herb / Shrub / Tree — Tree (small)

Annual / Biennial / Perennial — Perennial

Used as — Fodder (leaves and pods fed to cattle / sheep / goats)

1. *Climatic requirements* — Very hot climate
2. *Soil requirements* — Sandy, sandy loam and loamy soils
3. *Required land situations* — High land
4. *Importance of crop* — Vegetable / fodder
5. *Fertilizer management* — 40 kg N, 60 kg P_2O_5 and 30 kg K_2O/ha, annually
6. *Water management* — 2-3 irrigations
7. *Duration of crop* — 5-10 years
8. *Parts used* — Leaves and pods used as fodder
9. *Where grown in India* — In drier tracts of India
10. *Export possibility* — High
11. *Economic yields (q / ha)* — 100-150 (green) annually
12. *Economic value (Rs. / ha)* — 10,000 to 30,000 annually
13. *By-product (If any)* — Trunks / branches

14. *Use of by-product* — Implements (agricultural) / poles / fuel / organic matter

469. ***CROP* — CHILD-LIFE TREE/Indian amulet plant / Putranjiva/*Jiaput***

Scientific Name* — *Putranjiva roxburghii Wall

Family — Euphorbiaceae

Herb / Shrub / Tree — Tree

Annual / Biennial / Perennial — Perennial

Used as — Fodder (leaves and twigs) / Timber (furniture)

1. *Climatic requirements* — Warm humid climate
2. *Soil requirements* — Loamy, sandy loam and alluvial soils
3. *Required land situations* — High lands
4. *Importance of crop* — Avenue plant with ever green leaves; resistant to flood
5. *Fertilizer management* — 80 kg N, 50 kg P_2O_5 and 30 kg K_2O/ha, annually
6. *Water management* — 2-3 irrigations annually
7. *Duration of crop* — 10-15 years
8. *Parts used* — Leaves and soft wigs
9. *Where grown in India* — All over India under humid conditions
10. *Export possibility* — High
11. *Economic yields (q / ha)* — 100-150 (green) annually
12. *Economic value (Rs. / ha)* — 40,000 to 50,000 annually
13. *By-product (If any)* — Trunk / branches
14. *Use of by-product* — Wood for furniture / poles / fuels / organic matter

470. ***CROP* — JUJUBE / CHINESE DATE / Ber / Badara / *Kul* / *Kuvala*/ *Karkandhu***

Scientific Name* — *Ziziphus jujuba (L.) Lam / ***Z. mauritiana*** Lamk

Family — Rhamnaceae

Herb / Shrub / Tree — Tree

Annual / Biennial / Perennial — Perennial

Used as — Fodder (leaves and soft twigs) / fruits

1. *Climatic requirements* — Warm and humid climate
2. *Soil requirements* — Loamy, sandy loam and alluvial soils
3. *Required land situations* — High lands

4. *Importance of crop* — Thorny branches
5. *Fertilizer management* — 80 kg N, 50 kg P_2O_5 and 30 kg K_2O/ha, annually
6. *Water management* — 2-3 irrigations annually
7. *Duration of crop* — 10-15 years
8. *Parts used* — Leaves and soft twigs
9. *Where grown in India* — All over India under humid conditions
10. *Export possibility* — High
11. *Economic yields (q / ha)* — 80-100 (green) annually
12. *Economic value (Rs. / ha)* — 30,000 to 40,000 annually
13. *By-product (If any)* — Trunk / branches
14. *Use of by-product* — Furnitures / poles / fuel / organic matter.

Chapter - 9

BEVERAGE AND NARCOTICS

INTRODUCTION

The beverage crops are those crops, which give the delightful mood or fresh energy after the consumption. Sometimes these crops give some intoxications or strong feelings also. Again some narcotic crops, if consumed in different forms, produce torpor or sleep, affecting the central nervous systems so as to produce dizziness emphasis, loss of memory and of neuromuscular co-ordination and ultimately, unconsciousness, if consumed more or for a long time.

Tea, coffee, *paan* etc. are the 'beverage crops'; whereas tobacco, hemp / *bhang / charas / ganja, dhatura* etc. are the 'narcotics crops'. Narcotics should always be avoided for leading normal life. But the business with these narcotics is very strong all over the world, though illegal.

The salient points on identification, management, crop economics, exportability etc. of some of the narcotic and beverage crops are as follows -

471. *CROP*—TEA / *Cha* / Chaye / Chai

Scientific Name — Camellia sinensis (L) Kuntze

Family — Theaceae

Herb / Shrub / Tree — Shrub (ever green)

Annual / Biennial / Perennial — Perennial

Used as — Beverage (cured leaves)

1. *Climatic requirements* — Moderately cool weather
2. *Soil requirements* — Rocky, loamy, sandy loam and acidic soils
3. *Required land situations* — High altitude / plains with hillocks
4. *Importance of crop* — Mostly top two leaves give the tea / Tea the big plant is pruned to shape

5. *Fertilizer Management*— 50 kg N, 50 kg P_2O_5 and 30 kg K_2O/ha, annually
6. *Water management* — 4-5 irrigations annually
7. *Duration of crop* — 15-20 years
8. *Parts used* — Leaves
9. *Where grown in India* — Assam, Darjeeling (W.B), Nilgiris (T.N.), Malabar, Travancore, Deharadun, Kumaon etc.
10. *Export possibility* — Very high
11. *Economic yields (q / ha)* — 2-3 (dry tea) annually
12. *Economic value (Rs. / ha)* — 1,50,000 to 2,00,000 annually
13. *By-product (If any)* — Trunk / branches
14. *Use of by-product* — Furniture / fuel / organic matter

472. *CROP* — COFFEE / Arabian coffee / Kafi / *Cafi*

Scientific Name* — *Coffea arabica L.

Family — Rubiaceae

Herb / Shrub / Tree — Shrub (ever green) / tree

Annual / Biennial / Perennial — Perennial

Used as — Beverage (powdered dried seeds, roasted)

1. *Climatic requirements* — Moderately cool climate
2. *Soil requirements* — Loamy, soils with organic matter and sandy loam soils
3. *Required land situations* — High lands
4. *Importance of crop* — Stimulant beverage
5. *Fertilizer management* — 80 kg N, 50 kg P_2O_5 and 30 kg K_2O/ha, annually
6. *Water management* — 4-5 irrigation annually
7. *Duration of crop* — 15-20 years
8. *Parts used* — Seeds
9. *Where grown in India* — Nilgiris, Karnataka, Travancore etc.
10. *Export possibility* — High
11. *Economic yields (q / ha)* — 10-15 (dried seeds) annually
12. *Economic value (Rs. / ha)* — 1,50,000 to 2,00,000 annually
13. *By-product (If any)* — Trunks and branches
14. *Use of by-product* — Furniture / implements / fuel

473. *CROP* — CACAO / COCOA / Coco

Scientific Name* — *Theobroma cacao L.

Family — Sterculiaceae

Herb / Shrub / Tree — Tree (small)

Annual / Biennial / Perennial — Perennial

Used as — Beverage (seeds) / confectionery / cosmetics

1. *Climatic requirements* — Warm climate
2. *Soil requirements* — Loamy, sandy loam and alluvial soils
3. *Required land situations* — High lands
4. *Importance of crop* — Seeds are source of non-alcoholic beverage
5. *Fertilizer Management*— 80 kg N, 50 kg P_2O_5 and 30 kg K_2O/ha, annually
6. *Water management* — 4-5 irrigations annually
7. *Duration of crop* — 5-10 years
8. *Parts used* — Seeds
9. *Where grown in India* — South India, Orissa etc.
10. *Export possibility* — High (very)
11. *Economic yields (q / ha)* — 10-12 (dry) annually
12. *Economic value (Rs. / ha)* — 1,00,000 to 2,00,000 annually
13. *By-product (If any)* — Branches / trunks
14. *Use of by-product* — Furniture / implements / fuel timber / organic matter

474. *CROP* — ARECANUT / Betel-nut palm / *Supari / Guasupari*

Scientific Name* — *Areca catechu L.

Family — Palmae

Herb / Shrub / Tree — Palm (erect long) / shrub

Annual / Biennial / Perennial — Perennial

Used as — Beverage (used in 'Paan' and fermented dough nuts are also used)

1. *Climatic requirements* — Warm (moderate) climate
2. *Soil requirements* — Alluvial, loamy and sandy loam soils
3. *Required land situations* — High lands
4. *Importance of crop* — Nuts, (within the husk of the fruits) chewed along with '*paan*' (betel leaves), *katha (Khair)* and spices and also with tobacco.
5. *Fertilizer management* — 80 kg N, 50 kg P_2O_5 and 30 kg K_2O/ha, annually
6. *Water management* — 2-3 irrigations annually
7. *Duration of crop* — 5-10 years

8. *Parts used* — Nuts from the fruits
9. *Where grown in India* — Karnataka, Kerala, T.N., West Bengal, Assam, Maharashtra, Orissa etc.
10. *Export possibility* — High
11. *Economic yields (q / ha)* — 100-150 (green) annually
12. *Economic value (Rs. / ha)* — 1,20,000 to 2,00,000 annually
13. *By-product (If any)* — Long slender stem and leaves
14. *Use of by-product* — Poles / fuel / cattle feed / thatching of *boroj* (pandal betelvine)

475. *CROP*— BETELVINE / *Paan* / Vetel pepper / *Pan* / Tambul patra

Scientific Name* — *Piper betle L.

Family — Piperaceae

Herb / Shrub / Tree — Shrub (climbing)

Annual / Biennial / Perennial — Perennial

Used as — Beverage (leaves with betel nut, *khair* and lime) / climbing vine

1. *Climatic requirements* — Warm climate
2. *Soil requirements* — Alluvial and loamy soils
3. *Required land situations* — High lands (mostly near river bank, *Jhills, ponds, bills* etc.)
4. *Importance of crop* — Alternate leaves on vine, which grip the supports firmly with adventitious roots; susceptible to flood water.
5. *Fertilizer management* — 80 kg N, 50 kg P_2O_5 and 30 kg K_2O/ha, annually (mostly from organic sources)
6. *Water management* — 4-5 irrigations annually
7. *Duration of crop* — 10-15 years
8. *Parts used* — Leaves (as such)
9. *Where grown in India* — West Bengal, Assam, Orissa, T.N., Karnataka, M.P. etc.
10. *Export possibility* — Very high
11. *Economic yields (q / ha)* — 100-150 (green leaves) annually
12. *Economic value (Rs. / ha)* — 1,20,000 to 2,00,000 annually
13. *By-product (If any)* — Vines / roots
14. *Use of by-product* — Fuel / cattle feed / organic matter / medicine

476. *CROP — DHATURA* / Thorn apple / Jimson weed / *Dhutura* / *Stramonium*

Scientific Name* — *Datura stramonium L.

Family — Solanaceae

Herb / Shrub / Tree — Shrub

Annual / Biennial / Perennial — Annual

Used as — Narcotics (leaves and seeds) / poisoning / medicine (asthma)

1. *Climatic requirements* — Warm climate
2. *Soil requirements* — Alluvial, loamy and sandy loam soils
3. *Required land situations* — High lands
4. *Importance of crop* — Fruits / seed / leaves are dangerously poisonous
5. *Fertilizer management* — 80 kg N, 50 kg P_2O_5 and 30 kg K_2O/ha, annually
6. *Water management* — 2-3 irrigations
7. *Duration of crop* — 120-160 days
8. *Parts used* — Leaves / seeds / fruits
9. *Where grown in India* — West Bengal, Assam, Tripura, Orissa, Maharashtra, A.P., M.P., Bihar, Karnataka etc.
10. *Export possibility* — High
11. *Economic yields (q / ha)* — 10-12 (dry fruits)
12. *Economic value (Rs. / ha)* — 50,000 to 80,000
13. *By-product (If any)* — Branches / main plants
14. *Use of by-product* — Poles / fuel / for fencing / organic matter

477. *CROP* — HEMP / Bhang / charas / ganja

Scientific Name — Cannabis sativa L.

Family — Cannabinaceae

Herb / Shrub / Tree — Herb (tall and smelling)

Annual / Biennial / Perennial — Annual

Used as — Narcotics (different forms of *bhang, ganja* and *charas* from leaves and flowers) / bast fibre (stem) / oil (seeds)

1. *Climatic requirements* — Warm climate
2. *Soil requirements* — Loamy and sandy loam soils
3. *Required land situations* — High land (in road side and even in irrigation channel bunds)
4. *Importance of crop* — Very strong narcotics
5. *Fertilizer management* — 80 kg N, 50 kg P_2O_5 and 30 kg K_2O/ha
6. *Water management* — 3-4 irrigations
7. *Duration of crop* — 120-160 days

8. *Parts used* — Leaves and flowers
9. *Where grown in India* — All over India specially West Bengal, Assam, Orissa, Bihar, Kerala etc.
10. *Export possibility* — High
11. *Economic yields (q / ha)* — 5-7 (dried laves and flowers)
12. *Economic value (Rs. / ha)* — 1,50,000 to 2,00,000
13. *By-product (If any)* — Plants and branches
14. *Use of by-product* — Fuel / fencing / organic matter

Note — *Bhang* is composed of dried leaves and flowering shoots of male and female plants; *ganja* consists of dried unfertilized inflorescence and *charas* is a crude resin. *Ganja* and *Charas* are used through smoking, while *bhang* is taken in the form of drink.

Chapter - 10

SPICES

INTRODUCTION

The crops (any part like, bark, seed, leaves, flower etc.), which give aroma, good odour, flavour or pungent vegetable substances, to our food materials, are called the 'Spices'. Here only the perennial spices have been dealt with (annual spices have been dealt with under field crops).

India has been recognized the world over as the 'home of some spices', since time immemorial.

Once considered a luxury, spices have been today become an integral part of our daily diets. Spices, all over the world, are used to flavour the foods and beverages. Spices also are added to meat preservation; medicinal preparation, cosmetics, perfumery, backery goods and various other products.

Therefore, the salient points on identification, management, crop economics, exportability etc. of some spice crops are given as below.

478. *CROP*—CARDAMOM / Chhoti elaichi / *Choto alach*

Scientific Name — Elettaria cardamomum Maton.

Family — Zingiberaceae

Herb / Shrub / Tree — Herb (tall)

Annual / Biennial / Perennial — Perennial

Used as — Spices [dried fruits (capsules)]

1. *Climatic requirements* — Cool climate
2. *Soil requirements* — Organic, loamy and sandy loam soils
3. *Required land situations* — High land but continuous water flow needed
4. *Importance of crop* — Seeds are hot and pungent smelling

5. *Fertilizer Management*— 80 kg N, 50 kg P_2O_5 and 30 kg K_2O/ha, annually
6. *Water management* — Continuous water flow
7. *Duration of crop* — 2-3 years
8. *Parts used* — Fruits
9. *Where grown in India* — West ghats / Karnataka and Kerala, Madurai, Nilgiri and Tirunelveli (T.N), Darjeeling (West Bengal) etc.
10. *Export possibility* — High
11. *Economic yields (q / ha)* — 5-10 (dried) annually
12. *Economic value (Rs. / ha)* — 1,00,000 to 2,00,000 annually
13. *By-product (If any)* — Stems and leaves
14. *Use of by-product* — Fuel / cattle feed / organic matter

479. *CROP*— PEPPER / Black pepper / Kalimirch / Gol mirch / Golmarich

Scientific Name — *Piper nigrum* L.

Family — Piperaceae

Herb / Shrub / Tree — Herb / Shrub (vine)

Annual / Biennial / Perennial — Perennial

Used as — Spices / Condiment / Medicine (Stimulant)

1. *Climatic requirements* — Warm climate with high rainfall
2. *Soil requirements* — Alluvial, loamy, sandy loam soils
3. *Required land situations* — High
4. *Importance of crop* — Climbing vine where fruits (small bids) are available
5. *Fertilizer management* — 80 kg N, 50 kg P_2O_5 and 30 kg K_2O/ha, annually
6. *Water management* — 3-4 irrigations annually
7. *Duration of crop* — 3-4 years
8. *Parts used* — Fruits
9. *Where grown in India* — Western Ghats, Karnataka, Maharashtra, Kerala, Andaman, W.B. etc.
10. *Export possibility* — Very high
11. *Economic yields (q / ha)* — 5-10 (dry fruits) annually
12. *Economic value (Rs. / ha)* — 1,50,000 to 2,00,000 annually
13. *By-product (If any)* — Vines and leaves
14. *Use of by-product* — Fuel / cattle feed / organic matter / medicine

480. ***CROP* — CINNAMON / Dalchini / *Darchini***

Scientific Name — Cinnamomum zeylanicum Blume

Family — Lauraceae

Herb / Shrub / Tree — Trees

Annual / Biennial / Perennial — Perennial

Used as — Spice / bark / condiments / medicine (stimulant)

1. *Climatic requirements* — Warm climate
2. *Soil requirements* — Loamy, sandy loam and sandy soils
3. *Required land situations* — High land
4. *Importance of crop* — Delicate fragrance and warm sweet taste
5. *Fertilizer Management*— 80 kg N, 50 kg P_2O_5 and 30 kg K_2O/ha, annually
6. *Water management* — 3-4 irrigations annually
7. *Duration of crop* — 10-15 years
8. *Parts used* — Barks
9. *Where grown in India* — W. Ghats and confined to Kerala, only
10. *Export possibility* — Very high
11. *Economic yields (q / ha)* — 5-10 (dry bark) annually
12. *Economic value (Rs. / ha)* — 50,000 to 80,000 annually
13. *By-product (If any)* — Trunk / branch / leaves
14. *Use of by-product* — Pole / furniture / fuel / cattle feed / organic matter

481. ***CROP* — LAUNG / CLOVE / *Labanga***

Scientific Name — Syzygium aromaticum (L) Merri. & Perry / ***Eugenia aromatica*** O. Kuntze and ***E. caryophyllata*** Thunb

Family — Myrtaceae

Herb / Shrub / Tree — Tree (evergreen)

Annual / Biennial / Perennial — Perennial

Used as — Spice (clover flower buds) / medicine / industrial uses

1. *Climatic requirements* — Moderately cool climate in medium range of altitude of the hills
2. *Soil requirements* — Loamy and sandy loam soils
3. *Required land situations* — High lands
4. *Importance of crop* — Dried flower bud (*labanga*) is used as spice.
5. *Fertilizer management* — 80 kg N, 50 kg P_2O_5 and 30 kg K_2O/ha, annually

6. *Water management* — 3-4 irrigations annually
7. *Duration of crop* — 10-20 years
8. *Parts used* — Flower buds
9. *Where grown in India* — Tamil Nadu and Kerala
10. *Export possibility* — Very high
11. *Economic yields (q / ha)* — 3-8 (dry flower buds) annually
12. *Economic value (Rs. / ha)* — 80,000 to 1,20,000 annually
13. *By-product (If any)* — Trunk / branches / leaves
14. *Use of by-product* — Timber / fuel / cattle feed / organic matter

482. ***CROP*— TEJPAT / *Tejpata* / INDIAN CASSIA**

Scientific Name — Cinnamomum tamala (Buch-Ham) T. Eberm Nees

Family — Lauraceae

Herb / Shrub / Tree — Tree

Annual / Biennial / Perennial — Perennial

Used as — Spice / condiments (leaves)

1. *Climatic requirements* — Moderately cool (of subtropical Himalayas, Khasi of Jaintia hills) climate
2. *Soil requirements* — Loamy, sandy, loam soils
3. *Required land situations* — High land
4. *Importance of crop* — Aromatic leaves used in curry
5. *Fertilizer management* — 80 kg N, 50 kg P_2O_5 and 30 kg K_2O/ha, annually
6. *Water management* — 3-4 irrigations annually
7. *Duration of crop* — 10-15 years
8. *Parts used* — Leaves
9. *Where grown in India* — Sikkim, Mizoram, Meghalaya, Assam, T.N., West Bengal, Karnataka etc.
10. *Export possibility* — High
11. *Economic yields (q / ha)* — 10-12 (dry leaves) annually
12. *Economic value (Rs. / ha)* — 60,000 to 80,000 annually
13. *By-product (If any)* — Trunk / branches
14. *Use of by-product* — Pole / fuel / timber / organic matter

483. ***CROP* — AJMUD / Ajmudi / *Radhuni / Saj***

Scientific Name — Trachyspermum roxburghianum (DC.) Craib

Family — Umbelliferae

Herb / Shrub / Tree — Herb

Annual / Biennial / Perennial — Annual

Used as — Spice (fruits) / medicine (dyspepsia / asthma)

1. *Climatic requirements* — Moderate cool weather
2. *Soil requirements* — Loamy, sandy loam and alluvial soils
3. *Required land situations* — High / medium / low lands
4. *Importance of crop* — White small cluster of flowers / small seeds
5. *Fertilizer management* — 60 kg N, 50 kg P_2O_5 and 30 kg K_2O/ha
6. *Water management* — 2-3 irrigations
7. *Duration of crop* — 120-140 days
8. *Parts used* — Seeds
9. *Where grown in India* — All over India (except South India)
10. *Export possibility* — High
11. *Economic yields (q / ha)* — 5-10 (dry seeds)
12. *Economic value (Rs. / ha)* — 30,000 to 60,000
13. *By-product (If any)* — Plants / branches / leaves
14. *Use of by-product* — Fuel / cattle feed / organic matter / spices (leaves)

484. *CROP*— CURRY LEAF/Mitha neem/Gandhela/Kadhineem/ *Karipata/ Curry patta*

Scientific Name — Murraya koenigii (L.) Spreng

Family — Rutaceae

Herb / Shrub / Tree — Shrub (large) / tree

Annual / Biennial / Perennial — Perennial

Used as — Spice and condiments

1. *Climatic requirements* — Warm weather
2. *Soil requirements* — Loamy, sandy loam and alluvial soils
3. *Required land situations* — High land
4. *Importance of crop* — Leaves are used for flavouring curries
5. *Fertilizer management* — 80 kg N, 50 kg P_2O_5 and 30 kg K_2O/ha, annually
6. *Water management* — 3-4 irrigations annually
7. *Duration of crop* — 3-4 years
8. *Parts used* — Leaves
9. *Where grown in India* — T.N., Maharashtra, Karnataka, North India, West Bengal, Orissa etc.

10. *Export possibility* — High
11. *Economic yields (q / ha)* — 5-10 (dry) annually
12. *Economic value (Rs. / ha)* — 30,000 to 60,000 annually
13. *By-product (If any)* — Plants and branches
14. *Use of by-product* — Fencing / fuel / organic matter / cattle feed

485. *CROP*— BARI CARDAMOM / Bengal cardamom / *Bari elaichi / Baraalach*

Scientific Name* — *Amomum aromaticum Roxb. / ***A. subulatum*** Roxb.

Family — Zingiberaceae

Herb / Shrub / Tree — Herb

Annual / Biennial / Perennial — Perennial

Used as — Spice (fruits) / medicine (seed oil for stomach)

1. *Climatic requirements* — Cool weather conditions
2. *Soil requirements* — Alluvial, loamy and sandy loam soil
3. *Required land situations* — Low land (swampy) under the trees
4. *Importance of crop* — Grows in swampy areas
5. *Fertilizer management* — 80 kg N, 50 kg P_2O_5 and 30 kg K_2O/ha, annually
6. *Water management* — No irrigation, as it grows in water flows
7. *Duration of crop* — 2-3 years
8. *Parts used* — Fruits
9. *Where grown in India* — Mostly in West Bengal, Sikkim, Assam etc.
10. *Export possibility* — High
11. *Economic yields (q / ha)* — 10-15 (dry) annually
12. *Economic value (Rs. / ha)* — 50,000 to 60,000 annually
13. *By-product (If any)* — Plants and leaves
14. *Use of by-product* — Fuel / cattle feed / organic matter / mulch.

Chapter - 11

INDUSTRIAL CROPS (Annual and Perennial)

INTRODUCTION

The plant, which is used in industry, for obtaining a particular product commercially on a mass scale, is known as industrial crop. These industrial crops have been classified as annual and perennial.

Groundnut (*Arachis hypogaea* L.) the annual plant is used as industrial crop for getting oil; whereas rubber (*Ficus elastica* Roxb.) the perennial plant is used as industrial crop for getting latex for type of vehicle or other rubber products.

The salient points on identification, management, production economics, exportability etc. of some of industrial crops, are indicated as below.

1. ANNUAL

486. CROP—EAST INDIAN TOBACCO / Tamak / Vilaiti Tambaku / Tamaki

Scientific Name* — *Nicotiana tabacum L. / ***N. rustica*** L.

Family — Solanaceae

Herb / Shrub / Tree — Herb

Annual / Biennial / Perennial — Annual

Used as — Narcotics industries (from leaves) production (like cigarette)

1. *Climatic requirements* — Cool weather
2. *Soil requirements* — Alluvial, loamy, sandy loam or black cotton soils
3. *Required land situations* — High / medium / low lands

4. *Importance of crop* — Nicotine content in leaves is high in *rustica*
5. *Fertilizer management* — 50 kg N, 50 kg P_2O_5 and 50 kg K_2O/ha
6. *Water management* — 4-5 irrigations
7. *Duration of crop* — 120-130 Days
8. *Parts used* — Leaves
9. *Where grown in India* — Punjab, Bihar, West Bengal, U.P., A.P., Maharashtra, Karnataka etc.
10. *Export possibility* — High
11. *Economic yields (q / ha)* — 5-10 (dry)
12. *Economic value (Rs. / ha)* — 40,000 to 50,000
13. *By-product (If any)* — Stem / branches / leaf ridge
14. *Use of by-product* — *Gutka* and other tobacco products / Snuff / *Khaini*

487. *CROP* — JUTE / Pat / White jute / Titapat / Mithapat / Pattam

***Scientific Name* — *Corchorus capsularis* / *C. olitorius* L.**

Family — Tiliaceae

Herb / Shrub / Tree — Herb

Annual / Biennial / Perennial — Annual

Used as — Jute mill factories (for sacks/carpet/mat etc.) / rope / clothings

1. *Climatic requirements* — Warm weather with high humidity
2. *Soil requirements* — Alluvial, loamy or sandy loam soils
3. *Required land situations* — High / medium / low lands
4. *Importance of crop* — A tall (nearly 2-3 m.) plant where from bast fibre is extracted; *capsularis* is resistant to flood water
5. *Fertilizer management* — 80 kg N, 50 kg P_2O_5 and 50 kg K_2O/ha
6. *Water management* — 1-2 irrigations (though rainfed in West Bengal)
7. *Duration of crop* — 120-130 days
8. *Parts used* — Bast fibre from the stem / *Capsularis* in low land
9. *Where grown in India* — West Bengal, Assam, Bihar, Orissa, Andhra Pradesh etc.
10. *Export possibility* — Very high
11. *Economic yields (q / ha)* — 15-20 (dry fibre)
12. *Economic value (Rs. / ha)* — 30,000 to 60,000
13. *By-product (If any)* — Stick / leaves

14. *Use of by-product* — Paper / board / manure / cattle feed / organic matter / vegetable

488. *CROP* — SUGARCANE / Noble cane / Ganna / Paunda / *Ikshu* / *Akh*

Scientific Name* — *Saccharum officinarum L.

Family — Gramineae

Herb / Shrub / Tree — Tall grass / herb

Annual / Biennial / Perennial — Perennial

Used as — Sugar factories / sugar industries (for sugar) / molasses / paper board / alcohol industry paper

1. *Climatic requirements* — Warm weather
2. *Soil requirements* — Alluvial / loamy / sandy loam and clayey soils
3. *Required land situations* — High / medium / low lands
4. *Importance of crop* — Sweet cane (node and inter node) for sugar
5. *Fertilizer management* — 120 kg N2, 60 kg P_2O_5 and 50 kg K_2O/ha (throughout life)
6. *Water management* — 5-6 irrigations
7. *Duration of crop* — 1-2 years (*Eksali* & *adhsali*)
8. *Parts used* — Cane (stems)
9. *Where grown in India* — Uttar Pradesh, Bihar, Punjab, Maharashtra, T.N., Karnataka etc.
10. *Export possibility* — High
11. *Economic yields (q / ha)* — 20-25 annually (sugar)
12. *Economic value (Rs. / ha)* — 40,000 to 50,000 annually
13. *By-product (If any)* — Leaves and bagasse (used materials)
14. *Use of by-product* — Fuel / cattle feed / paper / board / soil amendments

489. *CROP* — SUGAR BEET / Beet root / Garden beet / Chukandar/ Beet / Bit

Scientific Name* — *Beta vulgaris L.

Family — Chenopodiaceae

Herb / Shrub / Tree — Herb

Annual / Biennial / Perennial — Biennial

Used as — Sugar industries (with swollen roots) / vegetable

1. *Climatic requirements* — Cool weather conditions
2. *Soil requirements* — Loamy / sandy loam / saline and alluvial soils

3. *Required land situations* — High / medium / low lands
4. *Importance of crop* — Roots are swollen just below or above the soil levels
5. *Fertilizer management* — 80 kg N, 50 kg P_2O_5 and 40 kg K_2O/ha (at growing sage)
6. *Water management* — 3-4 irrigations
7. *Duration of crop* — 2 years
8. *Parts used* — Roots and leaves (for vegetables / sugar industries)
9. *Where grown in India* — All India except southern states
10. *Export possibility* — Very high
11. *Economic yields (q / ha)* — 300-350 (raw)
12. *Economic value (Rs. / ha)* — 80,000 to 1,00,000
13. *By-product (If any)* — Leaves and stems (negligible)
14. *Use of by-product* — Vegetable / cattle feed / organic matter / fuel mulch

490. ***CROP*—PEANUT / GROUNDNUT / Chinabadam / Badam / Mungphali**

Scientific Name* — *Arachis hypogaea L.

Family — Papilionaceae

Herb / Shrub / Tree — Herb

Annual / Biennial / Perennial — Annual

Used as — Oil or butter industries with peanut

1. *Climatic requirements* — Moderate warm weather
2. *Soil requirements* — Sandy loam, loamy or alluvial soils
3. *Required land situations* — High / medium / low lands
4. *Importance of crop* — Pods (peanut) are developed below the ground
5. *Fertilizer management* — 40 kg N, 80 kg P_2O_5 and 50 kg K_2O/ha
6. *Water management* — 3-4 irrigations
7. *Duration of crop* — 120-150 days
8. *Parts used* — Nut (for oils) / confectionery
9. *Where grown in India* — Gujarat, Maharashtra, Orissa, Karnataka, A.P., T.N., Kerala, West Bengal etc.
10. *Export possibility* — High
11. *Economic yields (q / ha)* — 15-25 (shelled pods)
12. *Economic value (Rs. / ha)* — 50,000 to 80,000
13. *By-product (If any)* — Vine with leaves and shell

14. *Use of by-product* — Fuel / cattle feed / paper / board / organic matter / green manure.

491. *CROP* — RAPE AND MUSTARD / Indian rape / Indian mustard / Toria / Rai, *Sarshe* / *Sarisha*

Scientific Name* — *Brassica campestries L. var. ***toria*** Duth & Full / ***B. juncea*** (L.) Czern and Coss

Family — Cruciferae

Herb / Shrub / Tree — Herb (branched)

Annual / Biennial / Perennial — Annual

Used as — Rape and mustard oil industries (with rape (*toria*) and mustard (*Rai*)

1. *Climatic requirements* — Cool weather
2. *Soil requirements* — Alluvial, loamy, sandy loam soils
3. *Required land situations* — High / medium / low lands
4. *Importance of crop* — Yellow flowers in long inflorescences
5. *Fertilizer management* — 80 kg N, 50 kg P_2O_5 and 40 kg K_2O/ha
6. *Water management* — 3-4 irrigations
7. *Duration of crop* — 120-130 days
8. *Parts used* — Seeds
9. *Where grown in India* — All over North India and East India (in winter)
10. *Export possibility* — High
11. *Economic yields (q / ha)* — 15-25 (dry seeds)
12. *Economic value (Rs. / ha)* — 40,000 to 80,000
13. *By-product (If any)* — Plants, leaves, silliqua and refuse materials
14. *Use of by-product* — Fuel / cattle feed / organic matter / oil cake

492. *CROP* — CASTOR / Castor seed / Castor bean / Castor oil / Arandi / *Ban Bheranda*

Scientific Name* — *Ricinus communis L.

Family — Euphorbiaceae

Herb / Shrub / Tree — Shrub/tree (small)

Annual / Biennial / Perennial — Perennial

Used as — 'Castor oil' manufacturing industry / Candle industry

1. *Climatic requirements* — Warm weather
2. *Soil requirements* — Loamy, sandy loam, sandy and alluvial soils
3. *Required land situations* — High / medium lands

4. *Importance of crop* — Latex comes out of stem and leaves/bio-fuel
5. *Fertilizer management* — 80 kg N, 50 kg P_2O_5 and 30 kg K_2O/ha, annually
6. *Water management* — 3-4 irrigations annually
7. *Duration of crop* — 3-4 years
8. *Parts used* — Seeds
9. *Where grown in India* — A.P., Maharashtra, Karnataka, Orissa, West Bengal, Bihar, M.P. etc.
10. *Export possibility* — Very high
11. *Economic yields (q / ha)* — 10-15(dry seeds) annually
12. *Economic value (Rs. / ha)* — 50,000 to 80,000 annually
13. *By-product (If any)* — Plants with leaves
14. *Use of by-product* — Fuel / cattle feed / organic matter / mulch

493. *CROP* — COTTON / Tula / Kapas / Upland cotton

Scientific Name — Gossypium hirsutum L*

Family — Malvaceae

Herb / Shrub / Tree — Shrub (much branched bush)

Annual / Biennial / Perennial — Annual

Used as — Textile industries (manufacturing of clothings)

1. *Climatic requirements* — Warm weather
2. *Soil requirements* — Black cotton soil, loamy and sandy loam soils
3. *Required land situations* — High / medium lands
4. *Importance of crop* — Woolly hairs on the bolls and stem / lint cotton
5. *Fertilizer management* — 80 kg N, 50 kg P_2O_5 and 30 kg K_2O/ha
6. *Water management* — 4-5 irrigations
7. *Duration of crop* — 120-140 days
8. *Parts used* — Bolls (sack of cotton fibre)
9. *Where grown in India* — Punjab, Rajasthan, Madhya Pradesh, Tamil Nadu, Maharashtra, U.P., A.P., Karnataka etc.
10. *Export possibility* — Very high
11. *Economic yields (q / ha)* — 10-25 (dry bolls)
12. *Economic value (Rs. / ha)* — 50,000 to 80,000
13. *By-product (If any)* — Bushy plants and leaves
14. *Use of by-product* — Fuel / paper / board / fencing / vine trailing.

* Other 3 types of cottons are *G. arboreum* (Tree cotton) *G. arboreum* var *nadam* (Egyptian cotton) and *G. herbaceum* (Asiatic cotton), which are also used in textile in different forms for different brands.

494. *CROP* — OIL PALM / Tel tal / *Tel gach*

Scientific Name* — *Elaeis guineensis Jacq.

Family — Palmae

Herb / Shrub / Tree — Tree

Annual / Biennial / Perennial — Perennial

Used as — Oil industry (for edible oils)

1. *Climatic requirements* — Warm weather
2. *Soil requirements* — Loamy and sandy loam soils
3. *Required land situations* — High lands
4. *Importance of crop* — Palm oil is obtained from fleshy mesocarp of fruit
5. *Fertilizer management* — 80 kg N, 50 kg P_2O_5 and 30 kg K_2O/ha,
6. *Water management* — 3-4 irrigations annually
7. *Duration of crop* — 10-15 years annually
8. *Parts used* — Mesocarp of fruits
9. *Where grown in India* — Maharashtra, Karnataka, Kerala etc.
10. *Export possibility* — High
11. *Economic yields (q / ha)* — 10-15
12. *Economic value (Rs. / ha)* — 40,000 to 50,000
13. *By-product (If any)* — Trunk / leaves
14. *Use of by-product* — Poles / furniture / cattle feed / organic matter / fuel

495. CROP — PINEAPPLE / Ananas / Anaras

Scientific Name* — *Ananas comosus (L.) Merr.

Family — Bromeliaceae

Herb / Shrub / Tree — Herb (leaves spiny)

Annual / Biennial / Perennial — Biennial

Used as — Pineapple soft drink industries (from fruit)

1. *Climatic requirements* — Warm weather conditions
2. *Soil requirements* — Alluvial, loamy or sandy loam or laterite soils
3. *Required land situations* — High land

4. *Importance of crop* — Used as canned fruit or fresh fruit other than soft drink
5. *Fertilizer management* — 80 kg N, 50 kg P_2O_5 and 50 kg K_2O/ha (Total)
6. *Water management* — 3-5 irrigations
7. *Duration of crop* — 1-2 years
8. *Parts used* — Fruits
9. *Where grown in India* — West Bengal, T.N., A.P. (Coastal), Assam, Kerala, Karnataka, Tripura and Orissa
10. *Export possibility* — Very high
11. *Economic yields (q / ha)* — 150-200 (fresh) total
12. *Economic value (Rs. / ha)* — 50,000 to 80,000
13. *By-product (If any)* — Plants and leaves
14. *Use of by-product* — Fuel / paper / board / cloth / oxalic acid (with waste materials) / organic matter

496. *CROP* — POTATO / Alu / Golalu

Scientific Name* — *Solanum tuberosum L.

Family — Solanaceae

Herb / Shrub / Tree — Herb

Annual / Biennial / Perennial — Annual

Used as — Potato product industries (chips / starch and alcohol)

1. *Climatic requirements* — Cool weather conditions
2. *Soil requirements* — Alluvial, loamy and sandy loam soils
3. *Required land situations* — High / medium / low lands
4. *Importance of crop* — 'Tubers' are formed just below the soil surface, on the stem of the plant
5. *Fertilizer management* — 120 kg N, 80 kg P_2O_5 and 80 kg K_2O/ha
6. *Water management* — 5-6 irrigations
7. *Duration of crop* — 110-130 days
8. *Parts used* — Tubers (i.e. potato)
9. *Where grown in India* — All over India except South
10. *Export possibility* — Very high
11. *Economic yields (q / ha)* — 250-300
12. *Economic value (Rs. / ha)* — 60,000 to 80,000
13. *By-product (If any)* — Haulm and leaves
14. *Use of by-product* — Cattle feed / fuel / organic matter / mulch

497. *CROP* — RAMIE / Kankhura / *Rami* / *Kund*

Scientific Name* — *Boehmeria nivea (L) Gaud

Family — Urticaceae

Herb / Shrub / Tree — Shrub (small)

Annual / Biennial / Perennial — Perennial

Used as — Fibre textile (thread / fabric / gas-mantel / packing material, rope and tow)

1. *Climatic requirements* — Moderate warm climate
2. *Soil requirements* — Alluvial, loamy and sandy loam soils
3. *Required land situations* — High
4. *Importance of crop* — Bast fibre from the stem / Grows in high rainfall areas
5. *Fertilizer management* — 60 kg N, 50 kg P_2O_5 and 50 kg K_2O/ha
6. *Water management* — 5-6 irrigations annually
7. *Duration of crop* — 2-3 years
8. *Parts used* — Stem (for fibres)
9. *Where grown in India* — Assam, West Bengal and Orissa (also grows in Himalayan regions and Nilgiri wildly)
10. *Export possibility* — Very high
11. *Economic yields (q / ha)* — 15-20 (fibre dry) annually
12. *Economic value (Rs. / ha)* — 30,000 to 50,000 annually
13. *By-product (If any)* — Stick and leaves
14. *Use of by-product* — Paper pulp / fuel / cattle feed / organic matter/ mulch

2. PERENNIAL CROPS

498. *CROP* — INDIAN RUBBER / *Rabar* /Assam rubber / Bor / Attah bar

Scientific Name* — *Ficus elastica Roxb

Family — Moraceae

Herb / Shrub / Tree — Tree (small)

Annual / Biennial / Perennial — Perennial

Used as — Rubber industries (tyres / rubbers etc.)

1. *Climatic requirements* — Warm climate conditions
2. *Soil requirements* — Loamy and sandy loam soils
3. *Required land situations* — High lands

4. *Importance of crop* — Ever green tree / Latex, obtained from bark, is used for preparing rubber; can withstand flood water
5. *Fertilizer management* — 80 kg N, 50 kg P_2O_5 and 30 kg K_2O/ha, annually
6. *Water management* — 4-5 irrigations annually
7. *Duration of crop* — 5-10 years
8. *Parts used* — Bark of plants/branches
9. *Where grown in India* — Assam, Khasi Hills, Tripura, Kerala etc.
10. *Export possibility* — Very high
11. *Economic yields (q / ha)* — 12-15 (raw latex) annually
12. *Economic value (Rs. / ha)* — 50,000 to 80,000 annually
13. *By-product (If any)* — Trunk / branches / leaves
14. *Use of by-product* — Fuel / timber / cattle feed / organic matter.

499. *CROP* — QUININE / Cinchona / Cincona / *Sinkona*

Scientific Name* — *Cinchona calisaya Wedd

Family — Rubiaceae

Herb / Shrub / Tree — Tree

Annual / Biennial / Perennial — Perennial

Used as — Drug industries (quinine for malaria treatment)

1. *Climatic requirements* — Cool weather conditions
2. *Soil requirements* — Loamy and sandy loam soils
3. *Required land situations* — High lands
4. *Importance of crop* — Bark is the source of quinine, a drug for malaria fever
5. *Fertilizer management* — 80 kg N, 50 kg P_2O_5 and 30 kg K_2O/ha, annually
6. *Water management* — 4-5 irrigations annually
7. *Duration of crop* — 10-15 years
8. *Parts used* — Bark from trunk / branches
9. *Where grown in India* — Nilgiris, Sikkim, W. Bengal (Darjeeling) etc.
10. *Export possibility* — Very high
11. *Economic yields (q / ha)* — 5-10 (bark, dry) annually
12. *Economic value (Rs. / ha)* — 80,000 to 1,20,000
13. *By-product (If any)* — Trunk / branches / leaves
14. *Use of by-product* — Fuel / timber / poles / fences / cattle feed / organic matter

500. ***CROP* — BAMBOO / Bansh / Peka / Chau / *Bangsa***

Scientific Name* — *Bambusa tulda Roxb

Family — Bambusaceae

Herb / Shrub / Tree — Shrub / culm

Annual / Biennial / Perennial — Perennial

Used as — Paper industries* / construction / housing

1. *Climatic requirements* — Moderately warm climate
2. *Soil requirements* — Alluvial, loamy and clayey soils
3. *Required land situations* — High / medium / low lands
4. *Importance of crop* — Mature culms are used for manufacturing paper / baskets / fans; can withstand flood water
5. *Fertilizer management* — 80 kg N, 50 kg P_2O_5 and 30 kg K_2O/ha, annually
6. *Water management* — 2-3 irrigations annually
7. *Duration of crop* — 3-4 years
8. *Parts used* — Culms of the stem
9. *Where grown in India* — Assam and Bengal, and other parts in India
10. *Export possibility* — High
11. *Economic yields (q / ha)* — 150-300 annually
12. *Economic value (Rs. / ha)* — 80,000 to 1,50,000
13. *By-product (If any)* — Branch and leaves
14. *Use of by-product* — Fences / cattle feed / stick / organic matter / fuel

* Notes : Different types of bamboos (under different names) are available in different places, but all of them are used in manufacturing papers / boards etc. These are as follows –

1. *Bambusa arundinaceae* (Retz) Willd (Thorny bamboo), 2. *B. balcoa* Roxb. (Plains bamboo, *Bhaluka*), 3. *B. falcata* Hort (Himalayan Bamboo), 4. *B. polymorph* Murno. (Betua) and 5. *B. vulgaris* Schrader, etc.

501. ***CROP* — SILK COTTON (White and red) / Semur / Safed simul**

Scientific Name* — *Bombax ceiba L. / ***B. malabaricum*** DC.

Family — Bombacaceae

Herb / Shrub / Tree — Tree

Annual / Biennial / Perennial — Perennial

Used as — Bedding industries (pillow / quilt / life belt / life jacket / cushions / surgical bandage and clothings) / box industries; can also withstand flood water

1. *Climatic requirements* — Warm climatic conditions
2. *Soil requirements* — Sandy, sandy loam and loamy soils
3. *Required land situations* — High / medium / low lands
4. *Importance of crop* — Bright big red / white flowers. Silk floss obtained from inner wall of fruit, is used for stuffing pillows and mattresses
5. *Fertilizer management* — 60 kg N, 50 kg P_2O_5 and 30 kg K_2O/ha annually
6. *Water management* — 3-4 irrigations annually
7. *Duration of crop* — 10-15 years
8. *Parts used* — Silk floss in fruits
9. *Where grown in India* — All over India, but more in Assam, West Bengal, Maharashtra, Orissa, A.P. etc.
10. *Export possibility* — Very high
11. *Economic yields (q / ha)* — 5-10 (floss) annually
12. *Economic value (Rs. / ha)* — 50,000 to 90,000 annually
13. *By-product (If any)* — Trunk / branches / leaves
14. *Use of by-product* — Packing box / match box / boat / furniture / cattle feed / organic matter / fuel

502. *CROP* — SILVER WATTLE / *Sada babla*

Scientific Name — Acacia dealbata Link

Family — Mimosaceae

Herb / Shrub / Tree — Tree

Annual / Biennial / Perennial — Perennial

Used as — Tanning / perfume and dyeing industries (in Palni and Nilgiri Hills)

1. *Climatic requirements* — Cool climatic conditions
2. *Soil requirements* — Loamy and sandy loam soils
3. *Required land situations* — High lands
4. *Importance of crop* — Flowers gives perfumes and bark gives tans and dyes
5. *Fertilizer management* — 40 kg N, 50 kg P_2O_5 and 30 kg K_2O/ha, annually
6. *Water management* — 3-4 irrigations annually
7. *Duration of crop* — 15-20 years

8. *Parts used* — Bark (tanning and dyeing) and flower (perfumes)
9. *Where grown in India* — Tamil Nadu (Palni and Nilgiri Hill), H.P. and Uttar Pradesh
10. *Export possibility* — Very high
11. *Economic yields (q/ha)* — 0.5-1.0 (flower) + 5-10 (dry bark), annually
12. *Economic value (Rs./ha)* — 80,000 to 1,50,000 annually
13. *By-product (If any)* — Trunk / branches / leaves
14. *Use of by-product* — Timber / fuel / cattle feed / organic matter

503. *CROP* — SWEET ACACIA / Cassie / *Gandh babul / Gandhababul* / Divana babul / Vilaiti babul / *Gandha babla*

Scientific Name — Acacia farnesiana (L.) Willd

Family — Mimosaceae

Herb / Shrub / Tree — Shrub / tree (thorny)

Annual / Biennial / Perennial — Perennial

Used as — Perfume, gum and tanning industries

1. *Climatic requirements* — Warm weather
2. *Soil requirements* — Loamy and sandy loam soils
3. *Required land situations* — High lands
4. *Importance of crop* — Golden yellow fragrant flowers (Cassie perfume)
5. *Fertilizer management* — 40 kg N, 50 kg P_2O_5 and 30 kg K_2O/ha, annually
6. *Water management* — 3-4 irrigations annually
7. *Duration of crop* — 5-10 years
8. *Parts used* — Flowers and bark
9. *Where grown in India* — All over India
10. *Export possibility* — Very high
11. *Economic yields (q/ha)* — Flower : 0.5-1.0 and Bark 5-10 (dry) annually
12. *Economic value (Rs./ha)* — 80,000 to 1,00,000 annually
13. *By-product (If any)* — Trunk / branches / leaves
14. *Use of by-product* — Timber implements / fuel / cattle feed / organic matter

504. *CROP* — BLACK WATTLE / *Kalobabla*

Scientific Name — Acacia mearnsii de Willd

Family — Mimosaceae

Herb / Shrub / Tree — Tree

Annual / Biennial / Perennial — Perennial

Used as — Tanning hide and skins industries

1. *Climatic requirements* — Cool weather
2. *Soil requirements* — Loamy and sandy loam soils
3. *Required land situations* — High lands
4. *Importance of crop* — Tanning hides / skin / cloth / printing etc.
5. *Fertilizer management* — 40 kg N, 50 kg P_2O_5 and 30 kg K_2O/ha annually
6. *Water management* — 3-4 irrigations annually
7. *Duration of crop* — 10-15 years
8. *Parts used* — Bark
9. *Where grown in India* — Nilgiris, Nagaland, Shillong and Travancore
10. *Export possibility* — Very high
11. *Economic yields (q / ha)* — 5-10 (Bark dry) annually
12. *Economic value (Rs. / ha)* — 80,000 to 1,00,000 annually
13. *By-product (If any)* — Wood / leaves
14. *Use of by-product* — Rayon-grade pulp / cattle feed / fuel / organic matter

505. *CROP* — GUM-ARABIC ACACIA / Kumta / Atha babla

Scientific Name — Acacia senegal (L) Willd

Family — Mimosaceae

Herb / Shrub / Tree — Shrub / tree (with hooked spines)

Annual / Biennial / Perennial — Perennial

Used as — Gum and textile industries (stem and bark)

1. *Climatic requirements* — Hot weather conditions
2. *Soil requirements* — Sandy, sandy loam and loamy soils
3. *Required land situations* — High lands
4. *Importance of crop* — White flowers with fragrance / good gum obtained
5. *Fertilizer management* — 40 kg N, 60 kg P_2O_5 and 30 kg K_2O/ha, annually
6. *Water management* — 2-3 irrigations annually
7. *Duration of crop* — 10-15 years
8. *Parts used* — Bark

9. *Where grown in India* — Dry areas of Punjab, Haryana, Rajasthan and Gujarat
10. *Export possibility* — Very high
11. *Economic yields (q / ha)* — 5-10 (dry bark) annually
12. *Economic value (Rs. / ha)* — 50,000 to 80,000 annually
13. *By-product (If any)* — Trunk / branch / leaves
14. *Use of by-product* — Timbers / implement / fuel / cattle feed / organic matter

506. CROP — ACACIA TORTILIS / Tortilis babla / Ritha

Scientific Name* — *Acacia tortilis (Forsk) Hayne

Family — Mimosaceae

Herb / Shrub / Tree — Tree

Annual / Biennial /Perennial — Perennial

Used as — Afforestation / industries programme / fodder in dry hot areas

1. *Climatic requirements* — Hot weather
2. *Soil requirements* — Sandy and sandy loam soil
3. *Required land situations* — High
4. *Importance of crop* — Highly drought and frost resistant in deserts
5. *Fertilizer management* — 40 kg N, 50 kg P_2O_5 and 30 kg K_2O/ha, annually
6. *Water management* — If water available, 2-3 irrigation, annually
7. *Duration of crop* — 10-15 years
8. *Parts used* — Saplings / whole plant
9. *Where grown in India* — Dry hot areas of Gujarat, Rajasthan and Punjab
10. *Export possibility* — High
11. *Economic yields (q / ha)* — 10-15 (dry wood) / fodder (twigs and leaves) annually
12. *Economic value (Rs. / ha)* — 10,000 to 15,000 annually
13. *By-product (If any)* — Leaves
14. *Use of by-product* — Cattle feed / fuel / organic matter

507. *CROP* — LEMON SCENTED EUCALYPTUS / Tasmanian blue-gum eucalyptus

Scientific Name* — *Eucalyptus citriodora Hook / ***E. globules*** Labill

Family — Myrtaceae

Herb / Shrub / Tree — Tree (big and handsome)

Annual / Biennial / Perennial — Perennial

Used as — Eucalyptus industries (for oil / perfumery / deodorant) / shade and wind break

1. *Climatic requirements* — Moderate warm / cool weather
2. *Soil requirements* — Loamy and sandy loam soils
3. *Required land situations* — High lands
4. *Importance of crop* — *Citriodora*, the source of 'citronellal' for manufacture of 'Citronellol' and menthol, from leaves and terminal buds, is famous for perfumery industries. From *E. globules*, deodorant is manufactured, besides utilization of oils for asthma / bronchitis / perfumery etc.
5. *Fertilizer management* — 80 kg N, 50 kg P_2O_5 and 30 kg K_2O/ha, annually
6. *Water management* — 3-4 irrigations annually
7. *Duration of crop* — 10-20 years
8. *Parts used* — Leaves and terminal buds along with trunk
9. *Where grown in India* — Punjab, U.P., A.P., hill stations of H.P. and T.N.
10. *Export possibility* — Very high
11. *Economic yields (q / ha)* — 1-2 (oil annually)
12. *Economic value (Rs. / ha)* — 80,000 to 1,20,000 annually
13. *By-product (If any)* — Wood pulp, used leaves
14. *Use of by-product* — Board / paper / fuel / cattle feed / organic matter

508. *CROP* — KARANJA / Pongam / *Karamja*

Scientific Name — Pongamia pinnata (L) Pierre

Family — Papilionaceae

Herb / Shrub / Tree — Tree

Annual / Biennial / Perennial — Perennial

Used as — Agro-industries [seed oil (for soap making) and treatment (of skin diseases) seeds and leaves are used for manures and cattle feeds]

1. *Climatic requirements* — Warm climate
2. *Soil requirements* — Alluvial, loamy and sandy loam soils
3. *Required land situations* — High / medium
4. *Importance of crop* — Tree is medium in size with profuse flowers

5. *Fertilizer management* — 40 kg N, 50 kg P_2O_5 and 30 kg K_2O/ha, annually
6. *Water management* — 2-3 irrigations annually
7. *Duration of crop* — 10-12 years
8. *Parts used* — Seed (as bio-fuel)
9. *Where grown in India* — Near the bank of the streams in peninsulas, West Bengal and Travancore
10. *Export possibility* — High
11. *Economic yields (q / ha)* — 20-30 (seeds) annually
12. *Economic value (Rs. / ha)* — 50,000 to 80,000 annually
13. *By-product (If any)* — Leaves, branches and trunk
14. *Use of by-product* — Manures / cattle feed / fuel / timber

509. *CROP* — CASHEWNUT / Kaju / Kajubadam

Scientific Name* — *Anacardium occidentale L.

Family — Anacardiaceae

Herb / Shrub / Tree — Tree

Annual / Biennial / Perennial — Perennial

Used as — Confectionery industries [cashew nut, dessert, edible oil, cashew apple, wine, shell and for industrial uses]

1. *Climatic requirements* — Warm climate with salinity
2. *Soil requirements* — Loamy, sandy loam and alluvial soils
3. *Required land situations* — High / medium
4. *Importance of crop* — Small evergreen tree
5. *Fertilizer management* — 50 kg N, 50 kg P_2O_5 and 30 kg K_2O/ha annually
6. *Water management* — 2-3 irrigations annually
7. *Duration of crop* — 10-15 years
8. *Parts used* — Seeds (Cashewnut)
9. *Where grown in India* — Malabar, Karnataka, A.P., W.B., T.N., Orissa, Maharashtra, Goa and Kerala
10. *Export possibility* — Very high
11. *Economic yields (q / ha)* — 6-12 (dry seeds) annually
12. *Economic value (Rs. / ha)* — 50,000 to 90,000 annually
13. *By-product (If any)* — Trunk / branches / leaves
14. *Use of by-product* — Timber / fuel / cattle feed / organic matter.

Chapter - 12

FUEL (Annual and Perennial)

INTRODUCTION

The residues of some crops either in the field or plantation may be used as 'fuel' in the rural areas. All the residues of small crops in both, *rabi* / pre-*kharif* / *kharif* seasons, may be used as fuel in rural areas, therefore, straw/stalk/ stick etc. of crops, which are left behind in the field, are used as fuel. In some cases some perennial plants are also used as fuel. Their fuel assessment may be done as per production of economic yields of crops. Their monetary values are also assessed according to the amount and market price, in different places.

One type of fuel is very costly in some places but the same materials may be cheaper in other places. So production and money value differ according to places.

Sometimes, under different situations, biomass for fuel or fodder is more valuable than the economic yields of that crop.

The salient points on identification, management, production economics, exportability etc. of some crops either annual or perennial, are stated as below.

1. ANNUAL CROPS

510. *CROP*—ARAHAR / CAJAN PEA / PIGEONPEA / Redgram / Tur / Rahar /Arhar

Scientific Name* — *Cajanus cajan (L) Millsp

Family — Papilionaceae

Herb / Shrub / Tree — Shrub

Annual / Biennial / Perennial — Annual

Used as — Fuel (plant / sticks / branches / leaves)

1. *Climatic requirements* — Warm (for growth) and cool (for flower)

2. *Soil requirements* — Alluvial, loamy and sandy loam soils
3. *Required land situations* — High lands
4. *Importance of crop* — Tall and bushy plants with yellow flowers
5. *Fertilizer management* — 40 kg N, 60 kg P_2O_5 and 30 kg K_2O/ha
6. *Water management* — 2-3 irrigations
7. *Duration of crop* — 140-180 days
8. *Parts used* — Total biomass
9. *Where grown in India* — All over India, mainly in M.P., Bihar, W.B., A.P., Maharashtra, Karnataka, U.P., Bihar, Orissa etc.
10. *Export possibility* — Less
11. *Economic yields (q / ha)* — 75-150 (dry plants)
12. *Economic value (Rs. / ha)* — 8,000 to 10,000
13. *By-product (If any)* — Seeds / pod shells
14. *Use of by-product* — Pulses / Bhusa

511. *CROP* — MAIZE / Corn / Indian corn / *Makai / Bhutta*

Scientific Name* — *Zea mays L.

Family — Gramineae

Herb / Shrub / Tree — A tall grass / herb

Annual / Biennial / Perennial — Annual

Used as — Fuel (whole plant without cobs)

1. *Climatic requirements* — Moderate warm weather
2. *Soil requirements* — Alluvial, loamy and sandy loam soils
3. *Required land situations* — High lands
4. *Importance of crop* — 'Cobs' (cluster of seeds are formed though silk is there on the head)
5. *Fertilizer management* — 120 kg N, 60 kg P_2O_5 and 50 kg K_2O/ha
6. *Water management* — 4-6 irrigations
7. *Duration of crop* — 120-140 days
8. *Parts used* — Whole plant (without seeds)
9. *Where grown in India* — All over India, specially in U.P., Punjab, M.P., Karnataka, Bihar, A.P., W.B., J.&K. etc.
10. *Export possibility* — High
11. *Economic yields (q / ha)* — 60-80
12. *Economic value (Rs. / ha)* — 7,000 to 8,000
13. *By-product (If any)* — Grains (from cobs)
14. *Use of by-product* — Carbohydrate food / starch / food product

512. *CROP* — SUGARCANE / Noble cane / Ganna / Paunda / *Akh* / *Ikshu*

Scientific Name* — *Saccharum officinarum L.

Family — Gramineae

Herb / Shrub / Tree — A tall grass / herb

Annual / Biennial / Perennial — Perennial

Used as — Fuel (bagasse and leaves)

1. *Climatic requirements* — Warm weather
2. *Soil requirements* — Loamy and alluvial soils
3. *Required land situations* — High / medium / low lands
4. *Importance of crop* — Sweet cane due to sugar
5. *Fertilizer management* — 120 kg N, 60 kg P_2O_5 and 50 kg K_2O/ha
6. *Water management* — 6-7 irrigations
7. *Duration of crop* — 2 years
8. *Parts used* — Bagasse and leaves
9. *Where grown in India* — All over India, specially in U.P., Bihar, Punjab, T.N. Karnataka, Maharashtra etc.
10. *Export possibility* — High
11. *Economic yields (q / ha)* — 90-150
12. *Economic value (Rs. / ha)* — 9,000 to 12,000
13. *By-product (If any)* — Juice
14. *Use of by-product* — For manufacturing sugar / *gur* / *alcohol*

513. *CROP* — JUTE / Pat / White jute / Titapat / Narcha / Nalicha / Mithapat

Scientific Name* — *Corchorus capsularis L. / ***C. olitorius*** L.

Family — Tiliaceae

Herb / Shrub / Tree — Shrub

Annual / Biennial / Perennial — Annual

Used as — Fuel (jute stick and refuse materials)

1. *Climatic requirements* — Warm weather with high humidity
2. *Soil requirements* — Alluvial, loamy and sandy loam soils
3. *Required land situations* — Low / medium lands
4. *Importance of crop* — Round / elongated cluster of pods / *Olitorius* for high land and *Capsularis* for low lands
5. *Fertilizer management* — 60 kg N, 60 kg P_2O_5 and 50 kg K_2O/ha
6. *Water management* — 2-3 irrigations (if necessary)

7. *Duration of crop* — 120-140 days
8. *Parts used* — Jute sticks and leaves
9. *Where grown in India* — West Bengal, Assam, Bihar, Orissa, U.P., A.P., Tripura etc.
10. *Export possibility* — Very high
11. *Economic yields (q / ha)* — 50-60 (stick)
12. *Economic value (Rs. / ha)* — 6,000 to 7,000
13. *By-product (If any)* — Fibre
14. *Use of by-product* — Sacks / mattress / carpet / garments

514. *CROP*—SUNNHEMP / *Shan* / *Sanpat* / Sunn / San

Scientific Name* — *Crotalaria juncea L.

Family — Papilionaceae

Herb / Shrub / Tree — Shrub (tall)

Annual / Biennial / Perennial — Annual

Used as — Fuel (from sticks and leaves) / paper

1. *Climatic requirements* — Warm weather (for seed, cool weather) with humidity
2. *Soil requirements* — Alluvial, loamy, sandy loam soils
3. *Required land situations* — High / medium lands
4. *Importance of crop* — Big red-yellow flowers
5. *Fertilizer management* — 40 kg N, 60 kg P_2O_5 and 50 kg K_2O/ha
6. *Water management* — 2-3 irrigations
7. *Duration of crop* — 120-130 days
8. *Parts used* — Stick and leaves
9. *Where grown in India* — West Bengal, Assam, Orissa, Bihar, U.P., and Tripura
10. *Export possibility* — High
11. *Economic yields (q / ha)* — 60-70 (Stick)
12. *Economic value (Rs. / ha)* — 7,000 to 8,000
13. *By-product (If any)* — Bast fibre
14. *Use of by-product* — For preparation of sacks / mat / carpet etc.

515. *CROP* — RICE / *dhan* / Paddy / Chauval / *Dhanya*

Scientific Name* — *Oryza sativa L.

Family — Gramineae

Herb / Shrub / Tree — Herb

Annual / Biennial / Perennial — Annual

Used as — Fuel (from rice straw / rice husk) / thatching material for house and *boroj*

1. *Climatic requirements* — Warm weather with plenty water
2. *Soil requirements* — Clayey and loamy soils
3. *Required land situations* — Low / medium
4. *Importance of crop* — It grows under marshy conditions, after puddling
5. *Fertilizer management* — 120 kg N, 60 kg P_2O_5 and 50 kg K_2O/ha
6. *Water management* — 2-3 irrigations (if needed, as it is grown in waterlogged conditions)
7. *Duration of crop* — 120-140 days
8. *Parts used* — All straws and rice husk
9. *Where grown in India* — All over low lying areas of India
10. *Export possibility* — Very high
11. *Economic yields (q / ha)* — 40-80
12. *Economic value (Rs. / ha)* — 6,000 to 8,000
13. *By-product (If any)* — Grains
14. *Use of by-product* — Rice after boiling / rice food products

516. *CROP* — MESTA / Kenaf / Bimlipatam jute / Patsan / *Masta* / *Ambari* / *Patwa* / *Roselle* / *Rama*

Scientific Name* — *Hibiscus sabadariffa L./ ***H. cannabinus*** L.

Family — Malvaceae

Herb / Shrub / Tree — Shrub

Annual / Biennial / Perennial — Annual

Used as — Fuel (with stick and leaves)

1. *Climatic requirements* — Warm weather with humidity
2. *Soil requirements* — Alluvial, clayey and loamy soils
3. *Required land situations* — Medium / low lands
4. *Importance of crop* — Fibre is obtained from its bark similar to jute, for industrial use
5. *Fertilizer management* — 60 kg N, 50 kg P_2O_5 and 30 kg K_2O/ha
6. *Water management* — 2-3 irrigations
7. *Duration of crop* — 120-140 days
8. *Parts used* — Sticks and leaves
9. *Where grown in India* — West Bengal, Assam, Orissa, Tripura, Bihar, T.N., U.P. etc.

10. *Export possibility* — Very high
11. *Economic yields (q / ha)* — 50-70
12. *Economic value (Rs. / ha)* — 7,000 to 8,000
13. *By-product (If any)* — Fibre
14. *Use of by-product* — Used like those of jute (bag / sack / carpet)

517. *CROP*— Dhencha / SESBANIA / *Dhaincha* / Jayanti / *Bagha-dhaincha*

Scientific Name* — *Sesbania aculeata (Willd.) Poir. / ***S. rostrata***

Family — Papilionaceae

Herb / Shrub / Tree — Shrub (tall)

Annual / Biennial / Perennial — Annual

Used as — Fuel (from plant and branches) / green manure / paper pulp

1. *Climatic requirements* — Warm and humid climatic conditions
2. *Soil requirements* — Alluvial, clayey and loamy soils
3. *Required land situations* — Low / medium / high lands
4. *Importance of crop* — Used as green manure crop at tender age
5. *Fertilizer management* — 40 kg N, 60 kg P_2O_5 and 30 kg K_2O/ha
6. *Water management* — 2-3 irrigations (if needed)
7. *Duration of crop* — 120-180 days
8. *Parts used* — Whole plant
9. *Where grown in India* — All areas in low lying plains of India
10. *Export possibility* — High
11. *Economic yields (q / ha)* — 80-90 (fuel)
12. *Economic value (Rs. / ha)* — 5,000 to 10,000
13. *By-product (If any)* — Seeds
14. *Use of by-product* — Seed for future use and multiplication

2. PERENNIAL

518. *CROP* — ACACIA Suma / *Suma Babla*

Scientific Name* — *Acacia suma Buch-Ham

Family — Bombacaceae

Herb / Shrub / Tree — Tree

Annual / Biennial / Perennial — Perennial

Used as — Fuel (wood-trunk, branch and leaves)

1. *Climatic requirements* — Warm weather

2. *Soil requirements* — Alluvial, loamy and sandy loam soils
3. *Required land situations* — High land
4. *Importance of crop* — Bark is used as tan
5. *Fertilizer management* —80 kg N, 60 kg P_2O_5 and 30 kg K_2O/ha annually
6. *Water management* — 2-3 irrigations annually
7. *Duration of crop* — 10-12 years
8. *Parts used* — Branch and leaves (as fuel)
9. *Where grown in India* — West Bengal, Bihar and W. Peninsula
10. *Export possibility* — High
11. *Economic yields (q / ha)* — 160-200
12. *Economic value (Rs. / ha)* — 16,000 to 32,000
13. *By-product (If any)* — Trunk
14. *Use of by-product* — Timber for furniture / Fuel / organic matter

519. *CROP* — SILK COTTON / *Simul* / Kopok / Safed simal

Scientific Name* — *Ceiba pentandra (L) Gaertn [Other cotton (***Bombax ceiba*** L. Red silk cotton) is in the same category]

Family — Bombacaceae

Herb / Shrub / Tree — Tree

Annual / Biennial / Perennial — Perennial

Used as — Fuel (trunk / branches / leaves)

1. *Climatic requirements* — Warm weather
2. *Soil requirements* — Alluvial, loamy or sandy loam soils
3. *Required land situations* — High land
4. *Importance of crop* — White / red big flowers (flame of the forest)
5. *Fertilizer management* — 80 kg N, 50 kg P_2O_5 and 30 kg K_2O/ha annually
6. *Water management* — 2-3 irrigation / year
7. *Duration of crop* — 10-20 years
8. *Parts used* — Some portion (except timbers) of trunk, branches and leaves
9. *Where grown in India* — All over the country, specially in Assam, West Bengal, Andaman etc.
10. *Export possibility* — High
11. *Economic yields (q / ha)* — 160-200
12. *Economic value (Rs. / ha)* — 16,000 to 20,000

13. *By-product (If any)* — Trunk
14. *Use of by-product* — Timber for furniture (like boxes)

520. CROP — MANGO / Am / Amra

Scientific Name — Mangifera indica L.

Family — Anacardiaceae

Herb / Shrub / Tree — Tree

Annual / Biennial / Perennial — Perennial

Used as — Fuel (from trunk / branch / leaves)

1. *Climatic requirements* — Warm weather
2. *Soil requirements* — Alluvial, loamy and sandy loam soils
3. *Required land situations* — High lands
4. *Importance of crop* — The leaves are needed for Hindu religious ceremonies other than fodder
5. *Fertilizer management* — 100 kg N, 50 kg P_2O_5 and 30 kg K_2O/ha annually
6. *Water management* — 2-3 irrigations annually
7. *Duration of crop* — 10-20 years
8. *Parts used* — Trunks / branches / leaves
9. *Where grown in India* — Uttar Pradesh, Punjab, Maharashtra, A.P., West Bengal, Bihar, Orissa, Karnataka, T.N. etc.
10. *Export possibility* — High
11. *Economic yields (q / ha)* — 160-200 annually
12. *Economic value (Rs. / ha)* — 50,000 to 60,000 annually
13. *By-product (If any)* — Fruits (Mangoes)
14. *Use of by-product* — Fruits / *chutney* / pickles / organic matter

521. CROP — JACK FRUIT / Kanthal / Kathal / Paras

Scientific Name — Artocarpus heterophyllus Lamk

Family — Moraceae

Herb / Shrub / Tree — Tree

Annual / Biennial / Perennial — Perennial

Used as — Fuels (residual / trunk / branches / leaves)

1. *Climatic requirements* — Warm weather conditions
2. *Soil requirements* — Alluvial, loamy and sandy loam soils
3. *Required land situations* — High lands
4. *Importance of crop* — Fruits come from stem / trunk. Heart wood is a good timber for furniture

5. *Fertilizer management* — 80 kg N, 50 kg P_2O_5 and 30 kg K_2O/ha annually
6. *Water management* — 2-3 irrigations annually
7. *Duration of crop* — 10-15 years
8. *Parts used* — Rest (other than timber) of the trunk / branches and leaves
9. *Where grown in India* — All over in India, specially Assam, Tripura, West Bengal, Orissa, Western and East Ghats
10. *Export possibility* — High
11. *Economic yields (q / ha)* — 80-1000 annually
12. *Economic value (Rs. / ha)* — 8,000 to 10,000 annually
13. *By-product (If any)* — Fruits / barks / branches
14. *Use of by-product* — Fruits / timber / dye / organic matter / fuel

522. *CROP* — MADAR / Akund / Swallow-wart / Safed ak / *Akanda*

Scientific Name* — *Calotropis procera (Willd) Dryand ex W. Ait

Family — Asclepiadaceae

Herb / Shrub / Tree — Shrub (evergreen)

Annual / Biennial / Perennial — Perennial

Used as — Fuel (stem) / branches / leaves

1. *Climatic requirements* — Warm weather
2. *Soil requirements* — Loamy, alluvial and sandy loam soils
3. *Required land situations* — High lands
4. *Importance of crop* — Ever green shrub; can withstand flood water
5. *Fertilizer management* — 80 kg N, 50 kg P_2O_5 and 30 kg K_2O/ha, annually
6. *Water management* — 2-3 irrigations annually
7. *Duration of crop* — 5-10 years
8. *Parts used* — Stem / branches / leaves
9. *Where grown in India* — All over India
10. *Export possibility* — High
11. *Economic yields (q / ha)* — 100-150 annually
12. *Economic value (Rs. / ha)* — 10,000 to 15,000 annually
13. *By-product (If any)* — Floss from seeds and powdered flowers
14. *Use of by-product* — Stuffing materials / cold / cough.

523. *CROP* — **BAMBOO / *Bansh* / Peka / Chau / *Bansha***

Scientific Name* — *Bambusa tulda Roxb

Family — Bambusaceae

Herb / Shrub / Tree — Shrub / culm

Annual / Biennial / Perennial — Perennial

Used as — Fuel (branches / leaves)

1. *Climatic requirements* — Moderate warm climate
2. *Soil requirements* — Alluvial, loamy and clayey soils
3. *Required land situations* — High / medium / low lands
4. *Importance of crop* — Mature culms are used for house construction / paper manufacture etc.
5. *Fertilizer management* — 80 kg N, 50 kg P_2O_5 and 30 kg K_2O/ha, annually
6. *Water management* — 2-3 irrigations annually
7. *Duration of crop* — 3-4 years
8. *Parts used* — Branches / leaves
9. *Where grown in India* — Assam, West Bengal, Sikkim, Meghalaya, Bihar, Orissa, M.P., Maharashtra, Karnataka etc.
10. *Export possibility* — High
11. *Economic yields (q / ha)* — 100-160 annually
12. *Economic value (Rs. / ha)* — 10,000 to 16,000 annually
13. *By-product (If any)* — Culm or stem
14. *Use of by-product* — Furniture / house construction / paper / pandal.

524. *CROP* — **SOLA PITH / *Sola* / *Saula***

Scientific Name* — *Aeschynomene aspera L

Family — Papilionaceae

Herb / Shrub / Tree — Shrub (tall)

Annual / Biennial / Perennial — Perennial

Used as — Fuels (branches / leaves)

1. *Climatic requirements* — Warm weather and waterlogging
2. *Soil requirements* — Clayey and loamy soils
3. *Required land situations* — Low / medium lands
4. *Importance of crop* — For insulating properly, the soft wood is used as packing material, hat / toys / decorations; floating type
5. *Fertilizer management* — 40 kg N, 60 kg P_2O_5 and 30 kg K_2O/ha, annually

6. *Water management* — No need of irrigation
7. *Duration of crop* — 2-3 years
8. *Parts used* — Spongy floating stem
9. *Where grown in India* — Waterlogged areas of Uttar Pradesh, Bihar, W. Bengal, Andhra Pradesh, Orissa, Maharashtra, Assam, Tamil Nadu etc.
10. *Export possibility* — High
11. *Economic yields (q / ha)* — 10-12 (residues of stem, branches and leaves) annually
12. *Economic value (Rs. / ha)* — 1,000 to 2,000
13. *By-product (If any)* — Stem (swollen) annually
14. *Use of by-product* — Packing material / preparation of hats / toys bottle cork / swimming-jackets / life belt

525. *CROP* — SIRIS / KOKKO / Lebek tree / *Sirisi*

Scientific Name* — *Albizia lebbeck (L) Benth

Family — Mimosaceae

Herb / Shrub / Tree — Tree

Annual / Biennial / Perennial — Perennial

Used as — Fuel (branches and leaves)

1. *Climatic requirements* — Warm climate
2. *Soil requirements* — Loamy, alluvial and sandy loam soils
3. *Required land situations* — High land
4. *Importance of crop* — Common road side / avenue planting. It is also used in tea and coffee plantation for shade
5. *Fertilizer management* — 60 kg N, 80 kg P_2O_5 and 30 kg K_2O/ha annually
6. *Water management* — 2-3 irrigations annually
7. *Duration of crop* — 10-15 years
8. *Parts used* — Branches / twig / leaves / residue materials
9. *Where grown in India* — All over India specially Assam, West Bengal, T.N., M.P., Punjab, U.P., Orissa etc.
10. *Export possibility* — High
11. *Economic yields (q / ha)* — 120-200 annually
12. *Economic value (Rs. / ha)* — 15,000 to 20,000 annually
13. *By-product (If any)* — Trunk
14. *Use of by-product* — Timber for furniture / poles / organic matter

526. ***CROP* — Hizzal / Hjal / Hijjal / Hijal**

Scientific Name — Barringtonia acutaugula (L) Gaertn

Family — Barringtoniaceae

Herb / Shrub / Tree — Tree

Annual / Biennial / Perennial — Perennial

Used as — Fuel (Branches / leaves)

1. *Climatic requirements* — Warm weather (can grow in waterlogged condition)
2. *Soil requirements* — Clayey, sandy loam and loamy soils
3. *Required land situations* — Low / medium / high lands
4. *Importance of crop* — Roadside plant for pretty flowers with pendulous spikes / Grows in deep water also
5. *Fertilizer management* — 80 kg N, 50 kg P_2O_5 and 30 kg K_2O /ha, annually
6. *Water management* — 2-3 irrigations (if necessary)
7. *Duration of crop* — 10-15 years
8. *Parts used* — Branches and leaves (some times trunk which is used for boat building etc.)
9. *Where grown in India* — Assam, West Bengal, Tripura, Orissa, Bihar, U.P., A.P., T.N., Kerala etc.
10. *Export possibility* — High
11. *Economic yields (q / ha)* — 40-50 annually
12. *Economic value (Rs. / ha)* — 40,000 to 10,000 annually
13. *By-product (If any)* — Trunk (stem)
14. *Use of by-product* — Timber for furniture / boat / poles

Notes : Besides the above mentioned crops under the heading 'FUEL', seeds of some of the crops (*Jatropha* spp. / Pongam (*Karamja* etc.) may be utilized for 'biofuels'. Different species of *Jatropha* (*Jatropha curcas* L.) / *J. glandulifera* Roxb. / *J. gossypifolia* L. etc.), could be used as strong 'bio-fuel' crops in India.

PART 3 :
WATER-CROPS

Chapter - 13

CROPS IN WATER (IN DIFFERENT SEASONS)

INTRODUCTION

Chapter 13 deals with different water-crops, which mostly have not yet been utilized commercially, but with the increase of population in the country, these crops that grow in waterbodies should be utilized for different purposes (such as foods / fodders / vegetables / medicines / flowers / spices / soil-binder / industrial uses / fruits / fuels / beautification / organic matter / mulch etc.) during different seasons.

The climatic conditions of the three main crop-seasons (e.g. *kharif*, *rabi* and pre-*kharif*) in the country are different in different parts of the country.

In India 'rainy / *kharif* season' varies from place-to-place. In the eastern part, the rainfall is high (ranging from 1,200 mm to 3,000 mm), whereas, in the central tract, rainfall is medium (from 800 mm to 1400 m). In the western tract, of course, rainfall is very low (ranges in between 600 mm to 900 mm). In the southern tract, of course, rainfall is high (except some areas), ranging from 1,000 mm to 2,500 mm. During this period, temperature is also more or less bearable in most of the areas (33°C to 41°C). The lowlying areas in the eastern parts of India are also inundated due to excessive rainfall and as a result, these lowlying areas accumulate water and farmers go for rice cultivation, as well as for most of the water-crops in the rainy season.

'Winter / *rabi* season' in India varies from place-to-place. In the northern parts winter is very severe (temperature varies from 0 to 8°C), whereas it is mild in the east, central and western tracts (7 to 14°C), while it is normal in the southern states. During winter season in India, rainfall is very low, in general, but in the southern parts, rain fall is received due to *'north-east monsoon'*, and from there slight amount of rainfall is obtained; also small amount of rainfall is received in other parts of the country particularly in the eastern parts. Here in the big waterbodies, some type of water-crops are grown in winter season also.

'Summer / pre-*kharif* season in India, varies from place-to-place. In the northern, central and western parts, summer is very severe (temperature ranges from 33 to 46°C), whereas it is mild in some eastern and southern parts of the country. In the eastern parts, rainfall is frequent with less amount of water; of course, water accumulation in waterbodies is higher in later part of the season. Hence some water-crops grow well.

Therefore, in some lowlying areas, particularly in the eastern parts of the country, huge number of water-crops are observed mostly in rainy/*kharif* season; less in winter or *rabi* season and moderate in summer / pre-*kharif* season. The water-crops are described below with the new format:

527. *CROP*—ASTERACANTHA/HYGROPHILA/ Talmakhana / Kulekhara / Asterakantha

Scientific Name — Asteracantha longifolia (L) Nees. / ***Hygrophila auriculata*** (Schum) Heine

Family — Acanthaceae

Herb / Shrub / Tree — Herb (tall)

Annual / Biennial / Perennial — Perennial

Used as — Vegetable / medicine / fuel / organic matter

1. *Climatic requirements* — Warm weather with normal temperature and rainfall
2. *Soil requirements* — Soils with organic matter in marshy land
3. *Required land situations* — Low land with marshy or shallow-water conditions
4. *Importance of the Crop* — When leaves are boiled, pinkish water is obtained. Some varieties are spinous, some are not
5. *Fertilizer management* — No fertilizer management, as it is in marshy land; but N, P_2O_5 and K_2O @ 100, 50 & 50 kg/ha respectively required annually
6. *Duration of crop* — Perennial, continuous plant-harvest
7. *Parts used* — Leaves, seeds and roots are used as medicine; vegetable, organic matter and fuel
8. *Where grown in India* — In eastern part of the country, in low lands
9. *Export possibility* — Processed medicines may be exported
10. *Economic estimated Yields (q / ha)* — 200 (green biomass)
11. *Economic value (Rs. / ha)* — 2500 (as estimated, because of non-commercial)
12. *By-product (If any)* — Stems and old leaves
13. *Use of by-product* — Fuel (after drying) / organic matter

14. *Peoples' response for use* — Good, though not commercially cultivated but locally utilized.

528. *CROP*—ARUM / TARO / COLOCASIA / Kachu / Arvi / Ghuiyan / Kachalu / Jal kachu

Scientific Name* — *Colocasia escutenta (L.) Schott. / ***Arum esculentum*** L.

Family — Araceae

Herb / Shrub / Tree — Herb

Annual / Biennial / Perennial — Perennial

Used as — Vegetable / fuel / organic matter / mulch

1. *Climatic requirements* — Warm weather with normal temperature and rainfall
2. *Soil requirements* — Soils with organic matter in water or marshy lands
3. *Required land situations* — Low lands having marshy lands / waterlogged conditions
4. *Importance of the Crop* — Its sap gives permanent black stains in cloths; gives yellow long flowers.
5. *Fertilizer management* — When in dry or wet conditions, it requires N, P_2O_5 and K_2O @ 100, 50 & 50 kg/ha respectively, annually. Normally no fertilizer is given
6. *Duration of the Crop* — Perennial in nature; harvested annually
7. *Parts used* — Its stolon, rhizomes, petiole and leaves are used as vegetable; starchy tuberous rhizomes are edible
8. *Where grown in India* — Mostly in low lying areas of the country
9. *Export possibility* — Dry processed rhizomes may be exported
10. *Economic estimated Yields (q / ha)* — 300 (green biomass)
11. *Economic value (Rs. / ha)* — 4000 to 5000 (Estimated)
12. *By-product (If any)* — Fuel and organic matter
13. *Use of by-product* — As a fuel or as organic matter
14. *Peoples' response for use* — Good to some persons, particularly for 'sag' and other vegetable-forms.

Notes : Taros or arums are of different types, both under dry or waterlogged conditions :

1. *Colocasia esculenta* (old cocoyam) is generally arums which grows in marshy lands
2. *Colocasia esculenta* var *esculenta* (known as 'dasheen taros'/ panikachu/ jalkachu/narikeli-kachu / sola kachu / saur–grow in shallow/deep water.

3. *Colocasia esculenta* var. *antiquorum* (known as 'eddoe taro' / gati kachu / mukhi kachu / chhara-kachu / arvi / panchamukhi kachu / ghat kachu/saru). Arums grow under different conditions – such as, (a) General *Colocasia esculenta* grows in dampy/wet / marshy lands, while (b) *Colocasia esculenta* var. *esculenta* (dasheen taro) grows in shallow/deep water, but (c) *Colocasia esculenta* var. antiquorum (eddoe taro) grows in dry soils

529. *CROP*—WATER SPINACH / SWAMP CABBAGE / WATER BIND-WEED / Kalmisak

Scientific Name — Ipomoea reptans (L.) Poir. / ***I. aquatica*** *Forsk*/ ***I. alba*** L.

Family — *Convolvulaceae*

Herb / Shrub / Tree — Herb (creeping)

Annual / Biennial / Perennial — Perennial

Used as — Vegetable / fodder/ fuel / organic matter

1. *Climatic requirements* — Warm weather with normal temperature and rainfall
2. *Soil requirements* — Soils with high organic matter in shallow / deep water
3. *Required land situations* — Low and marshy lands
4. *Importance of the Crop* — It runs over water surface or grows in marshy lands, though roots are in soils
5. *Fertilizer management* — N, P_2O_5 and K_2O @ 100, 50 & 50 kg/ha annually or no fertilizer in soils, with organic matter only
6. *Duration of crop* — Perennial but cut within the year
7. *Parts used* — Leaves and tips as vegetable; whole plant as fodder for cattle, goat and buffalos / organic matter / fuel
8. *Where grown in India* — Low land under humid conditions mostly in the eastern parts of the country.
9. *Export possibility* — If low temperature and containers are available for vegetable
10. *Economic estimated Yields (q / ha)* — 150-200 (only green vegetable)
11. *Economic value (Rs. / ha)* — 3,000-4,000
12. *By-product (If any)* — Stems and old leaves
13. *Use of by-product* — Fuels / organic matter / mulching
14. *Peoples' response for use* — It is a good leafy vegetable, people mostly in low lying areas, take it as a good food

530. *CROP*—ALTERNANTHERA / Senchisak / Kanchari / Madaranga / Nadarang

Scientific Name* — *Alternanthera amoena Voss. / ***A. sessilis*** (L) Br.

Family — *Amaranthaceae*

Herb / Shrub / Tree — Herb (creeping)

Annual / Biennial / Perennial — Perennial

Used as — Vegetable / fodder/ fuel / organic matter / mulch

1. *Climatic requirements* — Warm weather with normal temperature and rainfall
2. *Soil requirements* — Soils with organic matter (mostly clayey)
3. *Required land situations* — Low lands under marshy conditions
4. *Importance of the Crop* — It creeps with the rise of flood-water
5. *Fertilizer management* — Normally no fertilizer is used for, higher yield that should be given; N, P_2O_5 and K_2O @ 100, 50 & 50 kg/ha annually, respectively
6. *Duration of crop* — It is a perennial crop, but harvested within a year
7. *Parts used* — Leaves & tips are used as vegetable; entire plant is used as fodder and stems and old leaves are used as fuels / organic matter
8. *Where grown in India* — In marshy lands of low lying areas of the country
9. *Export possibility* — High, if low temperature and containers are available
10. *Economic estimated Yields (q / ha)* — 150-200 (green vegetable)
11. *Economic value (Rs. / ha)* — 4,000-5,000
12. *By-product (If any)* — Stems and old leaves of the plants
13. *Use of by-product* — Fuel / organic matter / mulch
14. *Peoples' response for use* — Good response mostly from the people of the lowlying areas of the country.

531. *CROP*— GIANT TARO / GIANT ARUM / LARGE-LEAVED CADIUM / Mankachu / Mankand / Bara mankachu

Scientific Name* — *Alocasia indica (Roxb.) Schott / ***A. cuculiata*** (Lour) Schott.

Family — *Araceae*

Herb / Shrub / Tree — Herb

Annual / Biennial / Perennial — Perennial

Used as — Vegetable / Medicine / fuel / organic matter

1. *Climatic requirements* — Warm weather with normal temperature and rainfall
2. *Soil requirements* — Clayey soils with organic matter
3. *Required land situations* — Lowlying land under marshy conditions
4. *Importance of the Crop* — Rhizome very long, big and thick, with heavy weight; big leaves with big petiole
5. *Fertilizer management* — N, P_2O_5 and K_2O @ 100, 50 & 50 kg/ha, respectively annually; normally no fertilizer is given
6. *Duration of crop* — Perennial but harvested each year
7. *Parts used* — Rhizome, petiole and leaf are used as vegetable; rhizome also used as medicine (for laxative and diuretic)
8. *Where grown in India* — In lowlying marshy land of eastern India or elsewhere in India
9. *Export possibility* — High, if lower temperature is controlled, for rhizomes
10. *Economic estimated Yields (q / ha)* — 300-400 (green / biomass)
11. *Economic value (Rs. / ha)* — 5,000-6,000
12. *By-product (If any)* — Refuse materials of the plants
13. *Use of by-product* — Organic matter / fuels / cattle food
14. *Peoples' response for use* — Good, particularly from people of lowlying areas

532. *CROP*—PES-TIGRIDIS / Senchi sak / Ghiabati / Langulilata

Scientific Name* — *Ipomoea pes-tigridis L.

Family — *Convolvulaceae*

Herb / Shrub / Tree — Herb (twining)

Annual / Biennial / Perennial — Perennial

Used as — Vegetable / fodder/ fuel / organic matters

1. *Climatic requirements* — Warm weather with normal temperature and rainfall
2. *Soil requirements* — Clayey soil with organic matter
3. *Required land situations* — Lowlying situation, with marshy conditions
4. *Importance of the Crop* — Twining herb, can run over water surface like creeper
5. *Fertilizer management* — Actually no fertilizer is given, but fertilizer is given with N, P_2O_5 and K_2O @ 100, 50 & 50 kg/ha, respectively annually when cultivated

6. *Duration of crop* — Perennial but harvested each year
7. *Parts used* — Leaf and tip as vegetable; whole plant as fodder, fuel and organic matter
8. *Where grown in India* — In lowlying areas in marshy lands of India
9. *Export possibility* — High for vegetable, if fridging container is available
10. *Economic estimated Yields (q / ha)* — 300 to 400 (green biomass)
11. *Economic value (Rs. / ha)* — 4,000-5,000
12. *By-product (If any)* — Stems (stalks) and old leaves
13. *Use of by-product* — Fuel, mulch and organic matter
14. *Peoples' response for use* — Good, particularly from the people of the lowlying areas of the country.

533. *CROP*—WATER FERN / MARSILEA / Susnisak

***Scientific Name* — *Marsilea quadrifolia* L./ *M. minuta* L.**

Family — *Marsileaceae*

Herb / Shrub / Tree — Herb

Annual / Biennial / Perennial — Perennial

Used as — Vegetable / medicine / organic matter / fuel

1. *Climatic requirements* — Warm weather with normal temperature and rainfall
2. *Soil requirements* — Clayey soils with organic matter
3. *Required land situations* — Lowlying situation in marshy lands
4. *Importance of the Crop* — Grows on soils by the sides of pond, *bill / jhill* & other lowlying water areas; plants run over water-surface
5. *Fertilizer management* — Normally no fertilizer is given, but N, P_2O_5 and K_2O @ 100, 50 & 50 kg/ha, respectively can be given as cultivated crop annually
6. *Duration of crop* — Perennial, but harvested each year
7. *Parts used* — Leaves and stalks are used as vegetable; Entire plants are used as medicine / organic matter / fuel.
8. *Where grown in India* — In marshy lands of lowlying areas of the country
9. *Export possibility* — High as medicine, when manufactured from it.
10. *Economic estimated Yields (q / ha)* — 150 to 200 (green biomass)
11. *Economic value (Rs. / ha)* — 4,000-6,000
12. *By-product (If any)* — Stalks and rhizomes

13. *Use of by-product* — Medicine (for urine-problems)
14. *Peoples' response for use* — Good, particularly from the people, residing in low lying areas of the country

534. *CROP*—PURSLANE / PORTULACA / Kulfa / Nuniasak / Khursa / Baralaniya

Scientific Name* — *Portulaca oleracea L. / ***P. grandiflora*** Hook.

Family — *Portulacaceae*

Herb / Shrub / Tree — Herb

Annual / Biennial / Perennial — Annual / Perennial

Used as — Vegetable / salad / soup / fuel / organic matters

1. *Climatic requirements* — Warm weather with normal temperature and rainfall
2. *Soil requirements* — Clayey soils, in marshy lowlying areas
3. *Required land situations* — Marshy lands in lowlying areas, with organic matter
4. *Importance of the Crop* — Leaves and tips used as vegetable / salads / soups, stems and old leaves as fuels and organic matter, grows better in saline soils
5. *Fertilizer management* — Normally no fertilizer is given but N, P_2O_5 and K_2O @ 100, 50 and 50 kg/ha may be given when cultivated, annually
6. *Duration of crop* — Perennial, harvested each year
7. *Parts used* — Tender leaves and tips are used as vegetable / salad / soup; stems or stalks and old leaves are used as fuel and organic matter
8. *Where grown in India* — In marshy lands of lowlying areas of the country
9. *Export possibility* — High, as vegetable, when packed and kept in fridging containers
10. *Economic estimated Yields (q / ha)* — 150 to 200 (green biomass)
11. *Economic value (Rs. / ha)* — 4,000-6,000
12. *By-product (If any)* — Stems / stalk / old leaves
13. *Use of by-product* — Fuels / organic matter
14. *Peoples' response for use* — Good, particularly from people of saline tract of lowlying marshy land areas of the country.

535. *CROP*—THATCH GRASS / Kansh / Kasa / Kans / Kansful

Scientific Name* — *Saccharum spontaneum L.

Family — *Gramineae*

Herb / Shrub / Tree — Herb (grass)

Annual / Biennial / Perennial — Perennial

Used as — Soil-binder / roofing / fencing / fodder / barrier / paper-pulp / fuel / organic matter

1. *Climatic requirements* — Warm weather with normal temperature and rainfall
2. *Soil requirements* — Any soil by the side of *bill / jhill* / pond, lake or other waterbodies
3. *Required land situations* — High / low lands, mostly in marshy lands
4. *Importance of the Crop* — It gives long white inflorescence, like horsetail; rough leaf-edge; flood resistant
5. *Fertilizer management* — Normally no fertilizer given, but for cultivation N, P_2O_5 and K_2O @ 100, 50 and 50 kg/ha is needed annually
6. *Duration of crop* — Perennial but harvested each year for different purposes
7. *Parts used* — Entire plant is used as soil-binder, roofing, barrier and paper pulp; stems are fencing; leaves are used as fodder. It is used as fuel and organic matter
8. *Where grown in India* — It grows in fallow lands / low lands in India
9. *Export possibility* — Not good for exportation
10. *Economic estimated Yields (q / ha)* — 400 to 500 (green biomass)
11. *Economic value (Rs. / ha)* — 4,000-5,000
12. *By-product (If any)* — Roots and stems
13. *Use of by-product* — Used as fuel and organic matter
14. *Peoples' response for use* — People use it as soil-binder / fodder / fuel / stick / thatching / organic matters / mulch / religious purpose etc.

536. *CROP*—WATER CRESS/NASTURTIUM / *Brahmisag*/ Piriyahalim

Scientific Name—Nasturtium officinale Br. / ***N. fontanum*** Aschers

Family — *Cruciferae*

Herb / Shrub / Tree — Herb

Annual / Biennial / Perennial — Perennial

Used as — Vegetable / medicine / fuel / organic matter

1. *Climatic requirements* — Warm weather with normal temperature and rainfall
2. *Soil requirements* — Clayey and other soils with organic matter

3. *Required land situations* — Marshy land or with shallow water in lowlying areas
4. *Importance of the Crop* — Small erect / trailed plant, with medicinal value
5. *Fertilizer management* — Normally no fertilizer, for cultivation, N, P_2O_5 and K_2O @ 100, 50 & 50 kg/ha, respectively should be given
6. *Duration of crop* — Perennial, but harvested each year
7. *Parts used* — Tender leaves and tips are used for vegetable / medicine; old leaves and stalks are used as fuel and organic matter
8. *Where grown in India* — In marshy lands / shallow water in lowlying areas of India
9. *Export possibility* — High, for medicine
10. *Economic estimated Yields (q / ha)* — 200 to 250 (green biomass)
11. *Economic value (Rs. / ha)* — 4,000-6,000
12. *By-product (If any)* — Stalks and old leaves
13. *Use of by-product* — Fuels and organic matter
14. *Peoples' response for use* — There is good response particularly from people of lowlying areas

537. *CROP*—COMMON REED-GRASS / Nal / Bajali / Sharkathi / Narkal/ Nal-khagra

Scientific Name* — *Phragmites karka (Retz.) Trin. / ***P. communis*** Trin./ ***P. maxima*** (Forsk) Blott.

Family — *Gramineae*

Herb / Shrub / Tree — Herb (tall grass)

Annual / Biennial / Perennial — Perennial

Used as — Rayon and paper making fibre and rope making / alcohol / lactic acid / basket / chairs / fodder (for elephant)

1. *Climatic requirements* — Warm weather with normal temperature and rainfall
2. *Soil requirements* — Mostly clayey soils, with organic matters
3. *Required land situations* — Low or medium lands or marshy lands
4. *Importance of the Crop* — It is a tall grass, leaves are rough and hairy in nature and acts as barrier and soil-binder
5. *Fertilizer management* — Normally no fertilizer is given, but for cultivation N, P_2O_5 and K_2O @ 100, 50 & 50 kg/ha respectively is given annually
6. *Duration of crop* — Perennial, but harvested annually

7. *Parts used* — Reeds are used for paper-pulp / rayon / absolute alcohol / feeding yeast for lactic acid / basket or chairs; fodder (for elephant)
8. *Where grown in India* — In marshy lands of lowlying areas in India
9. *Export possibility* — Reeds may be exported
10. *Economic estimated Yields (q / ha)* — 400 to 500 (green biomass)
11. *Economic value (Rs. / ha)* — 5,000-6,000
12. *By-product (If any)* — Old and new leaves
13. *Use of by-product* — Leaves (green) used as fodder; old leaves are used as fuels and organic matter
14. *Peoples' response for use* — Good response from people particularly from lowlying areas

538. *CROP*— COMMELINA / Kanshira / Kanteri / Kanchara / Kankawa

***Scientific Name* — *Commelina benghalensis* L.**

Family — *Commelinaceae*

Herb / Shrub / Tree — Herb

Annual / Biennial / Perennial — Perennial

Used as — Vegetable / fodder / fuel / organic matters / medicine

1. *Climatic requirements* — Warm weather with normal temperature and rainfall
2. *Soil requirements* — Mostly clayey soils with organic matter
3. *Required land situations* — Marshy lands in lowlying areas
4. *Importance of the Crop* — It gives blue flower inside soils; while in water roots and stems are modified and give blue flowers above water
5. *Fertilizer management* — Normally no fertilizer is given, but for cultivation N, P_2O_5 and K_2O @ 100, 50 & 50 kg/ha respectively, is given annually
6. *Duration of crop* — Perennial but harvested annually
7. *Parts used* — Tender leaves and tips are used as vegetable; entire plants are used as fodder / medicine / fuel and organic matters
8. *Where grown in India* — In marshy lands / shallow water in lowlying areas of the country
9. *Export possibility* — Less possibility
10. *Economic estimated Yields (q / ha)* — 300 to 400 (green biomass)
11. *Economic value (Rs. / ha)* — 4,000-5,000

12. *By-product (If any)* — Stalks and old leaves
13. *Use of by-product* — Fuels and organic matter
14. *Peoples' response for use* — Good, particularly from people in lowlying areas

539. *CROP*— INDIAN SPINACH / BASELLA / Pui / Poi / Puisak

***Scientific Name* — *Basella rubra* L./ *B. alba* L.**

Family — *Basellaceae*

Herb / Shrub / Tree — Under shrub / Herb

Annual / Biennial / Perennial — Perennial

Used as — Vegetable dyeing / Jam / Jellies / Sweets

1. *Climatic requirements* — Warm weather with normal temperature and rainfall
2. *Soil requirements* — Mostly clayey soils with organic matters
3. *Required land situations* — Marshy lands / medium lands in lowlying areas
4. *Importance of the Crop* — Larger heavy weight-leaves; absorbs heavy metal; trailing type; berries used in jellies
5. *Fertilizer management* — N, P_2O_5 and K_2O @ 100, 50 & 50 kg/ha, respectively annually
6. *Duration of crop* — Perennial, but harvested each year
7. *Parts used* — Stems, leaves and seeds are used as vegetable; spurious berries are reported to yield dye; berries are used for preparation of jam, jellies and sweets
8. *Where grown in India* — In dampy / marshyland in lowlying areas of the country
9. *Export possibility* — Less possibility
10. *Economic estimated Yields (q / ha)* — 400 to 500 (green biomass)
11. *Economic value (Rs. / ha)* — 4,000-6,000
12. *By-product (If any)* — Old leaves
13. *Use of by-product* — Organic matter / fuel
14. *Peoples' response for use* — Good, particularly people from lowlying areas; use it very much. It is used elsewhere also.

540. *CROP*— SITALPATI / CLINOGYNE / Motra / Pati / Chataigachh

Scientific Name* — *Clinogyne dichotoma Salisb. / ***C. forskalii*** Wall.

Family — *Marantaceae*

Herb / Shrub / Tree — Shrub (bushy)

Annual / Biennial / Perennial — Perennial

Used as — Sitalpati / mats / fodder / cattle-bedding / *atti* / soil-binder

1. *Climatic requirements* — Warm weather with normal temperature and rainfall
2. *Soil requirements* — Mostly clayey soils with organic matter
3. *Required land situations* — Marshy lands / in lowlying areas or by the sides of waterbodies
4. *Importance of the Crop* — Long uniform stem with branches at high, with white flowers and oval leaves
5. *Fertilizer management* — Needed N, P_2O_5 and K_2O @ 100, 50 & 50 kg/ha, respectively, annually
6. *Duration of crop* — Perenial; some stems are harvested annually
7. *Parts used* — Culms / stems and branches for sitalpati / *atti* / mats; leaves for fodder; pith for cattle-bedding; it is a soil-binder
8. *Where grown in India* — In marshy lands of low lying areas in India
9. *Export possibility* — High for sitalpati and mats or other products
10. *Economic estimated Yields (q / ha)* — 300 to 400 (green biomass)
11. *Economic value (Rs. / ha)* — 5,000-7,000
12. *By-product (If any)* — Piths & others
13. *Use of by-product* — Bedding for poultry / cattle / duck / or other insulators
14. *Peoples' response for use* — Good, particularly in lowlying areas.

541. *CROP*— SESBANIA / DHENCHA / Dhaincha / Sabujsar / Jayanti

Scientific Name — Sesbania aculeata (Willd) Poir. / ***S. bispinosa*** (Jacq.) Wright

Family — *Papilionaceae*

Herb / Shrub / Tree — Shrub (tall and bushy)

Annual / Biennial / Perennial — Annual

Used as — Green manure, fodder, fuel / seeds / mulch / stick

1. *Climatic requirements* — Warm weather with normal temperature and rainfall
2. *Soil requirements* — Clayey / alluvial soils with organic matter
3. *Required land situations* — Lowlying or medium lands
4. *Importance of the Crop* — It is grown for 'green manuring' for *kharif* rice for increasing soil fertility, at earlier stage
5. *Fertilizer management* — N, P_2O_5 and K_2O @ 40, 20 & 20 kg/ha, respectively

6. *Duration of crop* — Ploughed under at 45-50 days after sowing or kept under flooded conditions for seed for the year
7. *Parts used* — The entire plants at earlier stage is used as green manure; under flooded conditions it is kept for seeds/fuel/fencing etc.
8. *Where grown in India* — Mostly rice growing areas of the country for maintaining soil fertility.
9. *Export possibility* — Good, for seeds / seed-products
10. *Economic estimated Yields (q / ha)* — 1200 to 1400 green biomass; 15-20 q/ha seeds
11. *Economic value (Rs. / ha)* — 4,000-6,000
12. *By-product (If any)* — Leaves and branches (when in flooded condition)
13. *Use of by-product* — Fodder / fuel / organic matter / mulch
14. *Peoples' response for use* — Good, among the rice farmers in the country

542. *CROP*— BERMUDA GRASS / BAHAMA GRASS / DOG GRASS / Hariali / Doob / Durba

Scientific Name* — *Cynodon dactylon (L.) Pers./ ***Panicum dactylon*** L.

Family — *Gramineae*

Herb / Shrub / Tree — Herb (short when in dry and long in floods)

Annual / Biennial / Perennial — Perennial

Used as — Fodder / pasture / lawn / soil-binder

1. *Climatic requirements* — Warm weather with normal temperature and rainfall
2. *Soil requirements* — All soils, under different conditions
3. *Required land situations* — High, medium or low land both with or without water
4. *Importance of the Crop* — Very hardy in nature; small when in dry condition but long with big leaves, when flooded
5. *Fertilizer management* — Normally no fertilizer given; but when cultivated, it requires N, P_2O_5 and K_2O @ 100, 50 & 50 kg/ha, respectively annually
6. *Duration of crop* — Perennial, annually utilization for different purposes
7. *Parts used* — Entire plant for fodder / pasture / soil-binder (for soil erosion) / for preparation of lawn

8. *Where grown in India* — All over the country, under different conditions
9. *Export possibility* — Less possibility
10. *Economic estimated Yields (q / ha)* — 300 to 400 (green biomass)
11. *Economic value (Rs. / ha)* — 4,000-6,000
12. *By-product (If any)* — Nil (normally, but in flooded conditions, long stems are the by-products)
13. *Use of by-product* — Nil [normally, but they (stems) are used as organic matter]
14. *Peoples' response for use* — Good, particularly for checking soil erosion, lawn-making and for fodder when floods exist.

543. *CROP*— CANNA / INDIAN SHOT / QUEENSLAND ARROWROOT / Sarbajaya / Kalabati

Scientific Name* — *Canna indica L. / ***C. orientalis*** Rose.

Family — *Cannaceae*

Herb / Shrub / Tree — Herb

Annual / Biennial / Perennial — Perennial

Used as — Flower-plant /fodder / seeds / necklaces / fuels

1. *Climatic requirements* — Warm weather with normal temperature and rainfall
2. *Soil requirements* — Clayey soil with high organic matter
3. *Required land situations* — Marshylands under lowlying areas
4. *Importance of the Crop* — It gives beautiful coloured flowers; leaves like banana; improves beauty of the marshy lands
5. *Fertilizer management* — Under cultivated conditions, it requires, N, P_2O_5 and K_2O @ 100, 50 & 50 kg/ha, respectively annually
6. *Duration of crop* — Perennial, set flowers throughout the year
7. *Parts used* — Leaves and stems are used for fodder; flowers are for beautification in marshy lands; seeds for necklaces; it can be grown in gardens also
8. *Where grown in India* — In marshy lands of lowlying areas and gardens in India
9. *Export possibility* — Less possibility
10. *Economic estimated Yields (q / ha)* — 300 to 400 (green biomass)
11. *Economic value (Rs. / ha)* — 4,000-5,000
12. *By-product (If any)* — Nil (but when the plants are dried, by products are there)

13. *Use of by-product* — Nil (dried plants are used as fuel / organic matter)
14. *Peoples' response for use* — Good, particularly in lowlying areas of India

544. *CROP*— ASIATIC PENNYWORT / CENTELLA / Brahmi / Thankuni / Thankuri / Tholkuri

Scientific Name* — *Centella asiatica (L.) Urban / ***Hydrocotyle asiatica*** L.

Family — *Hydrocotylaceae*

Herb / Shrub / Tree — Herb

Annual / Biennial / Perennial — Perennial

Used as — Vegetable / medicine / fuel / organic matter

1. *Climatic requirements* — Warm weathers with normal temperature and rainfall
2. *Soil requirements* — Clayey soils with organic matters
3. *Required land situations* — Marshy lands in lowlying areas
4. *Importance of the Crop* — Curled leaves with long petiole, having good smell
5. *Fertilizer management* — Normally no fertilizer given; but for cultivation, N, P_2O_5 and K_2O @ 100, 50 and 50 kg/ha, respectively are given annually
6. *Duration of crop* — Perennial, but harvested annually
7. *Parts used* — Leaves and stems are used as vegetable / medicine; decoction of plants is used in treatment of leprosy
8. *Where grown in India* — Marshy lands in lowlying areas in India
9. *Export possibility* — High, for medicine
10. *Economic estimated Yields (q / ha)* — 150 to 200 (green biomass)
11. *Economic value (Rs. / ha)* — 4,000-10,000
12. *By-product (If any)* — Old leaves and rhizomes
13. *Use of by-product* — Fuels and organic matter
14. *Peoples' response for use* — Good, particularly people in lowlying areas use it

545. *CROP*— PALUDOSA / COMMELINA / Kanjuna / Kanjunaful

Scientific Name* — *Commelina paludosa BL. / ***C. obliqua*** Buch-Ham

Family — *Commelinaceae*

Herb / Shrub / Tree — Herb

Annual / Biennial / Perennial — Perennial

Used as — Vegetable / fodder / fuel / organic matter

1. *Climatic requirements* — Warm weather with normal temperature and rainfall
2. *Soil requirements* — Clayey soils with organic matter
3. *Required land situations* — Marshy lands in lowlying areas
4. *Importance of the Crop* — Can run over water-surface, with root-modification
5. *Fertilizer management* — Normally no fertilizer given; but for cultivation, N, P_2O_5 and K_2O @ 100, 50 and 50 kg/ha, respectively are given
6. *Duration of crop* — Perennial, but harvested annually
7. *Parts used* — Young shoots are eaten as vegetable; entire plant for fodder / fuels / organic matter
8. *Where grown in India* — In marshy lands in lowlying areas in India
9. *Export possibility* — Less possibility
10. *Economic estimated Yields (q / ha)* — 300-400 (green biomass)
11. *Economic value (Rs. / ha)* — 3,000-5,000 annually
12. *By-product (If any)* — Old leaves and stalks
13. *Use of by-product* — Organic matter
14. *Peoples' response for use* — Good, particularly for people in lowlying areas.

546. *CROP*— COMMON WOOD SORREL / INDIAN SORREL / Amrulsak / Susni sak / Chuka tripati

Scientific Name — ***Oxalis ripens*** L. / ***O. acetosella*** L. / ***O. corniculata*** L.

Family — *Oxalidaceae*

Herb / Shrub / Tree — Herb

Annual / Biennial / Perennial — Perennial

Used as — Vegetable / medicine / organic matter

1. *Climatic requirements* — Warm or low temperature with normal rainfall
2. *Soil requirements* — Clayey soil in lowlying areas, having organic matters
3. *Required land situations* — Marshy lands in lowlying areas
4. *Importance of the Crop* — Leaves possess a refreshing sub-acid flavour used in salad and rich in Vitamin C

5. *Fertilizer management* — Normally no fertilizer given; but for cultivation, N, P_2O_5 and K_2O @ 100, 50 & 50 kg/ha, respectively are given annually
6. *Duration of crop* — Perennial, but harvested annually
7. *Parts used* — Tender leaves and stalk are used as vegetable / medicine (good for urine-problem; entire plant for organic matter
8. *Where grown in India* — In marshy lands of lowlying areas
9. *Export possibility* — High, for medicine
10. *Economic estimated Yields (q / ha)* — 200 to 300 (green biomass)
11. *Economic value (Rs. / ha)* — 5,000-6,000
12. *By-product (If any)* — Old leaves / stalks
13. *Use of by-product* — Organic matter / fuel
14. *Peoples' response for use* — Good, to people of the lowlying areas.

547. *CROP*— KOVAI FRUIT / IVY GOURD / Kunduri / Kundru / Telakuch

Scientific Name* — *Coccinia indica Wight & Arn. / ***C. cordifolia*** Cogn. / ***C. grandis*** (L.) Voigt.

Family — *Cucurbitaceae*

Herb / Shrub / Tree — Twining or spreading shrub

Annual / Biennial / Perennial — Perennial

Used as — Vegetable / medicine / fuel / organic matter / fodder

1. *Climatic requirements* — Warm weather with normal temperature and rainfall
2. *Soil requirements* — Clayey soils, in marshy lands, in lowlying areas
3. *Required land situations* — Marshy / normal soils with organic matters in lowlying areas
4. *Importance of the Crop* — Can withstand waterlogging / water in floods; deep red or green fruits; bears white flowers
5. *Fertilizer management* — Normally no fertilizer given, but when cultivated, N, P_2O_5 and K_2O @ 100, 50 and 50 kg/ha, respectively are given annually
6. *Duration of crop* — Perennial, but harvested annually
7. *Parts used* — Green fruits are eaten as vegetable / medicine; tender leaves and tops are used as medicine. Juice from tender leaves and roots are used in diabetes; stem for fuel and organic matter
8. *Where grown in India* — Grows in any area, even in lowlying areas

9. *Export possibility* — Good for medicine for diabetic patients
10. *Economic estimated Yields (q / ha)* — 200 to 300 (green vegetable) i.e. 400-500 green biomass
11. *Economic value (Rs. / ha)* — 4,000-6,000
12. *By-product (If any)* — Stems and old leaves
13. *Use of by-product* — Fuel / fodder / organic matter
14. *Peoples' response for use* — Good, for tender leaves and tips for medicines particularly in medium / lowlying areas; it is also required for general people for diabetes.

548. *CROP*— BARNYARD GRASS / Samak / Sanwak / Shyama ghash / Shama

Scientific Name* — *Echinochloa crus-galli (L.) Beauv. / ***E. hispida*** (Retz.) Nees.

Family — *Gramineae*

Herb / Shrub / Tree — Herb (grass)

Annual / Biennial / Perennial — Annual

Used as — Fodder / Grain (but serious weed in rice crop) / Fuel / organic matter / mulch

1. *Climatic requirements* — Warm / cool weather with normal / low temperature with rainfall
2. *Soil requirements* — Clayey / clayey loam soils with organic matters
3. *Required land situations* — Waterlogged conditions in lowlying areas
4. *Importance of the Crop* — It is exactly like rice plant, and it is a weed in rice; seeds small and are eaten
5. *Fertilizer management* — Fertilizer N, P_2O_5 and K_2O @ 100, 50 & 50 kg/ha, respectively are given
6. *Duration of crop* — Annual crop, duration 3-4 months
7. *Parts used* — Entire plant is used as fodder / fuel / organic matter
8. *Where grown in India* — Waterlogged conditions in lowlying areas; It is an obnoxious weed in rice
9. *Export possibility* — Less possibility
10. *Economic estimated Yields (q / ha)* — 300 to 400 (green biomass)
11. *Economic value (Rs. / ha)* — 4,000 to 5,000 (when fodder is considered)
12. *By-product (If any)* — There is no by-product, straw (if used as fodder / fuel / organic matter) but if grains are taken, by-product is there

13. *Use of by-product* — Cattle feed / fuel / organic matter
14. *Peoples' response for use* — Good, for different purposes though it is a serious weed in rice field

549. *CROP*— CHINESE MAT GRASS / NUT GRASS / CORYMBOSUS / Sonali madurkathi

Scientific Name* — *Cyperus corymbosus Rottb. / ***C. tegetiformis*** Roxb.

Family — *Cyperaceae*

Herb / Shrub / Tree — Tall herb (sedge)

Annual / Biennial / Perennial — Perennial

Used as — Golden mat / fodder / fuel / organic matter / bedding

1. *Climatic requirements* — Warm weather with normal temperature and rainfall
2. *Soil requirements* — Clayey soils with organic matter
3. *Required land situations* — Marshylands in lowlying areas
4. *Importance of the Crop* — A tall rush-like plant, found in marshy lands, used for golden mats
5. *Fertilizer management* — Normally no fertilizer, but when cultivated, N, P_2O_5 and K_2O @ 100, 50 & 50 kg/ha, respectively are given annually
6. *Duration of crop* — Perennial, but harvested annually
7. *Parts used* — Culms / stems are used for preparation of golden mats; leaves are used as fodder / fuel / organic matter; piths are used as cattle / poultry beddings
8. *Where grown in India* — Grows in marshylands in lowlying areas
9. *Export possibility* — High, for golden mats
10. *Economic estimated Yields (q / ha)* — 300 to 400 (biomass)
11. *Economic value (Rs. / ha)* — 5,000-6,000
12. *By-product (If any)* — Refuse materials from pith, leaves or branches
13. *Use of by-product* — Fuel and organic matter
14. *Peoples' response for use* — Good, for mats (though it is an obnoxious weed in rice field).

550. *CROP*— DALLIS GRASS / KNOT GRASS / BAHIA GRASS / Delisghash

Scientific Name* — *Paspalum dilatatum Poir. / ***P. notatum*** Fluegge

Family — *Gramineae / Poaceae*

Herb / Shrub / Tree — Herb (tall grass)

Annual / Biennial / Perennial — Perennial

Used as — Fodder / soil-binder / fuel / organic matter

1. *Climatic requirements* — Warm weather with normal temperature and rainfall
2. *Soil requirements* — Clayey soils with high organic matter
3. *Required land situations* — Marshy lands in lowlying areas
4. *Importance of the Crop* — It is a tall grass (about 1.5 m) with heavy growth
5. *Fertilizer management* — Normally no fertilizer needed, but for cultivation N, P_2O_5 and K_2O @ 100, 50 & 50 kg/ha, respectively are needed annually
6. *Duration of crop* — Perennial but harvested annually
7. *Parts used* — Entire plant is used as fodder / soil-binder / fuel / organic matters
8. *Where grown in India* — In marshy lands in lowlying areas
9. *Export possibility* — Less possibility
10. *Economic estimated Yields (q / ha)* — 300 to 400 (green biomass)
11. *Economic value (Rs. / ha)* — 4,000-5,000 (when whole plant is considered)
12. *By-product (If any)* — Nil (other refuse materials, if any)
13. *Use of by-product* — Nil (fuel / organic matter / mulch)
14. *Peoples' response for use* — Good, particularly for people in lowlying areas

551. *CROP*— PARAGRASS / WATER GRASS / MAURITIUS GRASS / Paraghash

Scientific Name — Brachiaria mutica (Forsk) Stapf. / ***Panicum muticum*** Forsk.

Family — *Gramineae*

Herb / Shrub / Tree — Herb (grass)

Annual / Biennial / Perennial — Perennial

Used as — Fodder / soil-binder / cover crop / reclaims soils salinite

1. *Climatic requirements* — Warm weather with normal temperature and rainfall
2. *Soil requirements* — Clayey soils with organic matter
3. *Required land situations* — Marshy lands / shallow depth of water in lowlying areas
4. *Importance of the Crop* — Can be grown in marshy land/shallow / deep depth of water in lowlying areas

5. *Fertilizer management* — N, P_2O_5 and K_2O @ 100, 50 & 50 kg/ha, respectively when crop is in marshy land condition or in dry irrigated conditions
6. *Duration of crop* — Perennial but harvested annually
7. *Parts used* — Entire plants are used as fodder / soil-binder / fuel / organic matter
8. *Where grown in India* — In marshy lands or shallow / deep depth of water, in lowlying areas
9. *Export possibility* — Less possibility
10. *Economic estimated Yields (q / ha)* — 300 to 400 (green biomass)
11. *Economic value (Rs. / ha)* — 4,000-6,000 (when considered as fodder)
12. *By-product (If any)* — Nil (Refuse material)
13. *Use of by-product* — Nil (Fuel / organic matter)
14. *Peoples' response for use* — Good, particularly for the people from lowlying areas.

552. *CROP*— DEFLEXA / Dalghash / Natun paraghash

Scientific Name* — *Brachiaria deflexa (Schumach) Hubb. / ***B. regularis*** (Nees.) Stapf.

Family — *Gramineae*

Herb / Shrub / Tree — Herb (grass)

Annual / Biennial / Perennial — Perennial

Used as — Fodder / soil-binder / cover-crop / fuel / organic matter

1. *Climatic requirements* — Warm weather with normal temperature and rainfall
2. *Soil requirements* — Clayey soils with organic matters
3. *Required land situations* — Marshy lands/waterlogged conditions in lowlying areas
4. *Importance of the Crop* — It acts as soil-binder, gives higher biomass in marshy land
5. *Fertilizer management* —Required N, P_2O_5 and K_2O @ 100, 50 & 50 kg/ha, respectively annually
6. *Duration of crop* — Perennial but harvested annually
7. *Parts used* — Entire plants for fodder / cover-crop/fuel/organic matters
8. *Where grown in India* — In marshy lands / waterlogged conditions in lowlying areas of India
9. *Export possibility* — Less possibility

10. *Economic estimated Yields (q / ha)* — 300 to 400 (green biomass)
11. *Economic value (Rs. / ha)* — 4,000 to 6,000
12. *By-product (If any)* — Nil (Some refuse material)
13. *Use of by-product* — Nil (fuel / organic matter / mulch)
14. *Peoples' response for use* — Good, particularly in lowlying areas

553. *CROP*— RAGI / FINGER MILLET / AFRICAN MILLET / ELEUSINE / CYNOSURUS / Marua / Mandal / Marudhan

Scientific Name —***Eleusine coracana*** (L) Gaertn./ ***Cynosurus coracanus*** L.

Family — *Gramineae*

Herb / Shrub / Tree — Herb (grass)

Annual / Biennial / Perennial — Annual

Used as — Food / Fodder / cattle feed / fuel / organic matter

1. *Climatic requirements* — Warm weather with normal temperature and rainfall
2. *Soil requirements* — Clayey soils in lowlying areas
3. *Required land situations* — Waterlogged conditions, marshy lands in lowlying conditions
4. *Importance of the Crop* — It is a cereal crop, grows in some areas as a main food; dwarf in nature; inflorescence is club-like
5. *Fertilizer management* — N, P_2O_5 and K_2O @ 100, 50 & 50 kg/ha, respectively
6. *Duration of crop* — 100-120 days
7. *Parts used* — Grain for human food; straw for cattle feed / fuel / organic matter
8. *Where grown in India* — In waterlogged conditions, in marshy land in lowlying areas
9. *Export possibility* — Good (for grain)
10. *Economic estimated Yields (q / ha)* — 20-25 (dry grain); 250-350 fodder (green biomass)
11. *Economic value (Rs. / ha)* — 5,000-7,000
12. *By-product (If any)* — Straw (if grain is taken)
13. *Use of by-product* — Cattle feed / fuel / organic matter / mulch
14. *Peoples' response for use* — Good, it is a food crop in some areas

554. *CROP*— RICE*/PADDY / Dhan / Dhany / Chavul / Anna / Tandul / Chal / Bhat

Scientific Name — ***Oriza sativa*** L.

Family — *Gramineae*

Herb / Shrub / Tree — Herb (grass)

Annual / Biennial / Perennial — Annual

Used as — Grains for foods / Straw for cattle feed / board / paper; rice-bran for oil (for soap / cosmetics); husk for fuel / soil-conditioner/ bedding

1. *Climatic requirements* — Warm weather with normal temperature and rainfall
2. *Soil requirements* — Clayey soils, under waterlogged condition in lowlying areas
3. *Required land situations* — Lowlying areas under waterlogged conditions
4. *Importance of the Crop* — Cereal major crop in India, grows under different conditions (directed) transplanted waterlogged, shallow/ deep water in warm / cool weather
5. *Fertilizer management* — N, P_2O_5 and K_2O @ 100, 50 & 50 kg/ha, respectively
6. *Duration of crop* — 4-8 months (under different conditions)
7. *Parts used* — Grains are used as major cereal crop in India; straw is used as board/paper / matter cattle feed; rice-bran-oil is used as soap/cosmetic; rice-husk is used as fuel, soil-condition; from grains many ingredients are obtained.
8. *Where grown in India* — In waterlogged condition / deep water, in lowlying areas
9. *Export possibility* — Very high, particularly scented rice grains
10. *Economic estimated Yields (q / ha)* — 30 to 40 (dry seeds)
11. *Economic value (Rs. / ha)* — 7,000 to 10,000
12. *By-product (If any)* — Straw (dry)
13. *Use of by-product* — Fuel / cattle feed / bedding for cattle or poultry
14. *Peoples' response for use* — Good, for the people of both lowlying / medium or high lands of the country.

*Note : Rice is grown in all the seasons under different conditions, such as dry, waterlogged or in deep water conditions (i) *Aus / ahu* rice is grown under dry (i.e. rainfed in pre-*kharif* rice, (ii) *Aman* rice is grown under waterlogged condition in rainy season; (iii) Deep-water-rice is grown in very deep water in rainy season, (iv) *Boro* rice is grown in transplanted (waterlogged) conditions in winter season.

555. *CROP*— COMMON REED / PHRAGMITES / Baranal / Sarkathi / Khagra

Scientific Name* — *Phragmites communis Trin. / ***P. karka*** (Retz) Trin. / ***P. maxima*** (Forsk) Blatt.

Family — *Gramineae*

Herb / Shrub / Tree — Shrub (tall grass)

Annual / Biennial / Perennial — Perennial

Used as — Paper-pulp / paper / pipe / fodder / fuel / mulch/ organic matter

1. *Climatic requirements* — Warm weather with normal temperature and rainfall
2. *Soil requirements* — Clayey / clayey loam soils in lowlying areas
3. *Required land situations* — Lowlying areas or basin-like areas
4. *Importance of the Crop* — Long plants, having strong internodes and nodes
5. *Fertilizer management* — N, P_2O_5 and K_2O @ 100, 50 & 50 kg/ha, respectively annually
6. *Duration of crop* — Perennial, but harvested annually
7. *Parts used* — Entire plants are used as fodder; stems (culms) are used for paper-pulp / paper / pipe; leaves for fodder / bedding materials
8. *Where grown in India* — Lowlying areas
9. *Export possibility* — High (for paper / pipe / paper-pulp)
10. *Economic estimated Yields (q / ha)* — 300 to 400 (green biomass)
11. *Economic value (Rs. / ha)* — 4,000 to 5,000
12. *By-product (If any)* — Leaves (if culms are taken)
13. *Use of by-product* — Fodder / bedding for cattle / poultry
14. *Peoples' response for use* — Good, particularly for paper / paper-pulp.

556. *CROP*— VETIVER / VETIVERIA / Khas-khash / Khas / Gandar / Sugandhi sikar

Scientific Name —***Vetiveria zizanioides*** (L.) Nash. / ***Phalaris zizanioides*** L.

Family — *Gramineae*

Herb / Shrub / Tree — Shrub (tall grass)

Annual / Biennial / Perennial — Perennial

Used as — Roots for scented mat / fan / basket / curtain / oil perfumery / cosmetics / soap / paper-board / fodder / fuel

1. *Climatic requirements* — Warm weather with normal temperature and rainfall
2. *Soil requirements* — Clayey / clayey loam soils
3. *Required land situations* — Lowlying areas or medium lands
4. *Importance of the Crop* — Its dried roots are used for good scent, particularly in dry season, for cooling the air
5. *Fertilizer management* — N, P_2O_5 and K_2O @ 100, 50 & 50 kg/ha, respectively annually
6. *Duration of crop* — Perennial, but harvested partially annually
7. *Parts used* — Dry roots for aromatic scent in mats / fans basket / curtain which moistened, cool / scent the air; for manufacturing oil / perfumery / cosmetics soap; culms are used for paper-board; leaves are used for fodder / fuels
8. *Where grown in India* — Lowlying areas of dry wet region of Rajasthan
9. *Export possibility* — High, particularly for its products
10. *Economic estimated Yields (q / ha)* — 400 to 500 (green biomass)
11. *Economic value (Rs. / ha)* — 6,000 to 10,000
12. *By-product (If any)* — Leaves (green after removing culms)
13. *Use of by-product* — Fodder / fuel / organic matter / mulch
14. *Peoples' response for use* — Good, for all products from it.

557. *CROP*—SAGITTARIA / ARROW-HEAD / Tesraballam pata / Baratirpata

Scientific Name* — *Sagittaria trifolia L.

Family — *Alismataceae*

Herb / Shrub / Tree — Herb

Annual / Biennial / Perennial — Perennial

Used as — Vegetable / fuels / organic matter

1. *Climatic requirements* — Warm weather with normal temperature and rainfall
2. *Soil requirements* — Clayey / clayey loam soils with organic matters, of lowlying areas
3. *Required land situations* — Lowlying areas / marshylands
4. *Importance of the Crop* — Leaves are like arrows with three ends and some portions of the plant is within water and some is over water surface
5. *Fertilizer management* — N, P_2O_5 and K_2O @ 100, 50 & 50 kg/ha, respectively though normally no fertilizer is used

6. *Duration of crop* — Perennial, but harvested annually
7. *Parts used* — Tubers (at root points) are used as vegetable; leaves and stems are used as fuels / organic matters
8. *Where grown in India* — In marshy lands of lowlying areas
9. *Export possibility* — Less possibility
10. *Economic estimated Yields (q / ha)* — 300 to 400 (green biomass)
11. *Economic value (Rs. / ha)* — 3,000 to 4,000
12. *By-product (If any)* — Stems and other refuse materials
13. *Use of by-product* — Fuels / organic matter
14. *Peoples' response for use* — Good, for people of the lowlying areas particularly in eastern India

558. *CROP*— FIMBRISTYLIS / GLOBULOSA / LITTORALIS

Scientific Name* — *Fimbristylis littoralis Gaud. / ***F. globulosa*** (Retz) Kuntze

Family — *Cyperaceae*

Herb / Shrub / Tree — Herb

Annual / Biennial / Perennial — Perennial

Used as — Mat . basket / fuel / organic matter

1. *Climatic requirements* — Warm weather with normal temperature and rainfall
2. *Soil requirements* — Clayey / clayey loam soils with organic matter
3. *Required land situations* — Marshylands in lowlying areas
4. *Importance of the Crop* — It is a tall sedge with strong body, used for mat
5. *Fertilizer management* — Normally no fertilizer given, but for cultivation, N, P_2O_5 and K_2O @ 100, 50 & 50 kg/ha, respectively is applied
6. *Duration of crop* — Perennial, but harvested annually
7. *Parts used* — With culms, mat / basket are prepared. Leaves are used as fuel / organic matter
8. *Where grown in India* — In marshy lands of lowlying areas
9. *Export possibility* — Good, particularly the mats
10. *Economic estimated Yields (q / ha)* — 300 to 400 (green biomass)
11. *Economic value (Rs. / ha)* — 4,000 to 10,000
12. *By-product (If any)* — Leaves (green biomass)
13. *Use of by-product* — Fuel / organic matter
14. *Peoples' response for use* — Good, for people of lowlying areas

559. *CROP*— **WATERLILY / INDIAN WATERLILLY / Bhenght / Saluk / Sapla / Kanval / Koka / Koi**

Scientific Name* — *Nymphaea lotus auct. / ***N. nauchali*** Burm.f.

Family — *Nymphaeaceae*

Herb / Shrub / Tree — Herb

Annual / Biennial / Perennial — Perennial

Used as — Vegetable / medicine / fuel / organic matter

1. *Climatic Requirements*— Warm weather with normal temperature and rainfall
2. *Soil requirements* — Clayey / clayey loam soils with organic matter
3. *Required land situations* — Shallow / deep water in lowlying ponds, *bills* / *jhills* etc.
4. *Importance of the Crop* — It gives good fragrant flowers above water
5. *Fertilizer management* — Normally no fertilizer is given, but for cultivation, N, P_2O_5 and K_2O @ 100, 50 & 50 kg/ha, respectively is given annually
6. *Duration of crop* — Perennial, but harvested annually
7. *Parts used* — All parts (stems / leaves / rhizomes) are used as vegetable; flowers are astringent and cardiotonic; seeds are used as cooling medicines
8. *Where grown in India* — In lowlying areas, having shallow or deep clean water in India
9. *Export possibility* — Good, particularly medicine from it
10. *Economic estimated Yields (q / ha)* — 200 to 400 (green biomass)
11. *Economic value (Rs. / ha)* — 3,000 to 6,000
12. *By-product (If any)* — Leaves and other refuse materials
13. *Use of by-product* — Fuel / organic matter
14. *Peoples' response for use* — Good, for people living in lowlying areas.

560. *CROP*— **LOTUS / EAST INDIAN WATERLILY / Padma / Kamal / Kanwal**

Scientific Name* — *Nelumbium speciosum Willd. / ***Nelumbo nucifera*** Gaertn

Family — *Nelumbonaceae*

Herb / Shrub / Tree — Herb

Annual / Biennial / Perennial — Perennial

Used as — Vegetable / arrowroot / medicine / scent / decoration

1. *Climatic requirements* — Warm weather with normal temperature and rainfall
2. *Soil requirements* — Clayey / clayey loam soils with organic matter
3. *Required land situations* — Shallow / deep water, in lowlying areas
4. *Importance of the Crop* — It is planted in the gardens for fragrant flowers
5. *Fertilizer management* — Generally no fertilizer given but for cultivation N, P_2O_5 and K_2O @ 100, 50 & 50 kg/ha, respectively is given annually
6. *Duration of crop* — Perennial, but harvested annually
7. *Parts used* — Rhizomes / flower-stalks / leaf (with air pocket) are used as vegetable; rhizomous yields a kind of arrowroot / medicine (for dysentery / diarrhoea; garlands from flowers
8. *Where grown in India* — In shallow / deep water in lowlying areas
9. *Export possibility* — High, for arrowroot and medicines
10. *Economic estimated Yields (q / ha)* — 300 to 400 (green biomass)
11. *Economic value (Rs. / ha)* — 6,000 to 7,000
12. *By-product (If any)* — Leaf-stalks and other refuse materials
13. *Use of by-product* — Fuels / organic matter/bedding for cattle and poultry (after drying)
14. *Peoples' response for use* — Good, particularly for people of lowlying areas.

561. *CROP*— INDIAN BLUE WATERLILY / Nilpadma / Nilkamal

Scientific Name — Nymphaea stellata Willd

Family — *Nymphaeaceae*

Herb / Shrub / Tree — Herb

Annual / Biennial / Perennial — Perennial

Used as — Vegetable / medicine / fuel / organic matter

1. *Climatic requirements* — Warm / cool weather with high / low temperature and rainfall
2. *Soil requirements* — Clayey / clayey loam soils with organic matters
3. *Required land situations* — Shallow / deep water, in lowlying areas
4. *Importance of the Crop* — It has beautiful flowers, planted in water – gardens

5. *Fertilizer management* — Normally no fertilizer is given but for cultivation N, P_2O_5 and K_2O @ 100, 50 & 50 kg/ha, respectively, is given annually
6. *Duration of crop* — Perennial, but harvested annually
7. *Parts used* — Rhizomes / tender leaves / flowers / peduncles are used as vegetable; powdered rhizome is given in dyspepsia, diarrhoea and pile / medicine / fuel / organic matter
8. *Where grown in India* — In shallow / deep water in lowlying areas
9. *Export possibility* — High, with rhizomes for medicines
10. *Economic estimated Yields (q / ha)* — 300 to 400 (green biomass)
11. *Economic value (Rs. / ha)* — 6,000 to 10,000
12. *By-product (If any)* — Leaf-stalks/flower stalks and others
13. *Use of by-product* — Fuels and organic matters
14. *Peoples' response for use* — Good, for people of lowlying areas

562. *CROP*— INDIAN WHITE WATERLILY / NYMPHAEA / Chhoto sundi / Sundi

Scientific Name* — *Nymphaea esculenta Roxb.

Family — *Nymphaeaceae*

Herb / Shrub / Tree — Herb

Annual / Biennial / Perennial — Perennial

Used as — Vegetable / food / fuel / organic matters

1. *Climatic Requirements*— Warm weather with normal temperature and rainfall
2. *Soil requirements* — Clayey / clayey loam soils with organic matters
3. *Required land situations* — Shallow / deep water, in lowlying areas
4. *Importance of the Crop* — Beautiful small flowers, above the water surface
5. *Fertilizer management* — Normally no fertilizer is given, but for cultivation, N, P_2O_5 and K_2O @ 100, 50 & 50 kg/ha, respectively is given annually
6. *Duration of crop* — Perennial, but harvested annually
7. *Parts used* — Rhizomes and flower-stalks are used as vegetable; seeds as food (after frying)
8. *Where grown in India* — In shallow / deep water in lowlying areas of India
9. *Export possibility* — Less possibility

10. *Economic estimated Yields (q / ha)* — 200 to 300 (green biomass)
11. *Economic value (Rs. / ha)* — 3,000 to 4,000
12. *By-product (If any)* — Leaf-stalks and other refuse materials
13. *Use of by-product* — Fuel / organic matter / cattle feeds
14. *Peoples' response for use* — Good, for people of lowlying areas

563. *CROP*— WHITE WATERLILY / Pandharen kamal / Sapla

Scientific Name — ***Nymphaea alba*** L.

Family — *Nymphaeaceae*

Herb / Shrub / Tree — Herb

Annual / Biennial / Perennial — Perennial

Used as — Vegetable / tanning / fuel / organic matter

1. *Climatic requirements* — Warm weather with normal temperature and rainfall
2. *Soil requirements* — Clayey / clayey loam soils with organic matters
3. *Required land situations* — Shallow / deep water, in lowlying areas
4. *Importance of the Crop* — White good big fragrant flowers on water surface
5. *Fertilizer management* — Normally no fertilizer is given, but for cultivation, N, P_2O_5 and K_2O @ 100, 50 & 50 kg/ha, respectively, is given annually
6. *Duration of crop* — Perennial, but harvested annually
7. *Parts used* — Rhizomes / flower-stalks are used as vegetable; rhizomes are also used for tanning
8. *Where grown in India* — Shallow / deep water in lowlying areas in India
9. *Export possibility* — Good for tanning from rhizomes
10. *Economic estimated Yields (q / ha)* — 300 to 400 (green biomass)
11. *Economic value (Rs. / ha)* — 4,000 to 6,000
12. *By-product (If any)* — Leaf-stalks and other refuse materials
13. *Use of by-product* — Fuel and organic matter
14. *Peoples' response for use* — Good, for people of lowlying areas

564. *CROP*— MONOCHORIA / PICKEREL WEED / Nokal kachuripana

Scientific Name — ***Monochoria vaginalis*** (Burm.f.) Presl.

Family — *Pontederiaceae*

Herb / Shrub / Tree — Herb

Annual / Biennial / Perennial — Perennial

Used as — Vegetable / fodder / fuel / organic matter

1. *Climatic requirements* — Warm weather with normal temperature and rainfall
2. *Soil requirements* — Clayey and clayey loam soils with organic matter
3. *Required land situations* — Marshy lands or shallow / deep water in lowlying areas
4. *Importance of the Crop* — While in soils, it strikes roots, but while on water surface, its roots are modified
5. *Fertilizer management* — Normally no fertilizer is given, but when cultivated it requires N, P_2O_5 and K_2O @ 100, 50 & 50 kg/ha, respectively, is given annually
6. *Duration of crop* — Perennial, but harvested annually
7. *Parts used* — Rhizomes are used as vegetable, while entire plants are used as fodder / fuel / organic matter / mulch
8. *Where grown in India* — In marshy lands, shallow / deep water in lowlying areas of India
9. *Export possibility* — Less possibility
10. *Economic estimated Yields (q / ha)* — 300 to 400 (green biomass)
11. *Economic value (Rs. / ha)* — 3,000 to 4,000
12. *By-product (If any)* — Refuse materials after using as fodder
13. *Use of by-product* — Fuels / organic matter / mulch
14. *Peoples' response for use* — Good, particularly for the cattle of the farmers of the lowlying areas of the country, though it is an obnoxious weed in lowlands

565. *CROP*— WATER FENNEL / WATER CROW-FOOT / Jalmauri

Scientific Name — ***Ranunculus trichophyllus*** Chaix. / ***R. aquatilis*** L.

Family — *Ranunculaceae*

Herb / Shrub / Tree — Herb

Annual / Biennial / Perennial — Perennial

Used as — Fodder / medicine / spice / fuel / organic matter

1. *Climatic Requirements*— Warm weather with normal temperature and rainfall
2. *Soil requirements* — Clayey / clayey loam soils with organic matter
3. *Required land situations* — Marshy lands in lowlying aeas

4. *Importance of the Crop* — Its scented seed like 'fennel' (*mauri*) is a spice
5. *Fertilizer management* — Normally no fertilizer is given, but for cultivation it requires, N, P_2O_5 and K_2O @ 100, 50 & 50 kg/ha, respectively, is given annually
6. *Duration of crop* — Perennial, but harvested annually
7. *Parts used* — Entire plants are used as fodder / fuel / organic matter; seeds are used as medicine / spice
8. *Where grown in India* — In marshy lands in lowlying areas
9. *Export possibility* — High, particularly with the seeds (spice / medicine)
10. *Economic estimated Yields (q / ha)* — 250 to 300 (green biomass)
11. *Economic value (Rs. / ha)* — 3,000 to 4,000
12. *By-product (If any)* — Refuse material
13. *Use of by-product* — Fuel / organic matter
14. *Peoples' response for use* — Good, particularly for people living in lowlying areas of the country

566. *CROP*— BULRUSH / CAT'S TAIL / ELEPHANT GRASS / TYPHA / INDIAN REED-MACE / Hogla / Pater / Bora

Scientific Name— Typha elephantina Roxb. / ***T. angustifolia*** Watt.

Family — *Typhaceae*

Herb / Shrub / Tree — Shrub (tall)

Annual / Biennial / Perennial — Perennial

Used as — Vegetable / conservation of soils / ornamental / fodder / fuel / bedding / organic matter

1. *Climatic requirements* — Warm weather with normal temperature and rainfall
2. *Soil requirements* — Clayey / clayey loam, with organic matter
3. *Required land situations* — Marshy lands in lowlying areas
4. *Importance of the Crop* — Tall plant with long soft leaves, grows in marshy lands
5. *Fertilizer management* — Normally no fertilizer is given, but for cultivation, N, P_2O_5 and K_2O @ 100, 50 & 50 kg/ha, respectively, is given annually
6. *Duration of crop* — Perennial, but harvested annually
7. *Parts used* — Young shoots and rhizomes are used as vegetable. Good fodder for elephant; leaves for thatching, screen soft mats (cattle / human bedding) / rope / fuel / organic matter; used in conservation of soils / ornamental plants

8. *Where grown in India* — In marshy lands of lowlying areas of India
9. *Export possibility* — High, with the long soft leaves for bedding / mats
10. *Economic estimated Yields (q / ha)* — 400 to 500 (green biomass)
11. *Economic value (Rs. / ha)* — 6,000 to 7,000
12. *By-product (If any)* — Refuse materials like flower stalks
13. *Use of by-product* — Fuel / organic matter / mulch
14. *Peoples' response for use* — Good, for people living in lowlying areas

567. *CROP*— CHINA MAT-GRASS / TEGETUM / Baramadurkathi

Scientific Name* — *Cyperus tegetum Roxb.

***Family* —** *Cyperaceae*

***Herb / Shrub / Tree* —** Herb (tall sedge)

***Annual / Biennial / Perennial* —** Perennial

***Used as* —** Mats / fodder / fuel / organic matter / mulch

1. *Climatic Requirements*— Warm weather with normal temperature and rainfall
2. *Soil requirements* — Clayey / clayey loam soils with organic matter
3. *Required land situations* — Marshy lands, in lowlying areas
4. *Importance of the Crop* — Small perennial plant, gives long stems
5. *Fertilizer management* — Normally no fertilizer is given, but for cultivation it requires, N, P_2O_5 and K_2O @ 100, 50 & 50 kg/ha, respectively, annually
6. *Duration of crop* — Perennial, but harvested annually
7. *Parts used* — Culms (stems) are used for mats; leaves for fodder / fuel / organic matter
8. *Where grown in India* — In marshy lands in lowlying areas
9. *Export possibility* — High, with mats (prepared from it)
10. *Economic estimated Yields (q / ha)* — 300 to 400 (green biomass)
11. *Economic value (Rs. / ha)* — 5,000 to 7,000
12. *By-product (If any)* — Refuse materials after mat preparation
13. *Use of by-product* — Fuels / organic matter / mulch
14. *Peoples' response for use* — Good, particularly for people residing in lowlying areas

568. ***CROP*— JOB'S TEARS / COIX / Gurlu / Sankru / Kaiz / Kaiz dana**

Scientific Name —***Coix lachryma*** L. / ***C. lachryma–job*** L. / ***C. stenocarpa*** Balansa

Family — *Gramineae*

Herb / Shrub / Tree — Herb (tall sedge)

Annual / Biennial / Perennial — Perennial

Used as — Beer / porridge / fodder / fuel / ornaments / organic matter

1. *Climatic Requirements*— Warm weather with normal temperature and rain fall
2. *Soil requirements* — Clayey and clayey loam soils with organic matter
3. *Required land situations* — Lowlying areas, can withstand salinity
4. *Importance of the Crop* — Tall plant that gives good coloured seeds for necklace
5. *Fertilizer management* — Normally no fertilizer is given, but for cultivated crop, N, P_2O_5 and K_2O @ 100, 50 & 50 kg/ha, respectively, needed annually
6. *Duration of crop* — Perennial, but harvested annually
7. *Parts used* — Grains give beer / porridge; leaves and seems for fodder; seed for ornaments (necklace)
8. *Where grown in India* — Lowlying saline tracts of India
9. *Export possibility* — Good for grains, for preparation of beer
10. *Economic estimated Yields (q / ha)* — 300 to 400 (green biomass)
11. *Economic value (Rs. / ha)* — 4,000 to 7,000
12. *By-product (If any)* — Refuse materials after use
13. *Use of by-product* — Fuel / organic matter
14. *Peoples' response for use* — Good, particularly for people in saline tracts of lowlying areas

569. ***CROP*— WATER FERN / ADIANTUM / Jalpharn / Dhekisak**

Scientific Name — ***Adiantum candatum*** L. / ***Capillus veneris*** L.

Family — *Adiantaceae*

Herb / Shrub / Tree — Herb

Annual / Biennial / Perennial — Perennial

Used as — Medicine / fish-food / fuel / organic matter /vegetable; weed in deep-water-paddy

1. *Climatic requirements* — Warm weather with normal temperature and rainfall

2. *Soil requirements* — Clayey and clayey loam soils with organic matters
3. *Required land situations* — Shallow / deep water in lowlying areas
4. *Importance of the Crop* — Plants in moist lands, can withstand low temperature
5. *Fertilizer management* — Normally no fertilizer is given, but for cultivated crop, N, P_2O_5 and K_2O @ 100, 50 & 50 kg/ha, respectively is given
6. *Duration of crop* — Perennial, but vigorous growth in rainy season
7. *Parts used* — Entire plants are used as medicine / vegetable / fuel / fish-food / organic matter
8. *Where grown in India* — In shallow/deep water, with deep-water-paddy or in marshy lands, in lowlying areas of India
9. *Export possibility* — Good, particularly for medicines
10. *Economic estimated Yields (q / ha)* — 200 to 300 (green biomass)
11. *Economic value (Rs. / ha)* — 3,000 to 5,000
12. *By-product (If any)* — Refuse materials after use of main products
13. *Use of by-product* — Fuel / organic matter
14. *Peoples' response for use* — Good, particularly for people residing in lowlying areas.

570. *CROP*— GUINEA RUSH / ARTIFCULATUS / Kesur

Scientific Name — Scirpus artifculatus L.

Family — *Cyperaceae*

Herb / Shrub / Tree — Herb

Annual / Biennial / Perennial — Annual

Used as — Perfumery / mat (weed in rice) / fuel / organic matter

1. *Climatic Requirements*— Warm weather with normal temperature and rainfall
2. *Soil requirements* — Clayey and clayey loam with organic matter
3. *Required land situations* — Marshy lands / shallow or deep water in lowlying areas
4. *Importance of the Crop* — Annual sedge grows in waterlogged paddy soil
5. *Fertilizer management* — Normally no fertilizer is given, but for cultivation N, P_2O_5 and K_2O @ 100, 50 & 50 kg/ha, respectively, is given
6. *Duration of crop* — Annual, 4-5 months

7. *Parts used* — Tuberous roots are used as perfumery; stems are used as mats (poor quality)
8. *Where grown in India* — In marshy lands in lowlying areas in India
9. *Export possibility* — High, particularly for perfumes
10. *Economic estimated Yields (q / ha)* — 300 to 400 (green biomass)
11. *Economic value (Rs. / ha)* — 4,000 to 6,000
12. *By-product (If any)* — Leaves and refuse materials
13. *Use of by-product* — Fuel / organic matter
14. *Peoples' response for use* — Good, particularly for people of lowlying areas

571. *CROP*— GORGAN NUT / FOX NUT / EURYALE / Makhana / Makhna

Scientific Name* — *Euryale ferox Salisb.

Family — *Euryalaceae*

Herb / Shrub / Tree — Herb

Annual / Biennial / Perennial — Perennial

Used as — Food / fodder / fuel / organic matter

1. *Climatic requirements* — Warm weather with normal temperature and rainfall
2. *Soil requirements* — Clayey and clayey loam soils with organic matters
3. *Required land situations* — Shallow or deep / stagnant water in lowlying areas
4. *Importance of the Crop* — A prickly aquatic perennial herb, cultivated for its seeds which are very tasty after frying. Tasty '*moya*' is produced.
5. *Fertilizer management* — N, P_2O_5 and K_2O @ 100, 50 & 50 kg/ha, respectively, annually
6. *Duration of crop* — Perennial, but harvested annually
7. *Parts used* — Seeds are roasted and made of ball-like tasty food (*moya*) with good flavour; stem, leaves and others are used as fuel / organic matter
8. *Where grown in India* — In shallow / deep stagnant water in lowlying areas of India
9. *Export possibility* — Good, for roasted seeds / food with good flavour
10. *Economic estimated Yields (q / ha)* — 10-15 (dry seeds)
11. *Economic value (Rs. / ha)* — 5,000 to 8,000

12. *By-product (If any)* — Stem / leaves / other refuse materials
13. *Use of by-product* — Fuel / organic matter / fodder
14. *Peoples' response for use* — Good, particularly for people living in lowlying areas (good for business men also).

572. *CROP*— GIANT WATERLILY / ROYAL WATERLILY / Bara padma / Bara jalpadma

Scientific Name — Victoria amazonica Sow / ***V. regia*** Lindl.

Family — *Nymphaeaceae / Euryalaceae*

Herb / Shrub / Tree — Herb (with big fragrant flower)

Annual / Biennial / Perennial — Perennial

Used as — Vegetable / medicine / fuel / organic matter

1. *Climatic requirements* — Warm weather with normal temperature and rain fall
2. *Soil requirements* — Clayey and clayey loam soils with organic matter
3. *Required land situations* — Deep water, in lowlying areas
4. *Importance of the Crop* — It is a big lotus, having big leaves and big fragrant flowers above the water surface
5. *Fertilizer management* — Normally no fertilizer is given, but for cultivation N, P_2O_5 and K_2O @ 100, 50 & 50 kg/ha, respectively, is given annually
6. *Duration of crop* — Perennial, but harvested annually
7. *Parts used* — Rhizomes / flower – stalks are used as vegetable; rhizomes are used as medicine; seeds are edible after frying
8. *Where grown in India* — In lowlying areas of India
9. *Export possibility* — High, particularly rhizomes which are used as medicines
10. *Economic estimated Yields (q / ha)* — 300 to 400 (green biomass)
11. *Economic value (Rs. / ha)* — 5,000 to 8,000
12. *By-product (If any)* — Leaves / leaf-stalk and other refuse materials
13. *Use of by-product* — Fuel / organic matter / fodder
14. *Peoples' response for use* — Good, particularly for people residing in lowlying areas

573. *CROP*— JUTE (bitter) / CAPSULARIS JUTE / Titapat / Nalita / Jalkosh / Pat / Kosta

Scientific Name — Corchorus capsularis L.

Family — *Tiliaceae*

***Herb / Shrub / Tree* —** Herb / Shrub

***Annual / Biennial / Perennial* —** Annual

***Used as* —** Fibres / stick / paper-pulp / green manure / vegetable

1. *Climatic requirements* — Warm weather with normal temperature and rainfall
2. *Soil requirements* — Alluvial / clayey / clayey loam soils
3. *Required land situations* — Medium or lowlands
4. *Importance of the Crop* — It is a fibre crop that grows in very low areas and harvested with pegging 2 poles, keeping other two pieces, crossed tied, in deep water in floodings
5. *Fertilizer management* — N, P_2O_5 and K_2O @ 80, 40 & 40 kg/ha, respectively, while in dry conditions (at later stage it is in water)
6. *Duration of crop* — 3-4 months, harvested at dry or wet conditions
7. *Parts used* — Stem-bark as fibre; stem as stick for paper-pulp / fuel; leaves as green manure/vegetable
8. *Where grown in India* — In lowlying areas / in clayey / clayey loam soils of India
9. *Export possibility* — High, particularly fibre or fibre-products
10. *Economic estimated Yields (q / ha)* — 15-20 q/ha of dry fibre or 300-400 (green biomass)
11. *Economic value (Rs. / ha)* — 10,000 to 15,000/-
12. *By-product (If any)* — Refuse materials (after using all parts)
13. *Use of by-product* — Fuel / organic matter / mulch
14. *Peoples' response for use* — Good, for people in lowlying areas

574. *CROP*— COMMON WATER FERN / MARSILEA / Susnisak

Scientific Name* — *Marsilea minuta L. / ***M. quadrifolia*** L.

***Family* —** *Marsileaceae*

***Herb / Shrub / Tree* —** Herb

***Annual / Biennial / Perennial* —** Perennial

***Used as* —** Vegetable / medicine / fuel / organic matter

1. *Climatic requirements* — Warm weather with normal rainfall and temperature
2. *Soil requirements* — Clayey / clayey loam soils with organic matter
3. *Required land situations* — Marshy lands in lowlying areas
4. *Importance of the Crop* — A common fern-type plant, grows in marshy lands

5. *Fertilizer management* — Normally no fertilizer is given but for cultivation N, P_2O_5 and K_2O @ 100, 50 & 50 kg/ha, respectively, is given annually
6. *Duration of crop* — Perennial, but harvested annually
7. *Parts used* — Leaves and plants are used for sedation and insomnia; leaves are used as vegetable and medicine (for urinary problems)
8. *Where grown in India* — In marshy lands in lowlying areas of the country
9. *Export possibility* — High, particularly for medicine
10. *Economic estimated Yields (q / ha)* — 200 to 300 (green biomass)
11. *Economic value (Rs. / ha)* — 5,000 to 7,000/-
12. *By-product (If any)* — Stalks and other refuse materials
13. *Use of by-product* — Fuel / organic matter / fodder
14. *Peoples' response for use* — Good, particularly for people from lowlying areas.

575. *CROP*— WATER DROPWORT / OENANTHE

Scientific Name* — *Oenanthe javanica (Blume) DC. / ***O. stolonifera*** Wall

Family — *Umbelliferae*

Herb / Shrub / Tree — Herb (Stoloniferous)

Annual / Biennial / Perennial — Perennial

Used as — Vegetable / condiments / medicine / spice / organic matter

1. *Climatic requirements* — Warm weather with normal temperature and rainfall
2. *Soil requirements* — Clayey and clayey loam soils with organic matter
3. *Required land situations* — Marshy lands/waterlogged conditions in lowlying situations
4. *Importance of the Crop* — Perennial stoloniferous herb, used as vegetable and eaten as such
5. *Fertilizer management* — Normally no fertilizer is given but for cultivation N, P_2O_5 and K_2O @ 100, 50 & 50 kg/ha, respectively, is required annually
6. *Duration of crop* — Perennial, but harvested annually
7. *Parts used* — Young shoots, stolons are used as vegetable / condiment / spice
8. *Where grown in India* — Marshy and waterlogged conditions in lowlying areas

9. *Export possibility* — High, because of use of spice and condiments
10. *Economic estimated Yields (q / ha)* — 200 to 300 (green biomass)
11. *Economic value (Rs. / ha)* — 5,000 to 6,000/-
12. *By-product (If any)* — Refuse materials (particularly with leaves and stems)
13. *Use of by-product* — Fuel and organic matter
14. *Peoples' response for use* — Good, particularly with people living in lowlying areas

576. *CROP*— WATER PLANTAIN / ALISMA / Panikala

Scientific Name — Alisma plantago - aquatica L.

Family — *Alismaceae*

Herb / Shrub / Tree — Herb

Annual / Biennial / Perennial — Perennial

Used as — Vegetable (tuberous rhizomes) / fruit / fuel / organic matter

1. *Climatic Requirements*— Warm weather with normal temperature and rainfall
2. *Soil requirements* — Clayey and clayey loam soils with organic matter
3. *Required land situations* — Marshy lands / shallow water in lowlying areas
4. *Importance of the Crop* — It is a tuberous rhizomes, used as vegetable; fruit is like plantain and children like it very much
5. *Fertilizer management* — Normally no fertilizer is given but for cultivation N, P_2O_5 and K_2O @ 100, 50 & 50 kg/ha, respectively, is given annually
6. *Duration of crop* — Perennial, but harvested annually
7. *Parts used* — Tuberous rhizomes are used as vegetable; scented fruits are edible (hence children like it); used in aquarium
8. *Where grown in India* — Marshy lands / shallow water in lowlying areas
9. *Export possibility* — Less possibility
10. *Economic estimated Yields (q / ha)* — 250 to 300 (green biomass)
11. *Economic value (Rs. / ha)* — 4,000 to 5,000/-
12. *By-product (If any)* — Stem / leaves / other refuse material
13. *Use of by-product* — Fuel / organic matter
14. *Peoples' response for use* — Good, local people in lowlying use it.

577. ***CROP***— **ROXBURGHII / Kathsola / Kasthasola / Sola**

Scientific Name — Sesbania roxburghii Merrill. / ***S. aculeata*** var. ***paludosa*** Baker

Family — *Papilionaceae*

Herb / Shrub / Tree — Herb

Annual / Biennial / Perennial — Perennial

Used as — Toy / fine deity ornaments / other decorations in festivals / fodder / poultices

1. *Climatic Requirements*— Warm weather with normal temperature and rainfall
2. *Soil requirements* — Clayey and clayey loam soils with organic matter
3. *Required land situations* — Marshy lands / shallow water in lowlying areas
4. *Importance of the Crop* — A perennial herb with yellow flowers grows in marshy / shallow water; grown in garden. Stems (soft) are used for decoration
5. *Fertilizer management* —N, P_2O_5 and K_2O @ 100, 50 & 50 kg/ha, respectively, annually
6. *Duration of crop* — Perennial, but harvested annually
7. *Parts used* — Leaves and small branches are used as cattle feed; soft woods (stems) are used for making toys / decoration in deity or other festivals; leaves are also used as poultice.
8. *Where grown in India* — In marshy lands / shallow water in lowlying areas in India
9. *Export possibility* — High, for decoration purpose / toy with soft woods
10. *Economic estimated Yields (q / ha)* — 150 to 250 (green biomass)
11. *Economic value (Rs. / ha)* — 5,000 to 6,000/-
12. *By-product (If any)* — Refuse materials after using soft woods
13. *Use of by-product* — Fuel / fodder / organic matter
14. *Peoples' response for use* — Good, particularly for the people of lowlying areas

578. ***CROP***— **NYMPHOIDES /Pansewali/Chandmala / Barachandmala**

Scientific Name — Nymphoides hydrophylla (Loureiro) Kuntze. / ***N. indicum*** (L.) Kuntze.

Family — *Gentianaceae*

Herb / Shrub / Tree — Herb

Annual / Biennial / Perennial — Perennial

Used as — Vegetable / medicine / fuel / fodder / organic matters

1. *Climatic Requirements*— Warm weather with normal temperature and rainfall
2. *Soil requirements* — Clayey and clayey loam soils with organic matter
3. *Required land situations* — Shallow water in lowlying areas
4. *Importance of the Crop* — Perennial herb is like big moon; leaves with flowers are floating on water surface, though roots are in soils in shallow depth of water
5. *Fertilizer management* —Normally no fertilizer is given but for cultivation, it requires N, P_2O_5 and K_2O @ 100, 50 & 50 kg/ha, respectively, annually
6. *Duration of crop* — Perennial, but harvested annually
7. *Parts used* — Both flower and leaf stalks / fruits / rhizomes are used as vegetable; leaves are used as medicine (for reducing body-temperature of human beings)
8. *Where grown in India* — Shallow water-depth in lowlying areas
9. *Export possibility* — High, for medicines
10. *Economic estimated Yields (q / ha)* — 250 to 300 (green biomass)
11. *Economic value (Rs. / ha)* — 5,000 to 7,000/-
12. *By-product (If any)* — Old leaves and other refuse material
13. *Use of by-product* — Fuel / fodder / organic matter
14. *Peoples' response for use* — Good, particularly for the people of lowlying areas

579. *CROP*— NATUNS / APONOGETON / Ghenchu

Scientific Name — Aponogeton natans (L.) Engler.

Family — *Aponogetonaceae*

Herb / Shrub / Tree — Herb

Annual / Biennial / Perennial — Perennial

Used as — Vegetable / food / fodder / fuel / organic matter

1. *Climatic Requirements*— Warm weather with normal temperature and rainfall
2. *Soil requirements* — Clayey and clayey loam soils with organic matter
3. *Required land situations* — Marshy / waterlogged soils in lowlying areas
4. *Importance of the Crop* — Perennial herb, surviving in water, with tubers at roots

5. *Fertilizer management* —Normally no fertilizer is given, but for cultivation, N, P_2O_5 and K_2O @ 100, 50 & 50 kg/ha, respectively, annually is given annually
6. *Duration of crop* — Perennial, but harvested annually
7. *Parts used* — Plants having tuberous roots, are used for tuber-like bodies as vegetable; for tuber-like bodies stems and leaves are used as fodder
8. *Where grown in India* — Marshy lands / waterlogged soils in lowlying areas of India
9. *Export possibility* — High, for rhizomes/tuber-like bodies if low temperature is provided
10. *Economic estimated Yields (q / ha)* — 200 to 300 (green biomass)
11. *Economic value (Rs. / ha)* — 3,000 to 4,000/-
12. *By-product (If any)* — Refuse materials after using as vegetable
13. *Use of by-product* — Fuel / organic matter
14. *Peoples' response for use* — Good, particularly for the people of lowlying areas

580. *CROP*— WATER CHESTNUTS / CALTROPS / TRAPA / Panifal / Singhara / Singara

Scientific Name* — *Trapa natans L. var. ***bispinosa*** (Roxb.) Makino. / ***T. quadrispinosa*** Wall.

Family — *Trapaceae*

Herb / Shrub / Tree — Herb/shrub

Annual / Biennial / Perennial — Annual / Perennial

Used as — Food crop for fruits; flour is prepared by grinding dried fruits, used as substitute for cereal flour / fuel / organic matter

1. *Climatic Requirements*— Warm weather with normal temperature and rainfall
2. *Soil requirements* — Clayey and clayey loam soils with organic matter
3. *Required land situations* — Shallow water / stagnated water in lowlying areas
4. *Importance of the Crop* — An aquatic herb, having 2 types of roots and leaves, bears hard triangular fruits, cultivated in water
5. *Fertilizer management* —N, P_2O_5 and K_2O @ 100, 50 & 50 kg/ha annually
6. *Duration of crop* — Perennial, harvested annually
7. *Parts used* — Fruit as 'fruit'; flour is prepared (substitute for *atta*); plants are used as fodder / fuel / organic matter

8. *Where grown in India* — Shallow water / waterlogged conditions in lowlying areas
9. *Export possibility* — High, fruits or flour (in cool containers)
10. *Economic estimated Yields (q / ha)* — 300 to 400 (green biomass)
11. *Economic value (Rs. / ha)* — 4,000 to 10,000/-
12. *By-product (If any)* — Leaves and stems / with refuse materials
13. *Use of by-product* — Fuel / organic matter / mulch for crops
14. *Peoples' response for use* — Good, for people living in lowlying areas.

581. *CROP*— ARROW HEAD / SAGITTARIA / Chhotokut / Tirpata

Scientific Name* — *Sagittaria sagittifolia L.

Family — *Alismataceae*

Herb / Shrub / Tree — Herb

Annual / Biennial / Perennial — Perennial

Used as — Tubers as vegetable / fuel / organic matter

1. *Climatic requirements* — Warm weather with normal temperature and rainfall
2. *Soil requirements* — Clayey and clayey loam soils with organic matters
3. *Required land situations* — Marshy lands / waterlogged conditions in lowlying areas
4. *Importance of the Crop* — Aquatic herb, with arrow-type leaves, above water
5. *Fertilizer management* —Normally no fertilizer is given but for cultivation, P_2O_5 and K_2O @ 100, 50 & 50 kg/ha, respectively, annually is required
6. *Duration of crop* — Perennial, but harvested annually
7. *Parts used* — Tubers at root zone are used as vegetable; stems and leaves are used as fodder/fuel/organic matter
8. *Where grown in India* — Marshy lands / waterlogged conditions in lowlying areas
9. *Export possibility* — High (if low temperature container is provided)
10. *Economic estimated Yields (q / ha)* — 300 to 400 (green biomass)
11. *Economic value (Rs. / ha)* — 4,000 to 5,000/-
12. *By-product (If any)* — Refuse materials after using as vegetable
13. *Use of by-product* — Fodder / fuel / organic matter / mulch

14. *Peoples' response for use* — Good, particularly for the people living in lowlying areas

582. ***CROP***— **VALLISNERIA / EEL-GRASS/TAPE-GRASS / WILD CELERY / Shawla / Syala**

Scientific Name* — *Vallisneria spiralis L. / ***V. gigantean*** Graebn.

Family — *Hydrocharitaceae*

Herb / Shrub / Tree — Herb

Annual / Biennial / Perennial — Perennial

Used as — Salad (entire plant), for high content of calcium, phosphorus and iron / organic matter

1. *Climatic Requirements*— Warm weather with normal temperature and rainfall
2. *Soil requirements* — Clayey and clayey loam soils with organic matter
3. *Required land situations* — Shallow water / waterlogged conditions in lowlying areas
4. *Importance of the Crop* — Aquatic plants, having long belt-like leaves with long female stalks coming over water surface to receive pollens; used for salad for having more Ca, P and Fe.
5. *Fertilizer management* — N, P_2O_5 and K_2O @ 100, 50 & 50 kg/ha, respectively, annually
6. *Duration of crop* — Perennial, but harvested time-to-time annually
7. *Parts used* — Entire plants are used as salad, for having more 'calcium', phosphorus and iron
8. *Where grown in India* — Shallow water / waterlogged condition in lowlying areas
9. *Export possibility* — High, if containers with low temperature are available for export
10. *Economic estimated Yields (q / ha)* — 250 to 300 (green biomass)
11. *Economic value (Rs. / ha)* — 5,000 to 6,000/-
12. *By-product (If any)* — Refuse materials after using it as salad
13. *Use of by-product* — Fuel and organic matter
14. *Peoples' response for use* — Good, for people living in lowlying areas (or for all people in big hotels, for food in some countries)

583. ***CROP***— **ISCHAEMUM / Dhanua / Murchi**

Scientific Name* — *Ischaemum rugosum Salishb. / ***I. colladoa*** Spring

Family — *Gramineae*

Herb / Shrub / Tree — Herb

Annual / Biennial / Perennial — Annual

Used as — Fodder / food / fuel / organic matter / bedding

1. *Climatic requirements* — Warm weather with normal temperature and rainfall
2. *Soil requirements* — Clayey and clayey loam soils with organic matter
3. *Required land situations* — Marshylands / waterlogged conditions in lowlying areas
4. *Importance of the Crop* — Annual herb (grass), grows in shallow water, giving food and fodder mainly
5. *Fertilizer management* — N, P_2O_5 and K_2O @ 100, 50 & 50 kg/ha, respectively
6. *Duration of crop* — 3-5 months (mostly in wet conditions)
7. *Parts used* — Young plants are used as fodder for cattle / horse and grains are used as food
8. *Where grown in India* — Marshy lands / waterlogged condition in lowlying areas in the country
9. *Export possibility* — Less possibility
10. *Economic estimated Yields (q / ha)* — 300 to 400 (green biomass)
11. *Economic value (Rs. / ha)* — 4,000 to 5,000/-
12. *By-product (If any)* — Refuse materials after using fodder and food grain
13. *Use of by-product* — Fuel / bedding (for animal / poultry) / organic matter
14. *Peoples' response for use* — Good, particularly for people living in lowlying areas

584. *CROP*— STAGNINA / Chhoto shyama

Scientific Name — Echinochloa stagnina (Retz.) Beauv. ***Panicum crussgalli*** L.

Family — *Gramineae*

Herb / Shrub / Tree — Herb (grass)

Annual / Biennial / Perennial — Annual

Used as — Fodder / food / fuel / organic matter / mulch

1. *Climatic requirements* — Warm weather with normal temperature and rainfall

2. *Soil requirements* — Clayey and clayey loam soils with organic matter
3. *Required land situations* — Marshylands / waterlogged soils with organic matter
4. *Importance of the Crop* — It is an annual herb (grass), grows in water with good foliage
5. *Fertilizer management* — Normally no fertilizer is given but for cultivation, N, P_2O_5 and K_2O @ 100, 50 & 50 kg/ha, respectively is given annually
6. *Duration of crop* — 3-4 months
7. *Parts used* — Entire plants as fodders; grains are used as food : stems are sweet to taste to cattle
8. *Where grown in India* — Marshy land / shallow water in lowlying areas
9. *Export possibility* — Less possibility
10. *Economic estimated Yields (q / ha)* — 300 to 400 (green biomass)
11. *Economic value (Rs. / ha)* — 4,000 to 5,000/-
12. *By-product (If any)* — Refuse materials after use
13. *Use of by-product* — Fuel / organic matter / mulch
14. *Peoples' response for use* — Good, particularly for people of lowlying areas.

585. *CROP*— AESCHYNOMENE / SENSITIVA / Sola / Didhen / Phulan / Jal-lajjabati

Scientific Name* — *Aeschynomene sensitiva SW. / ***A. indica*** L.

Family — *Papilionaceae*

Herb / Shrub / Tree — Herb / shrub

Annual / Biennial / Perennial — Annual

Used as — Toys / hats / ornament for deity / fodder / fuel / industrial use (insulator) / organic matter

1. *Climatic requirements* — Warm weather with normal temperature and rainfall
2. *Soil requirements* — Clayey and clayey loam soils with organic matter
3. *Required land situations* — Shallow water / waterlogged conditions with organic matter
4. *Importance of the Crop* — An annual soft herb, floating on water from which hats / toys / deity-jewellery / fodder / fuel / organic matters are obtained; gives yellow flowers

5. *Fertilizer management* — Normally no fertilizer is given, but for cultivation it requires N, P_2O_5 and K_2O @ 40, 20 & 20 kg/ha, respectively
6. *Duration of crop* — 3-4 months in general but may vary
7. *Parts used* — Soft stems are used as insulator / hats / toys / deity-jewellery; leaves are used for fodder; refuse materials for fuels / organic matter
8. *Where grown in India* — Shallow water / waterlogged condition in lowlying areas of India
9. *Export possibility* — High, for insulator / toys and ornaments (for deity)
10. *Economic estimated Yields (q / ha)* — 150 to 250 (green biomass)
11. *Economic value (Rs. / ha)* — 4,000 to 6,000/-
12. *By-product (If any)* — Refuse materials after use
13. *Use of by-product* — Fuel / organic matter / mulch
14. *Peoples' response for use* — Good, for particularly people living in lowlying areas.

586. *CROP*— NEPTUNIA / Lajalu / Jal-lajjabati / Paninajak

Scientific Name* — *Neptunia oleraceae Lour. / ***N. prostrata*** Baill.

Family — *Mimosaceae*

Herb / Shrub / Tree — Herb

Annual / Biennial / Perennial — Annual

Used as — Vegetable / seeds as *dal* / medicine / fuel / organic matters

1. *Climatic Requirements*— Warm weather with normal temperature and rainfall
2. *Soil requirements* — Clayey and clayey loam soils, marshy lands / shallow water / waterlogged conditions
3. *Required land situations* — Waterlogged conditions / marshy land / shallow depth of water in lowlying areas
4. *Importance of the Crop* — Annual equatic herb found with sensitive leaves
5. *Fertilizer management* — Normally no fertilizer is given, but for cultivation N, P_2O_5 and K_2O @ 40, 20 & 20 kg/ha, respectively, is required
6. *Duration of crop* — 3-4 months in rainy season
7. *Parts used* — Young stems / leaves / pods are used as vegetable; juice from stems is used as ear-ache; roots are used for syphilis
8. *Where grown in India* — Marshy lands in lowlying areas of India
9. *Export possibility* — High, for medicinal purpose.

10. *Economic estimated Yields (q / ha)* — 200-300 (green biomass)
11. *Economic value (Rs. / ha)* — 4,000 to 6,000/-
12. *By-product (If any)* — Refuse materials
13. *Use of by-product* — Fuel / organic matter / mulch
14. *Peoples' response for use* — Good, particularly for people in lowlying areas

587. *CROP*— WATERHYACINTH / EICHHORNIA / Kachuripana

Scientific Name* — *Eichhornia cressipes (Mart) Solms.

Family — *Pontederiaceae*

Herb / Shrub / Tree — Herb

Annual / Biennial / Perennial — Perennial

Used as — Fodder / manure / fuel / organic matter / mulch

1. *Climatic requirements* — Warm weather with normal temperature and rainfall
2. *Soil requirements* — Clay and clayey loam soils with organic matter
3. *Required land situations* — Marshy lands / shallow or deep water in lowlying areas
4. *Importance of the Crop* — It is an aquatic perennial herb, growing under floating conditions above water, having big light-blue flowers. It is a bad weed, though some fishes (*Kai*) take shelter in it.
5. *Fertilizer management* — Normally no fertilizer is given, but for cultivation requirement will be N, P_2O_5 and K_2O @ 100, 50 & 50, respectively, annually
6. *Duration of crop* — Perennial but harvested annually / time-to-time
7. *Parts used* — Leaves and petiole are used as fodder; entire plants are used as manure / fuel / mulch / organic matter / *gobar* gas; an obnoxious weed in water; shelter of some fishes (*Kai*)
8. *Where grown in India* — In waterlogged / shallow or deep water in lowlying areas
9. *Export possibility* — Less possibility
10. *Economic estimated Yields (q / ha)* — 300-400 (green biomass)
11. *Economic value (Rs. / ha)* — 3,000 to 4,000/-
12. *By-product (If any)* — Stems and other refuse materials after using as fodder
13. *Use of by-product* — Fuels / organic matters / mulch / manure

14. *Peoples' response for use* — Not so good, though a fodder, but obnoxious weed in waterbodies, in lowlying areas

588. *CROP*— MORNING GLORY / Dholkalmi / Bankalmi / Kalmi

Scientific Name* — *Ipomoea sepiaria Roxb. / ***I. maxima*** SW

Family — *Convolvulaceae*

Herb / Shrub / Tree — Herb/Shrub (twining)

Annual / Biennial / Perennial — Perennial

Used as — Vegetable / fodder / flower / fuel / fencing / organic matter

1. *Climatic requirements* — Warm weather with normal temperature and rainfall
2. *Soil requirements* — Clayey and clayey loam soils with organic matter
3. *Required land situations* — Marshy lands in lowlying areas.
4. *Importance of the Crop* — A twining perennial herb, grows in marshy areas, can withstand deep water, having good big flowers
5. *Fertilizer management* — Normally no fertilizer is given, but for cultivation, requires N, P_2O_5 and K_2O @ 100, 50 & 50 kg/ha, respectively, annually
6. *Duration of crop* — Perennial, but harvested annually / time-to-time
7. *Parts used* — Soft twig and leaves are used as vegetable; entire plants are used as fodder; Sticks are used as fuel / fencing; flower for beautification
8. *Where grown in India* — Marshy lands / waterlogged condition in lowlying areas
9. *Export possibility* — Less possibility
10. *Economic estimated Yields (q / ha)* — 300-400 (green biomass)
11. *Economic value (Rs. / ha)* — 4,000 to 5,000/-
12. *By-product (If any)* — After using, refuse materials
13. *Use of by-product* — Fuel / organic matter / manure / mulch
14. *Peoples' response for use* — Good, particularly for the people of the lowlying areas.

589. *CROP*— GIANT REED / GREAT REED / ARUNDO / Baranal / Narhal / Narsal / Bajali

Scientific Name* — *Arundo donax L. / ***A. sativa*** Lamk.

Family — *Gramineae*

Herb / Shrub / Tree — Shrub (tall grain)

Annual / Biennial / Perennial — Perennial

Used as — Pipes/ baskets / paper / mat / used in *boroj* / medicine / musical instruments, fishing / walking sticks / fuel / fodder / organic matter / mulch

1. *Climatic requirements* — Warm weather with normal temperature and rainfall
2. *Soil requirements* — Clayey and clayey loam soils with organic matter
3. *Required land situations* — Marshy lands / waterlogged condition, in lowlying areas
4. *Importance of the Crop* — It is a hardy perennial reed, grows in marshy lands / shallow water. Hollow culms are used in manufacturing of baskets, pipes, betelvine pandals
5. *Fertilizer management* — Requirement of N, P_2O_5 and K_2O @ 100, 50 & 50 kg/ha, respectively annually
6. *Duration of crop* — Perennial but harvested annually
7. *Parts used* — Hollow culms are used as pipes / baskets / papers mats / pandal in *boroj* / rayon-grade-pulp; rhizomes are used for medicine; reeds are used as musical / fishing / walking stick / fodder fuel / organic matter
8. *Where grown in India* — Marshy lands / waterlogged soils in lowlying areas
9. *Export possibility* — High, for rhizomes (for medicine)
10. *Economic estimated Yields (q / ha)* — 400-500 (green biomass)
11. *Economic value (Rs. / ha)* — 6,000 to 8,000/-
12. *By-product (If any)* — Refuse materials after using stems / rhizomes, leaves are left
13. *Use of by-product* — Fodder / fuel / mulch / bedding for cattle / organic matter
14. *Peoples' response for use* — Good, for people of lowlying areas

590. *CROP*— INDIAN SALTWORT / SEA-BLITE / SUAEDA / Kharilani / Salsola

Scientific Name* — *Suaeda maritima (L.) Dum. / ***Salsola indica*** Wild

Family — *Chenopodiaceae*

Herb / Shrub / Tree — Herb

Annual / Biennial / Perennial — Annual

Used as — Vegetable / soil-binder in coastal areas / entire plants are used as fodder/fuel / organic matters

1. *Climatic requirements* — Warm weather with normal temperature and rainfall

2. *Soil requirements* — Sandy / sandy loam soils in coastal areas
3. *Required land situations* — More or less saline tracks in coastal areas
4. *Importance of the Crop* — It is a herb, grows in swamp / coastal areas only
5. *Fertilizer management* — Normally no fertilizer is given, but for cultivation N, P_2O_5 and K_2O @ 80, 40 & 40, respectively, is given
6. *Duration of crop* — 3-4 months
7. *Parts used* — Green leaves are used as vegetable; fixes sand / soil in aggregate form in sea-shores; entire plants are used as fodder / fuel / mulch / organic matter
8. *Where grown in India* — In sea-shore areas / swamp lands in India
9. *Export possibility* — Less possibility
10. *Economic estimated Yields (q / ha)* — 150-250 (total biomass))
11. *Economic value (Rs. / ha)* — 3,000 to 4,000/-
12. *By-product (If any)* — After using as vegetable, refuse material
13. *Use of by-product* — Fodder / fuel / mulch / organic matter
14. *Peoples' response for use* — Good, for the people of coastal areas in India

591. *CROP*— SALTY CAMEL FODDER / SALICORNIA / Nonta-gachh

Scientific Name — Salicornia brachiata Roxb.

Family — *Chemopodiaceae*

Herb / Shrub / Tree — Herb / shrub

Annual / Biennial / Perennial — Perennial

Used as — Vegetable (salad) / fodder / fuel / mulch / organic matters

1. *Climatic requirements* — Warm sultry weather with normal temperature and rainfall
2. *Soil requirements* — Sandy / sandy alluvial soils of coastal belts
3. *Required land situations* — Coastal areas / lowlying marshy areas
4. *Importance of the Crop* — It is a salty fodder (particularly for camel), grows in marshy land / shallow water in the sea-shores
5. *Fertilizer management* — Normally no fertilizer is given, but requires N, P_2O_5 and K_2O @ 100, 50 & 50 kg/ha, respectively, annually
6. *Duration of crop* — Perennial but harvested annually
7. *Parts used* — Young shoots and leaves are used as vegetable (particularly for 'salad'); it is used for pickle also; used as fodder particularly for camel

8. *Where grown in India* — In coastal areas / sea-shores of India
9. *Export possibility* — Less possibility
10. *Economic estimated Yields (q / ha)* — 150-300 (green biomass)
11. *Economic value (Rs. / ha)* — 3,000 to 4,000/-
12. *By-product (If any)* — After using as fodder, practically no by-product is available still refuse materials are there
13. *Use of by-product* — Fuel / mulch / organic matter
14. *Peoples' response for use* — Good, particularly for people in coastal areas.

592. *CROP*— YELLOW-BERRIED NIGHTSHADE / SURATTENSE/ Kantari / Kateli / Banbegun

Scientific Name — Solanum surattense Burn.f. / ***S. xanthocarpum*** Schrad & Wendi.

Family — *Solanaceae*

Herb / Shrub / Tree — Herb / Shrub

Annual / Biennial / Perennial — Perennial

Used as — Medicine / fuel / organic matter

1. *Climatic requirements* — Warm weather with normal temperature and rainfall
2. *Soil requirements* — Dampy clayey and clayey loam soils with organic matter
3. *Required land situations* — Dampy / marshy lands in lowlying areas
4. *Importance of the Crop* — Spiny perennial defuse herb that bears yellow berries and can withstand salinity
5. *Fertilizer management* — It requires N, P_2O_5 and K_2O @ 100, 50 & 50 kg/ha, respectively annually
6. *Duration of crop* — Perennial but harvested for the fruits annually
7. *Parts used* — Roots are used as an expectorant that form an ingredient of *ayurvedic* medicine (*Dasamula*); which are used in cough, asthma and pain in chest. Juice of fruits is used in soar-throat and juice of leaves, mixed with black-pepper is good for rheumatism. Fruits yield *glycoalkaloids* which can be source of materials for 'cortisone' and 'sex hormone preparation'
8. *Where grown in India* — Dampy area in lowlying conditions
9. *Export possibility* — High, particularly for medicine and sex hormone
10. *Economic estimated Yields (q / ha)* — 300-400 (green biomass)
11. *Economic value (Rs. / ha)* — 5,000 to 10,000/-

12. *By-product (If any)* — Stem, leaves and other refuse materials
13. *Use of by-product* — Fuels / organic matter
14. *Peoples' response for use* — Good, for medicinal values

593. *CROP*— GOOSE FOOT / LAMBS-QUARTERS / PIGWEED / Bathua / Bethosak

Scientific Name* — *Chenopodium album L.

Family — *Chenopodiaceae*

Herb / Shrub / Tree — Herb

Annual / Biennial / Perennial — Annual

Used as — Vegetable / fodder / fuels / organic matters / mulch

1. *Climatic requirements* — Warm / cool weather, having high / low temperature and rainfall
2. *Soil requirements* — Clayey and clayey loam soils in dampy areas
3. *Required land situations* — Dampy areas of lowlying conditions and situations
4. *Importance of the Crop* — An erect annual herb, grows under cool condition even in saline areas as vegetable and fodder
5. *Fertilizer management* — N, P_2O_5 and K_2O fertilizers require @ 100, 50 & 50 kg/ha, respectively
6. *Duration of crop* — 3-5 months (both in warm / cool conditions)
7. *Parts used* — Tender twigs and leaves are used as vegetable. Entire plants are used as fodder; dry plants are used as fuel / organic matter / mulch
8. *Where grown in India* — Under dampy conditions in lowlying areas
9. *Export possibility* — Less possibility
10. *Economic estimated Yields (q / ha)* — 300-400 (green biomass)
11. *Economic value (Rs. / ha)* — 3,000 to 5,000/-
12. *By-product (If any)* — After using it as vegetable, refuse materials
13. *Use of by-product* — Fuel / organic matter / mulch / fodder
14. *Peoples' response for use* — Good, particularly for vegetable in the dampy areas

594. *CROP*— NIPA / Golpata / Hat-pakha

Scientific Name* — *Nipa frutican L.

Family — *Palmae*

Herb / Shrub / Tree — Shrub (tall)

Annual / Biennial / Perennial — Perennial

Used as — Hand-fans / post / fodder / fuel / organic matters / beauatification

1. *Climatic requirements* — Warm weather with normal temperature and rainfall
2. *Soil requirements* — Clayey and clayey loam soils with organic matter
3. *Required land situations* — Dampy saline lands in low lying areas
4. *Importance of the Crop* — It is a perennial shrub, having round big beautiful leaves, used as fans; resistant to salinity.
5. *Fertilizer management* — N, P_2O_5 and K_2O requirement @ 100, 50 & 50 kg/ha, respectively, annually
6. *Duration of crop* — Perennial, but harvested (leaves) annually
7. *Parts used* — Leaves are used for hand-fans/fuel/fodder / thatching roof of house; stems are used as pole / posts; used for beautification in gardens
8. *Where grown in India* — Saline marshy/dampy lands in lowlying areas in India
9. *Export possibility* — Less possibility
10. *Economic estimated Yields (q / ha)* — 300-400 (green biomass)
11. *Economic value (Rs. / ha)* — 4,000 to 5,000/-
12. *By-product (If any)* — Materials after using main portion
13. *Use of by-product* — Fuel / organic matter / mulch
14. *Peoples' response for use* — Good, particularly in lowlying saline areas.

595. *CROP*— CANE / CALAMUS / Betra / Bet

Scientific Name* — *Calamus tenuis Roxb.

Family — *Palmae*

Herb / Shrub / Tree — Shrub (tall, with spines both in sheath and leaves)

Annual / Biennial / Perennial — Perennial

Used as — Furniture / mat / stick / screens/basket / canning (bad boys)

1. *Climatic Requirements*— Warm weather with normal temperature and rainfall
2. *Soil requirements* — Clayey and clayey loams in lowlying areas
3. *Required land situations* — Waterlogged / marshy / dampy in lowlying areas
4. *Importance of the Crop* — It is a spiny perennial tall / lanky plant grows in marshy / dampy condition; can withstand salinity.

5. *Fertilizer management* — Requires N, P_2O_5 and K_2O @ 100, 50 & 50 kg/ha, respectively, annually
6. *Duration of crop* — Perennial, some are used annually
7. *Parts used* — Stems are used for mat / walking sticks / furnitures / basket / canning bad boys; leaves and other for fuels / organic matter
8. *Where grown in India* — Dampy / marshy / waterlogged conditions in lowlying areas
9. *Export possibility* — High, for furniture / mat stick / baskets
10. *Economic estimated Yields (q / ha)* — 400-500 (green biomass)
11. *Economic value (Rs. / ha)* — 5,000 to 10,000/-
12. *By-product (If any)* — After using the stems, other portions
13. *Use of by-product* — Fuel / organic matter
14. *Peoples' response for use* — Good, both for people of main lands and lowlying areas.

Note : Because of spines on stem/sheath/leaves, it is very difficult to handle.

596. *CROP*— GREATER GALANGAL / ALPINIA / Borakulanjan

Scientific Name* — *Alpinia galanga (L.) SW / ***Langna galanga*** (L.) Stuntz.

Family — *Zingiberaceae*

Herb / Shrub / Tree — Shrub (robust)

Annual / Biennial / Perennial — Perennial

Used as — Spice and condiment / medicine / insecticide / fuel

1. *Climatic requirements* — Warm weather with normal temperature and rainfall
2. *Soil requirements* — Clayey / clayey loam soils with organic matter
3. *Required land situations* — Dampy lands in low lying areas
4. *Importance of the Crop* — It is a perennial robust herb, resistant to saline-water-flood; rhizomes are used as spice and flavouring agents; gives essential oils
5. *Fertilizer management* — Requirement of N, P_2O_5 and K_2O @ 100, 50 & 50 kg/ha, respectively, annually
6. *Duration of crop* — Perennial, but harvested annually
7. *Parts used* — Rhizomes are used as spice / flavouring agents for foods used as medicine / insecticides
8. *Where grown in India* — Dampy conditions of lowlying saline areas of India

9. *Export possibility* — High, for rhizomes (as such or dry)
10. *Economic estimated Yields (q / ha)* — 300-400 (green biomass)
11. *Economic value (Rs. / ha)* — 4,000 to 10,000/-
12. *By-product (If any)* — After using rhizomes, leaves and stems
13. *Use of by-product* — Fuel / fodder / organic matter
14. *Peoples' response for use* — Good for all types of people in the country

597. *CROP*— SCREW PINE / PANDANUS / Keya / Kewda / Keura

Scientific Name* — *Pandanus odoratissimus L.f. / ***P. fascicularis*** Lam.

Family — *Pandanaceae*

Herb / Shrub / Tree — Shrub (with spine on leaves)

Annual / Biennial / Perennial — Perennial

Used as — Scent / basket / cordage / paper-making / medicine / fuel / organic matter

1. *Climatic requirements* — Warm weather with normal temperature and rainfall
2. *Soil requirements* — Saline sandy clayey / clayey loam soils with organic matter
3. *Required land situations* — Marshy land / riverbed in lowlying areas
4. *Importance of the Crop* — Spiny perennial shrub, grown in saline lowlying areas; resistant to saline-water-floods; scent (*attar*) is obtained from spadice used in India and elsewhere, for scenting cloths
5. *Fertilizer management* — Requires N, P_2O_5 and K_2O @ 100, 50 & 50 kg/ha, respectively, annually
6. *Duration of crop* — Perennial, but harvested time-to-time as per need
7. *Parts used* — Flowers (spadice) are used as scent (as *Kewda-attar*) clothes and others; fruit is eaten; leaves are used for baskets / cordage/umbrellas paper-making; flowers are also used as medicine; others are used as organic matter / fuel / stick etc. (Some snakes reside in the bush of screwpine)
8. *Where grown in India* — Sea / river coast / lowlying areas
9. *Export possibility* — High, for scents / medicine
10. *Economic estimated Yields (q / ha)* — 300-400 (green biomass)
11. *Economic value (Rs. / ha)* — 5,000 to 10,000/-
12. *By-product (If any)* — After using as scent / basket etc.

13. *Use of by-product* — Fuel / organic matter / mulch
14. *Peoples' response for use* — Good, for all types of people.

598. *CROP*— DERRIS / TUBA ROOT / DERRIS POWDER / Bakalbip

Scientific Name — Derris sinuata* / *D. elliptica (Wall.) Benth.

Family — *Papilionaceae*

Herb / Shrub / Tree — Herb (traing / climbing)

Annual / Biennial / Perennial — Perennial

Used as — Fish-poison / insecticide / organic matter / fuel

1. *Climatic requirements* — Warm weather with normal temperature and rainfall
2. *Soil requirements* — Clayey / clayey loam / sandy clay with organic matter
3. *Required land situations* — Dampy / marshy conditions in lowlying areas
4. *Importance of the Crop* — A perennial herb/strub, climbing in mature trees, resistant to salinity in floods; Derris power (used for killing fish) is obtained from the roots of it.
5. *Fertilizer management* — It requires N, P_2O_5 and K_2O @ 100, 50 & 50 kg/ha, respectively, annually
6. *Duration of crop* — Perennial, but used from time-to-time
7. *Parts used* — Roots are used as insecticide / fish-poison. Leaves are used as fuel / organic matter; stems are used as fuel / sticks
8. *Where grown in India* — In dampy / marshy lowlying areas
9. *Export possibility* — High, for insecticidal / fish-poisoning
10. *Economic estimated Yields (q / ha)* — 300-400 (green biomass)
11. *Economic value (Rs. / ha)* — 4,000 to 6,000/-
12. *By-product (If any)* — Material after main use
13. *Use of by-product* — Fuel / organic matter / stick
14. *Peoples' response for use* — Good, by industries; but some naughty people misuse it by killing the fish from its roots.

599. *CROP*— COIX / Barakaiz

Scientific Name — Coix gigantea Koen. / ***C. lingulata*** Hack.

Family — Gramineae

Herb / Shrub / Tree — Shrub (tall grass)

Annual / Biennial / Perennial — Annual / Perennial

Used as — Fodder / fencing / fuel / necklace / rosaries / ornaments / furniture / organic matters

1. *Climatic requirements* — Warm sultry weather with normal temperature and rainfall
2. *Soil requirements* — Dampy / marshy saline clayey soils with organic matters
3. *Required land situations* — Marshy / dampy saline soils in lowlying areas
4. *Importance of the Crop* — It is a tall annual/perennial grass, resistant to saline-water-flood, having coloured beads
5. *Fertilizer management* — Required N, P_2O_5 and K_2O @ 100, 50 & 50 kg/ha, respectively, annually
6. *Duration of crop* — Harvested annually in earlier stage for fodder but for beads matured plants are needed
7. *Parts used* — Stony bracts are used for rosaries/beads ornaments / basket / boxes; as fodder it is used in the earlier periods; stems are used as fencing; leaves for fodder; entire plants as fuel / organic matter / mulch
8. *Where grown in India* — In dampy / marshy ands in lowlying areas in India
9. *Export possibility* — High, for beads for jewelaries
10. *Economic estimated Yields (q / ha)* — 300-400 (green biomass)
11. *Economic value (Rs. / ha)* — 4,000 to 6,000/-
12. *By-product (If any)* — Refuse materials after using for main products
13. *Use of by-product* — Fuel / organic matter / mulch
14. *Peoples' response for use* — Good, particularly for fodder and coloured seeds

600. *CROP*— **ENHYDRA / FLUCTUENS / Halencha / Hinche / Harkuch**

Scientific Name — Enhydra fluctuens Loureiro

Family — Compositiae

Herb / Shrub / Tree — Herb (climbing / trailing on water with a support)

Annual / Biennial / Perennial — Annual / Perennial

Used as — Vegetable / fuel / organic matter / mulch; resting place for fish

1. *Climatic requirements* — Cool weather with low temperature and less rainfall
2. *Soil requirements* — Clayey and clayey loam soils in lowlying areas

3. *Required land situations* — Marshy land / shallow water in lowlying areas
4. *Importance of the Crop* — This floating type of plant grows with the support, in water; it is a good vegetable; softmass on water surface nearer to adventitious roots
5. *Fertilizer management* — Requires N, P_2O_5 and K_2O @ 100, 50 & 50 kg/ha, respectively, annually
6. *Duration of crop* — Harvested in winter, for vegetable from time-to-time
7. *Parts used* — Tender leaves and stems are used as vegetable; entire plants are used as fuel / organic matter / mulch; some fishes take shelter under its bush
8. *Where grown in India* — Marshy lands / shallow water in lowlying areas
9. *Export possibility* — Less possibility
10. Economic estimated Yields (q/ha) — 200-300 (green biomass)
11. *Economic value (Rs. / ha)* — 3,000 to 4,000/-
12. *By-product (If any)* — Materials after using as vegetable
13. *Use of by-product* — Fuel / organic matter / mulch / fodder
14. *Peoples' response for use* — Good, particularly for people living in lowlying areas.

REFERENCES

Chatterjee, B.N.; Mondal, A.K.; Das, P.K. and Sengupta, Kajal (1999). *Moist and Wetland Agriculture,* Oxford & IBH Publishing Co. Pvt. Ltd., New Delhi, Calcutta, India.

Das, N.R. (2006). *Rainfed Paira / Utera Crops in India*, Kalyani Publishers, Ludhiana / New Delhi / Noida (U.P) / Hyderabad / Chennai / Calcutta / Cuttack, India.

Das, N.R. (2007). *Introduction to Crops of India*. Scientific Publishers (India), Jodhpur, India (First edition)

Das, N.R. (2008). *Agronomic Research Management*, Agrotech Publishing Academy, Udaipur, India.

Das, N.R. (2008). *Cropping System Research*, Kalyani Publishers, Ludhiana / New Delhi / Noida (U.P.) / Hyderabad / Chennai / Calcutta / Cuttack, India.

Das, N.R. (2008). *Tillage and Crop Production*, Scientific Publishers (India), Jodhpur, India.

Das, N.R. (2008). *Wheat Crop Management*, Scientific Publishers (India), Jodhpur, India.

Das, N.R. (2009). *Practical Manual on Basic Agronomy (with theory)*, Scientific Publishers (India), Jodhpur, India.

Das, N.R. (2012). *Introduction to Aquatic and Semi-aquatic Plants of India*, Kalyani Publishers, Ludhiana / New Delhi / Noida (U.P.) / Hyderabad / Chennai / Calcutta / Cuttack, India.

Das, N.R. (2013). *Different Names and Uses of Crops and Water-plants of India*, Kalyani Publishers, Ludhiana / New Delhi / Noida (U.P.) / Hyderabad / Chennai / Kolkata / Cuttack / Guwahati / Kochi / Bengaluru, India.

Das, N.R. (2013). *Glossary of Indian Crops*, Scientific Publishers (India), Jodhpur, India.

Das, N.R.; S.P. Bhattacharya and A.K. Das (1997a,b & c). Weeds in Crops of West Bengal, *World Weed*, (Winter / Summer / Rainy seasons), 4(3&4):195-226.

Datta, A.C. (1970). *Botany (for degree students)*. Third edition, Oxford University Press, Bombay / Calcutta / Madras, India.

FAO (1996). *Weed Management in Rice*, FAO plant production and protection paper 139, Food and Agriculture Organization of the United Nations, Rome.

Kirkpatrick, E.M.; C.M. Schwarz; G.W. Davidson; M.A. Seaton; J. Simpson and R.J. Sherrard (1983). *Chambers 20th Century dictionary*, Allied Publishers Private Limited, New Edition 1983, New Delhi, Calcutta, London, Bombay etc.

Narayanan, T.R. and Dabadghao, P.M. (1972). *Forage Crops of India*, Indian Council of Agricultural Research, New Delhi, India.

Naskar, K.R. (1990). Aquatic and Semi-aquatic Plants of Lower Ganga Delta, Daya Publishing House, New Delhi, India.

Palaniappan, S.P. and K. Annadurai (1999). *Organic Farming – Theory and Practices*. Scientific Publishers (India), Jodhpur, India.

Roy, A. (2007). *Banglar Jalar Gachh* (Bengali), West Bengal, Biodiversity Board, Calcutta, West Bengal, India.

Singh, S.S. (1988). *Crop Management*, Kalyani Publishers Ludhiana / New Delhi / Noida (U.P.) / Hyderabad / Chennai / Calcutta / Cuttack, First Edition, India.

Singh, Umrao, A.M. Wadhwani, B.M. Johri (1996). *Dictionary of Economic Plants in India,* Indian Council of Agricultural Research, New Delhi, (Reprinted) India.

Subbian, P.; K. Annadurai and S.P. Palaniappan (2000). *Agriculture : Facts and Figures*, Kalyani Publishers, Ludhiana / New Delhi / Noida (U.P) / Hyderabad / Chennai / Calcutta / Cuttack, Fist Edition, India.